Table of Integrals

$$\int x^n \, dx = \frac{x^{n+1}}{n+1} + c \quad \text{if } n \neq -1$$

$$\int \frac{dx}{x} = \ln |x| + c$$

$$\int e^x \, dx = e^x + c$$

$$\int \log x \, dx = x \log x - x + c$$

$$\int \frac{dx}{\sqrt{a^2 - x^2}} = \arcsin\left(\frac{x}{a}\right) + c \quad \text{or} \quad -\arccos\left(\frac{x}{a}\right) + c$$

$$\int \sin x \, dx = -\cos x + c$$

$$\int \cos x \, dx = \sin x + c$$

$$\int \tan x \, dx = \log \sec x + c$$

$$\int \cot x \, dx = \log \sin x + c$$

$$\int u(x) \frac{dv(x)}{dx} \, dx = u(x) \cdot v(x) - \int v(x) \frac{du(x)}{dx} \, dx$$

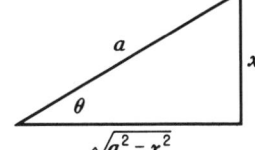 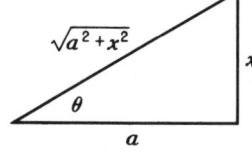 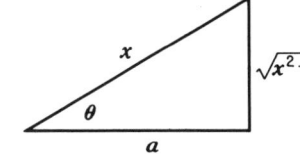

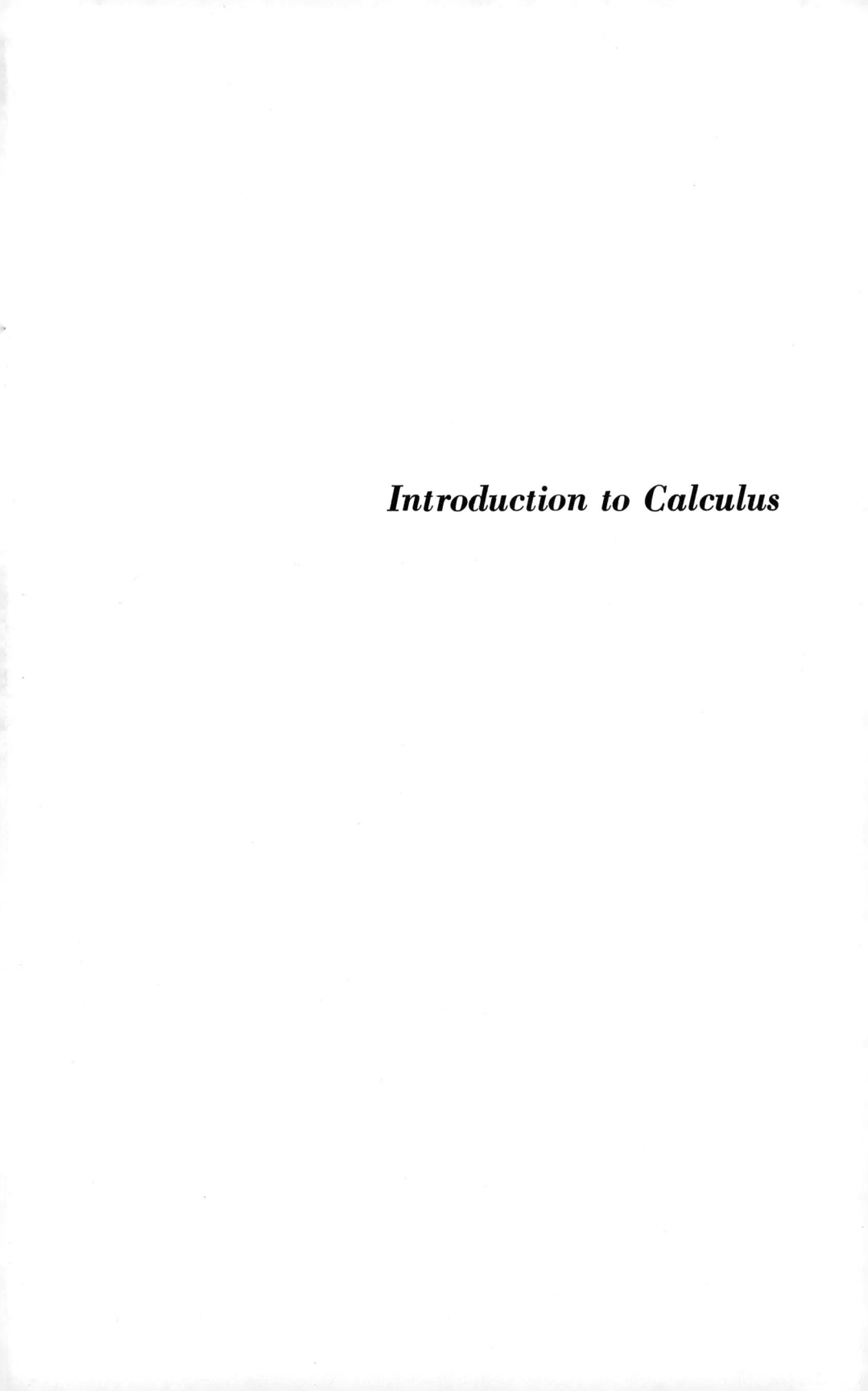

Introduction to Calculus

Introduction to *CALCULUS*
with analytic geometry

Richard V. Andree
THE UNIVERSITY OF OKLAHOMA

McGRAW-HILL BOOK COMPANY, INC.
New York   San Francisco   Toronto   London

Introduction to CALCULUS

Copyright © 1962 by the McGraw-Hill Book Company, Inc. All Rights Reserved. Printed in the United States of America. This book, or parts thereof, may not be reproduced in any form without permission of the publishers. *Library of Congress Catalog Card Number* 61-13427

Preface

This book is designed for the mature student who wishes to acquire the vital concepts of calculus. The emphasis is on basic ideas rather than on drill in manipulative techniques and tricks. Generalization is stressed throughout. The student is led to see that, for example, work under a variable force is a reasonable generalization of work under a constant force. It is this emphasis on fundamental ideas rather than on "formula juggling" which provides the background needed by a mature student who wishes to learn calculus as an aid to research in the social sciences. Secondary school teachers preparing students for the twelfth grade polynomial calculus course recommended by the Commission on Mathematics of the College Entrance Examination Board and by the Commission on the Undergraduate Program in Mathematics also need emphasis on theory, rather than "juggling."

Much of the text has been used in National Science Foundation Institutes for Secondary School Teachers, and in classes of Mathematics for Social Scientists. It has proved highly successful with both groups.

A social scientist who has a research problem which is stated in mathematical language can usually find a mathematician who is willing to help him solve it. However, very few mathematicians know enough social science to *formulate* the problem. The social scientist *must* learn enough mathematics to formulate his problems

Preface

in reasonable mathematical terms before he seeks aid. It can be done. Examine the literature in the field—it *is* being done.

This book is designed to provide the fundamentals of calculus needed to understand and formulate new problems.

It has proved reasonable to expect students to complete the text in four semester hours, especially if Chapter 1 can be covered rapidly. It seems inadvisable to skip Chapter 1 unless the student has had an adequate freshman course in mathematics during the last five years.

Richard V. Andree

Acknowledgements

Acknowledgements in a text traditionally thank people for assistance and absolve them from any possible shortcomings of the book. This book has developed so gradually that it would be impossible to name all of the friends, colleagues, and students who assisted in its preparation. However, it would be unfair not to acknowledge two outstanding contributors. Professor John C. Brixey, The University of Oklahoma, Norman, whose thought-provoking discussions during our coauthorship of several texts over the past ten years have helped formulate many of the ideas, and the majority of the problems, in this volume. The author expresses sincere thanks to Dr. Brixey for granting permission to borrow freely from earlier joint efforts, and even more for his penetrating analysis of many problems, both mathematical and pedagogical. Our second bouquet goes to the author's wife, Josephine Peet Andree, who possesses the outstanding qualities of patience and understanding which are so essential in an author's spouse. Since she, herself, is a competent mathematician and an excellent teacher, her contributions have been invaluable, not only on the thankless and arduous task of reading proof, but even more important in the actual formulation of the ideas and the expression of the text itself.

For any shortcomings which may appear, the author hereby blames the publisher, and the publisher blames the author.

R. V. A.

Contents

Preface		v
Chapter 1. Basic Algebraic Theory		1
1-1	Introduction	1
1-2	Functions and Relations	1
1-3	More General Functions	4
1-4	The Function Notation	5
1-5	Inequalities	8
1-6	Absolute Value	8
1-7	Two Basic Properties	10
1-8	The Structure of the Number System	10
1-9	Two Unusual Real Numbers	12
1-10	The Meaning of Division	12
1-11	Solution Set of an Equation	14
1-12	Auxiliary Equations	22
1-13	$\Delta f = f(x + \Delta x) - f(x)$	27
1-14	Self Test	30
Chapter 2. Basic Geometric Theory		31
2-1	Analytic (Coordinate) Geometry	31
2-2	The Distance between Two Points	38
2-3	Loci	43
2-4	Solution of Inequalities of First Degree	49
2-5	Slope of a Line	53
2-6	The Equation of a Line	57
2-7	Self Test	65
Chapter 3. Tangents and Limits		66
3-1	The Line Tangent to a Curve at a Point	66
3-2	Limit of a Function	69
3-3	Continuous Function	72

Contents

3-4	Slope of a Tangent Line	80
3-5	Increments	86
3-6	Self Test	90

Chapter 4. Differential Calculus — 91

4-1	The Derivative	91
4-2	The Delta Process	92
4-3	Generalization	97
4-4	Preliminary Theorems on Differentiation	97
4-5	The Derivative of a Polynomial	100
4-6	Maximum and Minimum Values of a Function	105
4-7	Applications Involving Maxima and Minima	118
4-8	Differentiation of a Product	123
4-9	Differentiation of a Power of a Function	126
4-10	Derivative of a Composite Function	129
4-11	Some Additional Applications	132
4-12	Self Test	136

Chapter 5. Extended Theorems of Differentiation — 138

5-1	Negative, Fractional, and Zero Exponents	138
5-2	Scientific Notation	142
5-3	An Extension of the Theorem $\dfrac{d(kx^n)}{dx} = knx^{n-1}$	147
5-4	A Further Extension of the Theorem $\dfrac{d(kx^m)}{dx} = kx^{m-1}$	149
5-5	Self Test	154

Chapter 6. Rates of Change — 156

6-1	Rate of Change	156
6-2	Average Rate of Change	156
6-3	Velocity	158
6-4	General Rate of Change	162
6-5	Self Test	166

Chapter 7. Integral Calculus — 167

7-1	Σ Notation	167
7-2	$\lim\limits_{x \to \infty} f(x)$	172
7-3	Area	173
7-4	The Spring Problem	180
7-5	Some Remarks on $\lim\limits_{n \to \infty} \sum\limits_{i=1}^{n} f(\xi_i)\,\Delta x_i$	184
7-6	The Definite Integral	184
7-7	Some Preliminary Theorems on Integrals	188
7-8	Fundamental Theorem of Integral Calculus	191
7-9	Integration Continued	198
7-10	Summary	206
7-11	Techniques of Integration	211
7-12	Self Test	213

Contents

Chapter 8. Applications of the Integral — 214

- 8-1 Motion — 214
- 8-2 On Setting Up Problems — 219
- 8-3 Further Applications of Integration — 227
- 8-4 Self Test — 237

Chapter 9. Logarithmic and Exponential Functions — 238

- 9-1 $\int \dfrac{dx}{x}$, ln x, and e — 238
- 9-2 Use of Tables of ln x — 247
- 9-3 The Inverse Function of $y = \ln x$, Namely, $y = e^x$ — 250
- 9-4 Self Test — 253

Chapter 10. Trigonometric Functions — 255

- 10-1 Trigonometric Definitions — 255
- 10-2 Limits of Trigonometric Functions — 260
- 10-3 Derivatives of cos u and sin u — 262
- 10-4 Derivatives of Other Trigonometric Functions — 268
- 10-5 Integration of Trigonometric Functions — 268
- 10-6 Self Test — 274

Chapter 11. Techniques of Integration — 276

- 11-1 Techniques — 276
- 11-2 Integration by Parts — 276
- 11-3 Trigonometric Substitution — 280
- 11-4 Self Test — 287

Chapter 12. Epilogue — 288

Reading List — 291

Answers and Hints — 303

List of Symbols — 355

Index — 357

Table of Integrals *Inside Front Cover*

1

Basic Algebraic Theory

1-1. Introduction

This book is designed for a mature reader who wishes a concise course in the fundamental ideas of the calculus. The applications and examples have been designed, specifically, to meet his needs. In addition to the calculus, certain other mathematical concepts which are vital to modern science and mathematics will be discussed. Preliminary concepts are presented in Chap. 1 for the benefit of students who may be unprepared in essential portions of elementary and intermediate algebra. Chapter 2 provides a similar preparation in analytic geometry. Those who feel familiar with these fundamentals may read these chapters rapidly but should not skip them entirely.

1-2. Functions and Relations

function

Possibly the most fundamental mathematical notion is that of a *function*. In brief, the statement "y is a function of x" means that if a suitable value of x is given, then one corresponding value of y is determined in some fashion. The word *function* is used to describe the *correspondence*.

Let X be a set of objects (numbers, points, ideas, or any other specific set). Let Y also be a set of objects. The objects in set Y need not be similar to objects in set X. For example, set X might consist of all the girls living in dormitories at Bryn Mawr College

1

Basic Algebraic Theory

this semester, and set Y might consist of the 10 telephone numbers (51000, 59142, 59143, 59138, 51473, 59176, 52801, 59145, 53544, 59158).

To say a functional relationship exists from set X into set Y means that for each element x in the set X (written $x \in X$),† a corresponding

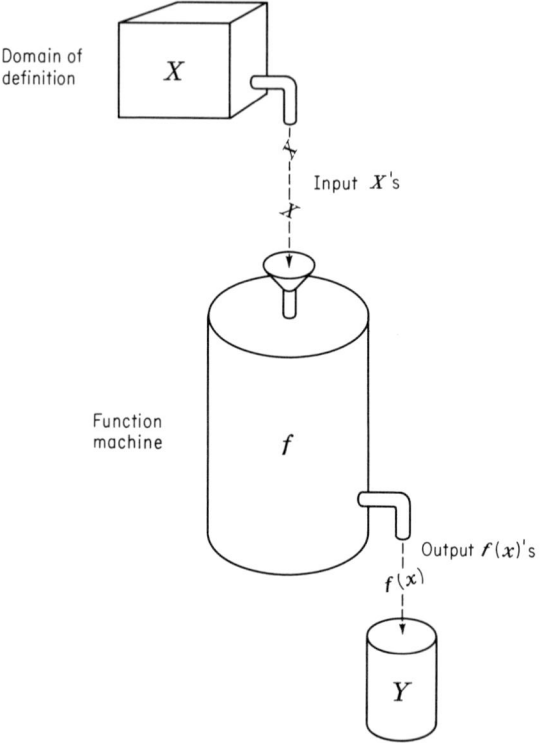

Figure 1-1

value $f(x)$ from the set Y (written $f(x) \in Y$) is determined. In more detail, a function consists of three things (see Fig. 1-1).

1. A set X of elements. This set may be called the *domain of definition* of the function.

function

2. A set Y of elements.
3. A rule f which assigns to each element x of X a unique element $f(x)$ of Y.

† The symbol ϵ is used in set theory. The statement $a \in S$ means that S is a set of elements, one of which is the specific element a. It is common practice to use small letters to represent elements and capital letters to represent a set of elements: thus $x \in X$.

Sec. 1-2. Functions and Relations

Another way of stating this is that a function is a set of *ordered pairs* $[x, f(x)]$ such that whenever two sets of pairs have the same first element they also have the same second element.

Note that neither the elements of X nor of Y need be numbers.

Illustration

Let X be the set of girls who attended the last major dance at your school and let Y be the set of dominant colors of their dresses. For each girl x in X there is a corresponding dominant dress color $f(x)$ in Y. Some possible examples are $f(\text{Suzanne}) = \text{white}$, $f(\text{Josie}) = \text{blue}$, $f(\text{Phyllis}) = \text{red}$, and $f(\text{Lois}) = \text{white}$.

If X is the collection of all names in the University of Oklahoma telephone directory and Y is the collection of telephone numbers with f being the obvious correspondence, then some of the busier offices have several telephone numbers which correspond to them. For example, if $x = \text{Psychology Department}$, then $f(x) = \frac{318}{319}$. The

relation word *relation* is currently used to describe such a "multiple-valued function."

The most serious difficulty students have is the feeling that this definition says more than it does. It is quite possible, for example, that the same element of Y may correspond to several elements of X (that is, change in x does not necessarily imply a change in y). There is no hint that the correspondence needs to be determined by an equation or formula. Indeed, none need exist. If $f(x)$ is given, there need not be a method of determining the x to which the $f(x)$ corresponds. All that the function concept states is that any

domain of time an object x is selected from a given set of objects X (the *domain*
definition *of definition* of the function), then a corresponding object $f(x)$ or y from a set of objects Y is determined. The *subset* of Y (consisting of those values actually assumed) is sometimes known as the *range* of the function, but we shall not make this distinction and shall refer to Y as the range. A function, therefore, is a matching of the objects in set X with objects in set Y such that every element in X is represented.

Letters such as p, g, or L may be used to represent the function in place of f, for example, $p(x)$, $g(x)$, or $L(x)$.

Example 1

Let the domain of definition X consist of those names listed in the current Los Angeles telephone directory. Let the range Y be the

Basic Algebraic Theory

telephone numbers listed in the same book. Let the function f be the correspondence (pairing) of the x's (names) and $f(x)$'s (telephone numbers) given in this book. For any name x in X, there is a corresponding telephone number $f(x)$ in Y, determined by the function f. It is consistent to use the notation f(John Magee) to denote the telephone number which the function f associates with the name John Magee.

Example 2

Let X be the set of integers (whole numbers) given below:

$$X = \{x\} = \{4, 5, 6, 7, 8, 9, 10, 11\}.$$

Y will also consist of a set of integers, although neither you nor the author knows exactly what integers. The value $B(x)$ of the function B is the number of scheduled buses which left the main New York City bus terminal in the 35-minute period immediately following x P.M. on September 25, 1961. Thus $B(7)$ is the number of scheduled buses leaving that terminal between 7:00 and 7:35 P.M., on the given date. The correspondence (function) B is determined by consulting the dispatcher's log for the given day.

Example 3

Let X be the set of all possible lengths of radii, that is, the positive real numbers. Let Y be the set of all possible areas of circles, that is, the positive real numbers. Let g be the function consisting of the correspondence between the radius and the area of a circle. In this case there exists a *formula* which also determines the function, namely, $y = \pi x^2$ and if $x = 5$, the corresponding y is 25π. However, the presence of such a formula is in no way essential to the existence of the function, as is illustrated by Examples 1 and 2. When a simple formula exists, it simplifies the description of the function considerably and will be used.

1-3. More General Functions

relation

Single-valued functions of *one variable* have exactly *one* $f(x)$ value which is determined for each permissible x value. The phrase "relation" or "multiple-valued function" is used to describe a correspondence in which a given x (input) may determine more than one corresponding $f(x)$ value (output).

Sec. 1-4. The Function Notation

Example 1

The relation C determined by $x^2 + y^2 = 25$ is multiple-valued since some of the permissible x values determine more than one corresponding y value. $C(3) = +4$ or -4, for example.

It is also possible to have a function of several variables in which the x's which are fed into the function "machine" are composites of several variables. The parcel post rate chart provides a familiar illustration of such a function. Both weight and zone must be given to determine the postage.

Example 2

For a given weight W and destination d (from the permissible values of W and d) a postal rate chart gives the corresponding cost of sending a package of weight W from Nashville to destination d for the given values of W and d. It is convenient to think of the x's in this case as consisting of pairs of values (W,d). The $f(x)$'s are the corresponding costs.

A Word of Caution: Note that there is nothing in the definition of function which asserts it is possible to "reason backward." If x is given, $f(x)$ is determined, but it does *not* say that if $f(x)$ is given, then x can be determined. If I tell you that the cost of sending a package is $1.36, it is not possible for you to tell me its weight or its destination.

1-4. The Function Notation

The notation bz means "the quantity b multiplied by the quantity z." Similar meanings are attached to combinations such as Ay, $x(x + 7)$, etc. However, by special agreement, when a scientist, a mathematician, or an engineer writes the Greek letter Δ (delta) next to a letter, as for example Δx, it does *not* mean Δ times that letter x. The symbol Δx (read "delta x") should be thought of as *one single element*, not as a product. In $\frac{\Delta y}{\Delta x}$ we may *not* divide out the deltas. The notation may seem peculiar at this point, but the powerful mathematical tools developed later using this notation will make you appreciate the need for its introduction. At present, just remember that Δx, Δy, Δt, and Δz are elements, not products. The symbol Δx is used to indicate a change in the value of x. For example, if x changes from 13 to 15, then $\Delta x = 2$. More generally, if the initial x value is $x = a$ and the terminal value is $x = b$, then the change in x is $\Delta x = b - a$.

The functional notations $f(x)$, $g(x)$, and $h(t)$ (read "f of x, g of x, and h of t") may also look like products but are not used as such. They represent the element $f(x)$ of Y which the correspondence (function) f associates with a given value of x in X. If there exists a formula which determines the function, such as $f(x) = (4x+1)(x-2)$ or $f(x) = 4x^2 - 7x - 2$, then $f(3)$ is determined by simply substituting 3 for x in the formula and combining terms to obtain $f(3) = (13) \cdot (1) = 13$ or $f(3) = 4(3^2) - 7(3) - 2 = 13$. If the function f is determined by means of a table, then $f(3)$ is found from the table.

Example 1

If $f(t) = t^2 - 7t + 2$, then $f(x)$ represents the same formula with t replaced by x. Thus $f(x) = x^2 - 7x + 2$, while

$$f\left(\frac{1}{y}\right) = \left(\frac{1}{y}\right)^2 - 7\left(\frac{1}{y}\right) + 2 = \frac{1 - 7y + 2y^2}{y^2}.$$

The symbol in parentheses need not be a letter; for example,

$$f(3) = (3)^2 - 7(3) + 2 = -10,$$
$$f(8) = (8)^2 - 7(8) + 2 = 10,$$
$$f(x+3) = (x+3)^2 - 7(x+3) + 2 = x^2 - x - 10,$$
$$f(y+\Delta y) = (y+\Delta y)^2 - 7(y+\Delta y) + 2 = y^2 + 2y\,\Delta y$$
$$+ (\Delta y)^2 - 7y - 7\Delta y + 2.$$

Example 2

Using $f(t) = t^2 - 7t + 2$ as before, find $f(y + \Delta y) - f(y)$.

$$f(y + \Delta y) - f(y) = [(y + \Delta y)^2 - 7(y + \Delta y) + 2]$$
$$- (y^2 - 7y + 2)$$
$$= 2y\,\Delta y + (\Delta y)^2 - 7\Delta y.$$

Example 3

If $f(t) = t/(t-3)$, determine $f(2)$, $f(2 + \Delta y) - f(2)$, and $f(2 + .01) - f(2)$. Clearly,

Sec. 1-4. The Function Notation

$$f(2) = \frac{2}{2-3} = -2.$$

$$f(2 + \Delta y) - f(2) = \frac{2 + \Delta y}{2 + \Delta y - 3} - \frac{2}{2-3}$$

$$= \frac{2 + \Delta y}{\Delta y - 1} + 2 = \frac{2 + \Delta y + 2\Delta y - 2}{\Delta y - 1}$$

$$= \frac{3\Delta y}{\Delta y - 1}.$$

The reader should note that both $\dfrac{3\Delta y}{\Delta y - 1}$ and $f(2 + \Delta y)$ are meaningless if $\Delta y = 1$. In a similar fashion, one obtains

$$f(2 + 0.01) - f(2) = \frac{3(0.01)}{0.01 - 1} = \frac{0.03}{-0.99} = \frac{-1}{33},$$

or approximately $f(2 + 0.01) = -0.03$. This may be obtained by setting $\Delta y = 0.01$ in $f(2 + \Delta y) - f(2)$, which was just computed. Before going further, show that if $g(z) = 3z^2 - 2z + 5$, then $g(5) = 70$ and $g(5 + \Delta x) - g(5) = 28\Delta x + 3(\Delta x)^2$.

Example 4

If $f(x) = 3/x$, find $f(x + \Delta x) - f(x)$.

$$f(x + \Delta x) - f(x) = \frac{3}{x + \Delta x} - \frac{3}{x} = \frac{3}{(x + \Delta x)} \cdot \frac{x}{x} - \frac{3}{x} \cdot \frac{(x + \Delta x)}{(x + \Delta x)}$$

$$= \frac{3x - 3x - 3\Delta x}{x(x + \Delta x)} = \frac{-3\Delta x}{x(x + \Delta x)}$$

Example 5

If $g(t) = 5/t^2$, find $\dfrac{g(t + \Delta t) - g(t)}{\Delta t}$.

$$\frac{g(t + \Delta t) - g(t)}{\Delta t} = \frac{\dfrac{5}{(t + \Delta t)^2} - \dfrac{5}{t^2}}{\Delta t} = \frac{5}{(t + \Delta t)^2 \Delta t} - \frac{5}{(t)^2 \Delta t}$$

$$= \cdots$$

$$= \frac{5t^2 - 5[t^2 + 2t\,\Delta t + (\Delta t)^2]}{(t + \Delta t)^2 t^2 \Delta t}$$

$$= \frac{-5(2t + \Delta t)\,\Delta t}{(t + \Delta t)^2 t^2 \Delta t} = \frac{-5(2t + \Delta t)}{(t + \Delta t)^2 t^2} \quad \text{if}$$

$$\Delta t \neq 0.\dagger$$

† The symbols $\Delta t \neq 0$ means "Δt is not equal to zero."

The division of numerator and denominator by Δt in the last step is valid only if Δt is not zero. If Δt is zero, the entire problem is without meaning since *division by zero is not defined*. (Why not? See Sec. 1-10 if in doubt.)

The expression $\dfrac{g(t + \Delta t) - g(t)}{\Delta t}$ is of great importance in the study of calculus, no matter what the function g happens to be (see Sec. 1-13).

1-5. Inequalities

Although the symbol $\neq$ means "not equal to," in some cases we may wish to say more than $x \neq 5$; we may wish to say "x is less than five." This is written $x < 5$. The statement "x is greater than minus one" can be written $x > -1$. The expression $z > t$ is read "z is greater than t." To include minus one as a possible value of x in the statement "x is greater than or equal to minus one," we write $x \geq -1$. The small ends of the symbols $<$ and $>$ point to the smaller quantity.

The symbols $-3 < t \leq 2$ indicate that t must lie between -3 and 2, including $t = 2$, but $t \neq -3$. The value of t must satisfy *both* of the relations $-3 < t$ and $t \leq 2$. The symbol $x > 5, x \leq 1$ indicates that *either* $x > 5$ or $x \leq 1$; x cannot satisfy both relations.

1-6. Absolute Value

The symbol $|b|$ is read "the absolute value of b." It refers to the numerical value of b without regard to its sign. The absolute value of a negative number $-b$ is the corresponding positive number $+b$. Using the notation of Sec. 1-5, we may say all this by writing, "If $b \geq 0$, $|b| = b$ and $|-b| = b$." In more general terms, $|b|$ is the distance *between* the point 0 and the point b without regard to sign. This latter concept generalizes to complex numbers, $x + iy$.

Clearly $|x| \geq 0$ for all x. Examples: $|3| = 3$, $|0| = 0$, $|-7| = 7$, $|\sqrt{5} + 2| = \sqrt{5} + 2$, and $|\sqrt{5} - 2| = \sqrt{5} - 2$, but $|-\sqrt{5} + 2| = |-(\sqrt{5} - 2)|$. (Why?)

The symbol $\sqrt{5}$ designates "the positive square root of five." If the negative root is desired, it must be indicated as $-\sqrt{5}$.

Sec. 1-6. Absolute Value

Hence $\sqrt{4} = 2$, not ± 2. Therefore $(\sqrt{5} - 2) > 0$, while $(-\sqrt{5} + 2) < 0$.

Problem Set 1-6

1. If $f(x) = x^2 - 4$, find $f(2), f(-7), f(1+x)$, and $f(2 + \Delta x)$.
2. If $h(x) = 3/x^2$, find $h(-5), h(2 + \Delta x)$, and $h(x + \Delta x)$.
3. Let A be the area of a circle and d its diameter. Then $A = f(d)$. Obtain a formula for $f(d)$.
4. Is the area of a circle a function of the circle's circumference? If possible, obtain a formula for the area of a circle in terms of its circumference.
5. Make up a single-valued function $y = f(x)$ in which it is impossible for you to find a unique x if y is given.
6. If $g(t) = (t^2 + 4)/3t$, find $g(2), g(3.1) - g(3)$, and also determine $g(x + \Delta x) - g(x)$.
7. Using $f(x) = (x^2/3) - x$ and $g(t)$ of Prob. 6, determine $f(7) - g(3)$ and $\dfrac{f(3)}{g(2) + 1}$.
8. Is it possible to express your age in years as a function of the number of months since you were five months old? State the domain of definition of this function.
9. Express your age, in years, as a function of the number of months since you were 15 months old. State the domain of definition of this function.
10. If $g(t) = 7/t^2$, find $\dfrac{g(x + \Delta x) - g(x)}{\Delta x}$. This expression is of considerable importance in the study of calculus.
11. Find $\dfrac{y(t + \Delta t) - y(t)}{\Delta t}$ where $y(t) = 3g(t) + 1$ for the $g(t)$ of Prob. 10.
12. If $g(x) = \dfrac{x}{x^2 + 1}$, find $\dfrac{g(y + \Delta y) - g(y)}{\Delta y}$.
13. If $p(x) = \dfrac{x^2 - 7}{3}$, find $\dfrac{p(t + \Delta t) - p(t)}{\Delta t}$.
14. If $q(x) = p(x) + g(x)$ for $p(x)$ of Prob. 13 and $g(x)$ of Prob. 10, find $q(2)$ and $q(x + 1) - q(x)$.
15. Is the total amount of rain which has fallen on your campus since the college was founded a function of the date? What is the

present domain of definition? What will the domain of definition be next year at this time?

16. Is the number of teeth T which a person has a function of his age? Is the function the same for all people? Is a person's age a function of the number of teeth T he has?

17. Make up a function which is not represented by a formula. State the domain of definition and the range.

18. It is quite possible to have a function of several variables. In some theaters the cost of admission is a function of several variables, $C = F(s, t, p, a, T)$ where s is the seat occupied, t is the time of attendance, p is the play or picture given, a is the age of the patron, and T is the amount of government tax imposed. In spite of the apparent complexity of this function, many people find it possible to compute the value of C with little effort. Write out a similar function for one of your local movie houses.

1-7. Two Basic Properties

Two basic properties of algebra are:

two basic properties

I. A product is zero if and only if at least one of its factors is zero. $a \cdot b = 0$ *if* and *only if* $a = 0$ or $b = 0$ or both.

II. A quotient is zero if and only if its numerator is zero and its denominator is not zero. $a/b = 0$ *if* and *only if* $a = 0$ and $b \neq 0$. If $b = 0$, the expression is undefined.

Simple as these two rules are, they form the heart of all equation solving in algebras based on the rational, real, or complex number systems. Every algebra classroom should have these two statements emblazoned on its walls; much algebraic folly would be avoided by the proper understanding and use of these two statements.

1-8. The Structure of the Number System

Before continuing, it seems desirable to examine briefly the numbers with which we shall work. Actually there is a hierarchy of number systems. One of the simplest systems is that of the integers (. . . , −6, −5, −4, −3, −2, −1, 0, 1, 2, 3, 4, 5, 6, 7, 8, 9, 10, 11, . . .). In this restricted system even such simple equations as

Sec. 1-8. The Structure of the Number System

$2x + 1 = 0$ and $5x = 3$ have no solutions, since neither $-\frac{1}{2}$ nor $\frac{3}{5}$ is an integer.

rational numbers
It is usual to extend this system to include those positive and negative fractions whose numerators and denominators are integers. The enlarged system is called the ***rational numbers***, since they are ***ratios*** of integers. This is the number system of commerce. It includes the positive and negative whole numbers and fractions (including decimal fractions). It suffices very well for ordinary trade, but even the ancient Greeks knew it was inadequate to express the length of the hypotenuse of a right triangle whose legs are each one unit long: it is impossible to solve $x^2 = 2$ in the rational number system. It is also impossible, using only rational numbers, to state the circumference of a circle of radius 1.

real numbers
The next step is to extend the rational number system to the real number system. The *real numbers* include numbers such as $\sqrt{2}\,\pi$, and $(13 + 4\sqrt[3]{17} - 3\pi\sqrt{5})$, as well as all the rational numbers. In general, a real number represents a distance or its negative. If a line (number axis) is given a zero point and a positive direction, then the directed distance from the zero point to any other point on the line is a real number. Furthermore, to every real number there corresponds one, and only one, point on the number axis. The real numbers are the measuring numbers. Calculus uses the real number system.

complex numbers
The equation $x^2 + 1 = 0$ has no real solution. The real number system is inadequate for today's science. Nevertheless, for more than a thousand years, the real numbers were considered adequate for the needs of mankind. It was not until about 1885 that Kronecker, using the complex plane of C. Wessel (1797), opened the door to a true understanding of the complex numbers. Much of the scientific progress of the last 75 years has been based on this concept. In the system of complex numbers ($a + bi$; a, b real and $i^2 = -1$), every polynomial equation with complex (or real) coefficients has as many solutions as its degree. The integers, rational numbers, and real number systems are each special subsets of the complex number system.

quaternions
The hierarchy of number systems does not stop with the complex numbers. They are a special case of the quaternions, upon which the modern "quantum electron spin" is based. The quaternions are themselves a very special case of one of the systems of matrices.

matrices   Even the matrices are not the final link in this chain, but that is another story for another course. Let it suffice to say that matrices play a vital role in psychology, aircraft design, quantum mechanics, chemical structure theory, game theory, statistics, economics, social science and, in fact, in the advanced stages of every science and near science known today. Nevertheless, for our purposes, we shall limit our attention primarily to the calculus of the real number system. It may occasionally be necessary to obtain the complex roots of an equation, merely to be sure that all the real roots have been discovered.

1-9. Two Unusual Real Numbers

There are two real numbers which have such important properties that they are worthy of special note. These numbers are one and zero.

The number one has the peculiar multiplicative property that

$b \cdot 1 = b$ for each real number b.

The number zero has a corresponding additive property:

$b + 0 = b$ for each real number b.

If a postulational approach for the development of the real numbers (or for the integers) were given, these two properties would be part of the postulates. The following property could then be derived as a theorem:

$0 \cdot b = 0$ for each real number b

as could

$(-1) \cdot (-1) = +1.$

We do not do this here, but feel that the reader should realize that these properties are theorems, derived from postulates and not obtained through any mystic insight.

1-10. The Meaning of Division

The meaning of the fraction N/D is best expressed as the *unique* solution of the equation $D \cdot x = N$. This simple but important

Sec. 1-10. The Meaning of Division

concept is often not made plain in beginning courses, with considerable confusion concerning "division by zero" being the result.

division

N/D is the unique solution of $D \cdot x = N$.

If $\frac{5}{0}$ is to be meaningful, it must be the unique solution of $0 \cdot x = 5$.

$\frac{5}{0}$ *meaningless*

However, $0 \cdot x = 5$ has no solution and therefore $\frac{5}{0}$ is meaningless.

$\frac{0}{0}$ *indeterminate*

If $\frac{0}{0}$ is meaningful, it must be the *unique* solution of $0 \cdot x = 0$. However, $0 \cdot x = 0$ is satisfied by *every* real value of x and no *unique* solution is determined. To distinguish the case represented by $\frac{5}{0}$ from that represented by $\frac{0}{0}$, the latter may be called *indeterminate*, although both $\frac{5}{0}$ and $\frac{0}{0}$ are without meaning.

$\frac{0}{5} = 0$

On the other hand, $\frac{0}{5}$ is the unique solution of the equation $5 \cdot x = 0$. Hence, by basic property I (Sec. 1-7), it follows that since $5 \neq 0$ and $5 \cdot x = 0$, x must equal zero. 0 is a solution of $5 \cdot x = 0$ and is a unique solution. Therefore, $\frac{0}{5} = 0$, a perfectly valid number.

Summary: $\frac{7}{0}$ is meaningless, $\frac{0}{0}$ is indeterminant, and $\frac{0}{7} = 0$. No mystic discussion of dividing zero apples among seven people is involved. Furthermore, the symbol ∞ is not used or *mis*used.

Problem Set 1-10

Use the two basic properties given in Sec. 1-7 to determine all *real* values of the variable which will make the expression given in each of Probs. 1 to 7 zero. Do *not* "solve equations" by some mysterious technique. Instead, apply the basic properties and elementary logic. You may also use the facts that $x^2 \geq 0$ for all real x and that the equation $x + 2 = 0$ has only one solution.

1. $(x - 7)(x + 5)$.

2. $\dfrac{x - 3}{x^2 + 4}$.

3. $\dfrac{x(x - 8)}{x^3 + 6}$.

4. $\dfrac{4}{x^2 + 7x + 10}$.

5. $\dfrac{7x - 3}{5x^2 + 31}$.

6. $\dfrac{x^2 + 1}{x - 3}$.

7. $\dfrac{x(x^2 + 4)}{2x - 3}$.

8. The distance s feet that a body falls in time t sec is given approximately as $s = 16t^2$.
(a) Find the change in s corresponding to a change Δt in the time t.
(b) Find the average rate of change in distance s with respect to the time t, as t changes from t to $t + \Delta t$; that is, find $\dfrac{s(t + \Delta t) - s(t)}{\Delta t}$.

Basic Algebraic Theory

(c) Apply the results of parts (a) and (b) when $t = 2$ and $\Delta t = 0.1$ so that $t + \Delta t = 2.1$.

9. If $S = -10t + 5$, find the average rate of change of S with respect to t as t changes from t to $t + \Delta t$; that is, find $\dfrac{s(t + \Delta t) - s(t)}{\Delta t}$.

10. Work Prob. 9 as t changes from 3 to 3.5 and from 6 to 6.5.

11. Show that $\dfrac{A}{B} \div \dfrac{C}{D} = \dfrac{A}{B} \cdot \dfrac{D}{C}$ if $BCD \neq 0$. HINT: $\dfrac{A/B}{C/D}$ may be multiplied above and below by BD. Discuss the restriction $BCD \neq 0$.

12. The acceleration of protons in a betatron (a type of atom smasher) gives remarkably high velocities in a short time. Speeds in excess of 99 per cent of the speed of light are possible in existing machines. If $v(t)$ is the velocity in miles per hour and t is time in seconds, where $v(t) = (100^{t/20})1.8$ for t between 0 and 45 sec, determine the velocity at $t = 30$ sec.

13.† The magnetic field H (in oersteds) of a solenoid containing N turns is given by the formula $H = 0.4\pi NI$, where I is the current in amperes. If a current of 2 amp produces a field of 300 oersteds, how many turns does the solenoid contain? Since there is some error in the meters used to measure H and I, if you had purchased the solenoid believing it contained 125 turns, would you feel justified in returning it?

14. If n cells of E volts and r ohms are connected in series to make a battery, a current of I amperes will be produced on a load of R ohms, where

$$I = \frac{nE}{R + nr}.$$

How many cells are needed in a battery which is to produce a current of 2 amp under a load of 5 ohms? Each cell produces 2.1 volts and has an internal resistance of 0.5 ohm.

1-11. Solution Set of an Equation

A number is said to satisfy an equation, or to be a *solution* of an equation, if and only if, upon substituting the number for the

---

† This book contains many problems involving the units of engineering, chemistry, and physics as well as those of economics, psychology, and finance. It will *not* be assumed that the student is familiar with these units. Each problem contains as much information about the units as is needed to work the problem.

Sec. 1-11. Solution Set of an Equation

variable in the equation, true arithmetical equality results. The word "root" is sometimes used instead of "solution." The set of all such solutions is called the *solution set* of the equation.

Example 1

Are 4 and 3 solutions of $P(x) = 4x^4 - 40x^3 + 95x^2 + 10x - 24 = 0$? Since $P(4) = 0$, 4 is a root of $P(x) = 0$. Since $P(3) \neq 0$, 3 is not a root.

identity An *identity* is an equation in which both members are equal for all values of the unknown for which both members are defined. For example, $\dfrac{x^2 - 4}{x^2 - x} = \dfrac{(x+2)(x-2)}{x(x-1)}$ is an identity, since both members are equal for all values of x except 0 and 1. Division by zero is not defined; consequently, neither member is defined for $x = 0$ nor for $x = 1$. (Why not?)

Equations which are not identities are called *conditional* equations, or simply equations. For example, $x^2 - 3x = 0$ is a conditional equation since 5, for example, does not satisfy the equation even though both members are defined. Can you find two solutions of $x^2 - 3x = 0$?

Two equations having the same solution set are called *equivalent*. In general, if one equation can be changed into the other by "permissible algebraic gyrations" such as adding like terms to both members, or multiplying both members by a *nonzero* constant, the equations are equivalent. This, however, is not a reasonable definition since we must define permissible algebraic gyrations as those which leave the solution set unaltered.

Example 2

The equations $x^2 + 4x - 7 = 2 + x$, $x^2 + 3x = 9$, and the equation $6 - x^2 = 3(x - 1)$ are equivalent equations. The inequalities $4x - 7 < 2x + 5$, $2x < 12$, $x < 6$, and $-x > -6$ are equivalent inequalities.†

A linear equation in one unknown is an equation involving only the first power of the unknown, say z, and constants. An example is $17z - 4 + 3z - 12 = 32z + 11$. The solution of this, or any other linear equation in one unknown, is obtained by arranging all

---

† The reader should recall that if an inequality is multiplied by a negative number, the inequality sign is reversed. For example, $3 < 5$, but $-3 > -5$.

Basic Algebraic Theory

terms involving the unknown z on one side of the equality sign and all other terms (constants) on the opposite side. This is accomplished, by addition, to obtain

$$17z + 3z - 32z = 11 + 4 + 12$$
$$-12z = 27.$$

Next, multiply both sides by the same number $(-\frac{1}{12})$ such that the coefficient of the unknown becomes 1:

$$z = \frac{27}{-12} = -\frac{9}{4}.$$

The reader should verify that $-\frac{9}{4}$ is a solution by direct substitution of $-\frac{9}{4}$ for z in the original equation.

Many quadratic equations, $Ax^2 + Bx + C = 0$, may be solved by factoring. The fundamental idea involved is that a product of factors is zero if and only if at least one of the factors is zero.†

Example 3

$$x^2 - 3x = 4.$$

An equivalent equation is $x^2 - 3x - 4 = 0$ or $(x - 4)(x + 1) = 0$. This is true if and only if either $(x - 4) = 0$ or $(x + 1) = 0$ (see Sec. 1-7). The two resulting linear equations may be solved to obtain $x = 4$ or $x = -1$. Both 4 and -1 will satisfy the original equation, as may be shown by direct substitution.

Example 4

$$2x^2 + 11x - 21 = 0$$
$$(2x - 3)(x + 7) = 0$$

Then either $(2x - 3) = 0$ or $(x + 7) = 0$.

Hence either $x = \dfrac{3}{2}$ or $x = -7$.

The equation $x^2 + 3x + 1 = 0$ cannot be solved directly by finding rational factors.‡ The quadratic formula $\dfrac{-B \pm \sqrt{B^2 - 4AC}}{2A}$

---

† The reader may be interested in knowing that there exist number systems in which a product of two elements may equal zero when neither element is zero. Although such systems have important applications, we shall not discuss them here.

‡ Note that we do *not* say that $x^2 + 3x + 1$ cannot be factored. Its factors are $x = \frac{3}{2} - \sqrt{5}/2$ and $x = \frac{3}{2} + \sqrt{5}/2$, but these factors are not rational since $\sqrt{5}$ cannot be expressed as a quotient of integers. They are perfectly valid *real* factors, but they are difficult to determine without first solving the given equation.

Sec. 1-11. Solution Set of an Equation

will be found helpful in this and similar cases. We derive this formula. A general quadratic equation, using literal coefficients, is $Ax^2 + Bx + C = 0$, with $A \neq 0$. Divide by A and subtract the constant term from both members.

$$x^2 + \frac{B}{A}x = -\frac{C}{A}.$$

Make the left member the square of $x + \frac{B}{2A}$ by adding $\frac{B^2}{4A^2}$ to each side.

$$x^2 + \frac{B}{A}x + \frac{B^2}{4A^2} = -\frac{C}{A} + \frac{B^2}{4A^2}$$

or

$$\left(x + \frac{B}{2A}\right)^2 = \frac{B^2 - 4AC}{4A^2}.$$

Find the square root of both sides.

$$x + \frac{B}{2A} = \pm\sqrt{\frac{B^2 - 4AC}{4A^2}}.$$

quadratic formula Whence $x = -\frac{B}{2A} \pm \frac{\sqrt{B^2 - 4AC}}{2A} = \frac{-B \pm \sqrt{B^2 - 4AC}}{2A}.$

All that has been shown is that *if* solutions exist, they must be either $\frac{-B + \sqrt{B^2 - 4AC}}{2A}$ or $\frac{-B - \sqrt{B^2 - 4AC}}{2A}$ or both. In the problems you will be asked to verify, by direct substitution into $Ax^2 + Bx + C = 0$, that each of these possible results is actually a solution (see Probs. 1, 2, and 3, Set 1-11). The solution, known as the *quadratic formula*, is useful enough to merit memorizing. Note that the formula applies only when the quadratic equation is written in the form $Ax^2 + Bx + C = 0$.

Example 5

Apply the quadratic formula to $x^2 + 3x + 1 = 0$.
Using $A = 1$, $B = 3$, $C = 1$, we obtain

$$x = \frac{-B \pm \sqrt{B^2 - 4AC}}{2A} = \frac{-3 \pm \sqrt{3^2 - 4 \cdot 1 \cdot 1}}{2 \cdot 1}$$

$$= \frac{-3 \pm \sqrt{5}}{2}.$$

Basic Algebraic Theory

Example 6

Solve $3x^2 + 7x + 2 = 0$. Using the quadratic formula with $A = 3$, $B = 7$, and $C = 2$, we obtain

$$x = \frac{-7 \pm \sqrt{49 - 4 \cdot 3 \cdot 2}}{2 \cdot 3} = \frac{-7 \pm \sqrt{25}}{6}$$

$$= \begin{cases} \dfrac{-7 + 5}{6} = \dfrac{-1}{3} \\ \text{or} \\ \dfrac{-7 - 5}{6} = -2. \end{cases}$$

This equation can be solved by factoring.

$$3x^2 + 7x + 2 = 0$$
$$(x + 2)(3x + 1) = 0$$
$$x + 2 = 0 \qquad 3x + 1 = 0$$
$$x = -2 \qquad x = -\tfrac{1}{3}.$$

Example 7

Solve $2z^2 - 5z + 4 = 0$. In this case $A = 2$, $B = -5$, $C = 4$. Hence

$$z = \frac{-(-5) \pm \sqrt{25 - 32}}{4} = \frac{5 \pm \sqrt{-7}}{4} = \frac{5 \pm i\sqrt{7}}{4}.$$

This equation has no solution in the set of real numbers. However, both $\tfrac{5}{4} + i\sqrt{7}/4$ and $\tfrac{5}{4} - i\sqrt{7}/4$ are complex numbers which satisfy the given equation. The reader should not think of the number i where $i^2 = -1$ as "a figment of the imagination" any more than he would the numbers $\sqrt{3}$, π, or $277/21{,}638$. Complex numbers are used in solving extremely practical problems in electrical circuit theory, television, vectors, and the study of atomic energy. For the present, consider the number i as another number like $\sqrt{5}$, which was introduced to enable you to solve equations. We can find a decimal approximation for $\sqrt{5}$ in tables but must wait for further interpretations of the number i.

Example 8

Solve for x in terms of y: $2x^2 - 3xy + 5y^2 - x + 7y - 6 = 0$. Write the equation in the form $Ax^2 + Bx + C = 0$.

$$2x^2 + (-3y - 1)x + (5y^2 + 7y - 6) = 0.$$

Sec. 1-11. Solution Set of an Equation

Then $A = 2 \quad B = -3y - 1 \quad C = 5y^2 + 7y - 6$

and $x = \dfrac{-(-3y - 1) \pm \sqrt{(-3y - 1)^2 - 4(2)(5y^2 + 7y - 6)}}{2 \cdot 2}$

$= \dfrac{3y + 1 \pm \sqrt{-31y^2 - 50y + 49}}{4}.$

Problem Set 1-11

1. Show by direct substitution that $\dfrac{-B + \sqrt{B^2 - 4AC}}{2A}$ satisfies the equation $Ax^2 + Bx + C = 0$.

2. Show by direct substitution that $\dfrac{-B - \sqrt{B^2 - 4AC}}{2A}$ satisfies the equation $Ax^2 + Bx + C = 0$.

3. Write a quadratic equation having roots $\dfrac{-B + \sqrt{B^2 - 4AC}}{2A}$ and $\dfrac{-B - \sqrt{B^2 - 4AC}}{2A}$.
Is this equation equivalent to $Ax^2 + Bx + C = 0$?

4. Solve for x: $3x^2 - 5x - 12 = 0$.

5. Solve for w: $7w^2 + 54w - 16 = 0$.

6. Show how the properties of Sec. 1-7 are used in the derivation of the quadratic formula.

7. Solve for t: $\dfrac{t^2 - 7t + 5}{t^4 + 8} = 0$. Use Sec. 1-7.

8. Solve for t: $1.4t^2 - 3.2t - 11 = 0$.

9. Solve for x in terms of z: $3x^2 - 4zx + 2x - z + 3z - 1 = 0$.

10. If $g(t) = t^2 - 7t + 6$, can you find a value of t such that $g(t) = 0$? Is it unique?

11. Determine z such that $g(z) = 5$ for the g of Prob. 10.

12. Solve for z: $3z^2 - 5iz + 2z + 3 - 2i = 0$, where $i^2 = -1$.

13. If $h(b) = b^2 + 5b - 7$, can you find values of b such that $h(b) = 3$? How many? What are they?

14. Is it possible to write a quadratic equation having equal complex roots? Would it be possible if the coefficients were restricted to real numbers?

15. Solve for x: $(x - 2)(x + 1) = 4$.

16. Discuss the fallacy of the following attempted solution of Prob. 15.

$(x - 2) = 2 \quad$ or $\quad (x + 1) = 2.$
Then $x = 4 \quad$ or $\quad x = 1.$

Show that neither 4 nor 1 is a root. Why is the procedure invalid?

17. An electrical circuit contains three resistances in parallel. If the three resistances are R_1, R_2, and R_3, then the total resistance is R where $\frac{1}{R_1} + \frac{1}{R_2} + \frac{1}{R_3} = \frac{1}{R}$. Determine R when $R_1 = 25$ ohms, $R_2 = 15$ ohms, and $R_3 = 42$ ohms.

18. The same as Prob. 17 where $R_1 = 10$ ohms, $R_2 = 60$ ohms, and $R_3 = 35$ ohms.

19. The length of a piece of tin is 3 in. longer than its width. The piece of tin costs $1.08 at 1 cent per square inch. Determine the dimensions of the piece of tin. Be sure to indicate the *units* of your answer on this and other verbal problems.

20. Express each of the following functions in a manner such that the quadratic expression in the denominator is a sum or a difference of squares, $A^2 \pm B^2$. For example, $\frac{3x + 2}{4x^2 + 3x - 1}$ is equal to

$$\frac{3x + 2}{(4x^2 + 3x + \frac{9}{16}) + (-\frac{9}{16} - 1)} = \frac{3x + 2}{(2x + \frac{3}{4})^2 - (\frac{5}{4})^2},$$

which is in the form requested. This technique is essential before tables can be used in certain calculus problems.

(a) $\dfrac{5x + 2}{4x^2 + 5x - 6}$. (b) $\dfrac{3}{\sqrt{6x - 2x^2}}$. (c) $\dfrac{3x + 1}{x^2 + 2x + 3}$.

21. A useful method of "completing a square," of importance in certain calculus problems, is shown in this example. The problem is to make the denominator of the fraction into a sum or difference of two squares. We begin by multiplying numerator and denominator by $4A$, where A is the coefficient of x^2 in the denominator, and then subtract 9 to make a perfect square $64x^2 + 48x + 9 = (8x + 3)^2$.

$$\frac{3x + 2}{4x^2 + 3x - 1} = \frac{4 \cdot 4(3x + 2)}{4 \cdot 4(4x^2 + 3x - 1)}$$
$$= \frac{48x + 32}{64x^2 + 48x + (9 - 9) - 16}$$
$$= \frac{48x + 32}{(8x + 3)^2 - (5)^2}.$$

Use this method of completing the square to express the quadratic functions as the sum or difference of two squares.

Sec. 1-11. Solution Set of an Equation

(a) $\dfrac{5x + 3}{x^2 + x + 1}$.

(b) $\dfrac{13t}{9t^2 - 12t + 20}$.

(c) $\dfrac{5}{2x - 3x^2}$.

(d) $\dfrac{49}{ax^2 + bx + c}$.

22. Prove that the sum Σ (Greek capital letter sigma) of the two roots of $Ax^2 + Bx + C = 0$ is $\Sigma = -B/A$.

23. Prove that the product Π (Greek capital letter pi) of the two roots of the above equation is $\Pi = C/A$. It is quite customary in science to use the Greek Σ and Π in the roles indicated in Probs. 22 and 23.

24. Problems 22 and 23 provide a quick check for the correctness of the roots of a quadratic equation. Use them to check some of the problems in this list.

25. Show that if $|k| < 10$, then $2t^2 - kt + 18 = 0$ has only complex roots. Can you set a larger bound on $|k|$ such that the roots remain complex?

26. Solve for t: $t^4 - 13t^2 + 36 = 0$. If you have trouble factoring this, you might let $t^2 = x$ and solve the resulting quadratic equation in x.

27. Solve for t: $2t^4 + 5t^2 - 12 = 0$ (see suggestion in Prob. 26).

28. Solve for z: $(z^4 - 4z^3 + 4z^2) - 22(z^2 - 2z) + 120 = 0$. HINT: let $x = z^2 - 2z$, then $x^2 = z^4 - 4z^3 + 4z^2$.

29. Solve for x: $3.6x^2 - 2.1x - 8.5 = 0$. Equations with decimal coefficients often appear in applied problems. The quadratic formula is almost always used to determine solutions.

30. Figure 1-2 shows an electric circuit containing a resistance of $R = 10$ ohms, an inductance coil of $L = 2$ henrys, and a condenser of capacitance $C = 1/10{,}050$ farads. To determine the current in the circuit, we must solve the equation $Lm^2 + Rm + 1/C = 0$ for m. Using the given values of L, R, and C, determine values of m. The complex value of m has an important interpretation in electrical theory. It is common practice among electrical technicians to use j rather than i to represent the number whose square is -1. Express your answer in this form.

31. Same as Prob. 30 for $R = 3$ ohms, $L = 2$ henrys, and $C = \frac{1}{200}$ farads.

32. For what integers (whole numbers) x is $3x^2 + 2x - 7 < 0$?

33. A ball bearing rolls down a track a distance s feet in t seconds where $s = 4t + t^2/2$. How long does it take the ball bearing to travel 24 ft? Obtain a general formula expressing t as a function of s.

34. A circuit contains two resistances in parallel, one of 20 ohms and one of 50 ohms. How large should a third resistance be if it is also to be connected in parallel with the other two and if the total resistance is to be 6 ohms? If resistances are available in multiples of 5 ohms only, what size should be used to come as near 6 ohms as possible (see Prob. 17)?

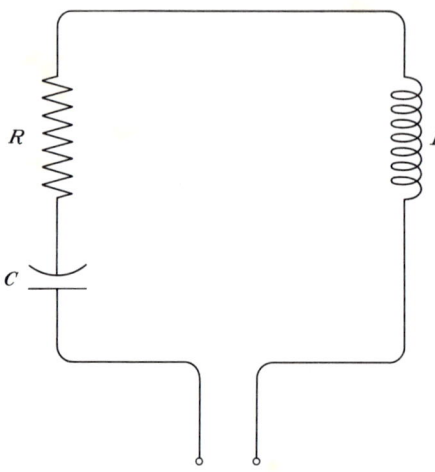

Figure 1-2

35. Calvin Butterball and Phoebe Small went to an amusement park on Calvin's day off. Calvin spent $7.65 plus 80 cents carfare on the date. Phoebe packed an excellent lunch. Calvin's new job in a nice air-conditioned onion packing plant pays 90 cents per hour plus 10 cents bonus for each box packed. Calvin averages 36 boxes per 8-hr day. Calvin is paid time and a half for overtime both on his base pay and on his bonus. If Calvin must finish packing a box once he starts it (that is, no fractional boxes permitted) how many hours overtime must he work to pay for the date?

1-12. Auxiliary Equations

The equations (1) $\dfrac{x^2 + 2x}{x - 3} = \dfrac{18 - x}{x - 3}$ and (2) $x^2 + 2x = 18 - x$ are *not* equivalent equations. (Why not?) Equation (2) is called an

Sec. 1-12. Auxiliary Equations

auxiliary equation

auxiliary equation to (1) since it is useful in determining the solutions of (1). Every root of (1) is also a root of (2), but not conversely. (What does this mean?) If the auxiliary equation can be solved, its roots may be checked to see which, if any, also satisfy the original equation. The solutions of $x^2 + 2x = 18 - x$ are $x = 3$ and $x = -6$. The first of these is *not* a root of Eq. (1) since the resulting members are undefined. (Why?) The value $x = -6$ is a root, and therefore is the only root of Eq. (1). In general, if the auxiliary equation is obtained by multiplying the original equation by an expression involving the unknown, the roots of the auxiliary equation which do *not* make that expression equal zero will be roots of the original equation.

Auxiliary equations may be obtained by squaring both members of an equation. If two numbers are equal, their squares are also equal. However, it is possible for the squares of unequal numbers to be equal: $-2 \neq 2$, but $(-2)^2 = (2)^2$. Thus a root of an auxiliary equation obtained by squaring may not be a root of the original equation. If the original equation has roots, they must be included among the roots of the auxiliary equation.

Example 1

Solve for x: $\sqrt{x - 2} = 3$.

$(\sqrt{x - 2})^2 = 9$ auxiliary equation
$x - 2 = 9$ $x = 11$.

The only root of the auxiliary equation is 11. Hence, 11 is the only possible root of the equation $\sqrt{x - 2} = 3$. Upon substitution, $\sqrt{11 - 2} = \sqrt{9} = 3$, we find that it is a root.

Example 2

Solve for x: $2x - \sqrt{2x - 3} - 9 = 0$.

First isolate the radical, then remove it by squaring.

$2x - 9 = \sqrt{2x - 3}$
$(2x - 9)^2 = (\sqrt{2x - 3})^2$ auxiliary equation
$4x^2 - 36x + 81 = 2x - 3$
$2x^2 - 19x + 42 = 0$ (Why?)
$(2x - 7)(x - 6) = 0$.

Therefore, $\tfrac{7}{2}$ and 6 are the only possible roots of the original. Each must be checked. Check $x = \tfrac{7}{2}$.

Basic Algebraic Theory

$$2\left(\frac{7}{2}\right) - \sqrt{2\left(\frac{7}{2}\right) - 3} - 9 \stackrel{?}{=} 0$$
$$7 - \sqrt{4}\dagger - 9 \stackrel{?}{=} 0$$
$$7 - 2 - 9 \neq 0.$$

Hence, $\frac{7}{2}$ is not a root of the original equation even though it satisfies the auxiliary equation.

Check $x = 6$.

$$2(6) - \sqrt{2(6) - 3} - 9 \stackrel{?}{=} 0$$
$$12 - 3 - 9 = 0.$$

Consequently, the original equation has 6 as its only solution.

extraneous value An extra value obtained in the process of solution, such as $\frac{7}{2}$ in the above problem, which is not a solution of the original equation, is called an *extraneous value*. The manner in which an extraneous value is introduced is more easily seen if $\frac{7}{2}$ is substituted for x in the equation in the form used just before it was squared.

$$2x - 9 = \sqrt{2x - 3}$$
$$2\left(\frac{7}{2}\right) - 9 \stackrel{?}{=} \sqrt{2\left(\frac{7}{2}\right) - 3}$$
$$-2 \stackrel{?}{=} \sqrt{4} = +2.\dagger$$

Example 3

Solve for x: $\sqrt{3x + 3} - \sqrt{2x} = \sqrt{x - 1}$. The given equation is equivalent to

$$\sqrt{3x + 3} = \sqrt{x - 1} + \sqrt{2x}.$$
Then $(\sqrt{3x + 3})^2 = (\sqrt{x - 1} + \sqrt{2x})^2$,
$$3x + 3 = x - 1 + 2\sqrt{x - 1} \cdot \sqrt{2x} + 2x.$$

Upon isolating the remaining radicals and dividing by 2, we obtain

$$2 = \sqrt{x - 1} \cdot \sqrt{2x}.$$
Then $\qquad 4 = (x - 1)(2x),$
$$x^2 - x - 2 = 0.$$

---

† In Sec. 1-6 it was pointed out that $-\sqrt{4} = -2$, not ± 2.

Sec. 1-12. Auxiliary Equations

Consequently, -1 and 2 are possible roots of the original equation. We must check each possible value.

Check $x = -1$.
$$\sqrt{-3+3} - \sqrt{-2} \stackrel{?}{=} \sqrt{-2}$$
$$0 - \sqrt{-2} \neq \sqrt{-2}$$
Hence -1 is extraneous.

Check $x = 2$.
$$\sqrt{6+3} - \sqrt{4} \stackrel{?}{=} \sqrt{1}$$
$$3 - 2 = 1$$
Hence 2 is a root.

Example 4

Determine values of S and C which satisfy both of the equations.

$$S^2 + C^2 = 1 \tag{1}$$
$$8S + C = 7. \tag{2}$$

Solve Eq. (1) for $C = \pm \sqrt{1 - S^2}$. If C is to be a value common to Eqs. (1) and (2), then C from Eq. (1) must also satisfy (2). Consequently

$$8S \pm \sqrt{1 - S^2} = 7$$
$$(8S - 7)^2 = (\mp \sqrt{1 - S^2})^2$$
$$64S^2 - 112S + 49 = 1 - S^2$$
$$65S^2 - 112S + 48 = 0$$
$$(13S - 12)(5S - 4) = 0.$$
$$(S, C) = \left(\frac{12}{13}, -\frac{5}{13}\right) \quad \text{and} \quad (S, C) = \left(\frac{4}{5}, \frac{3}{5}\right)$$

are two possible solutions. Both check in each given equation, hence both pairs are solutions.

A second and more desirable method of solution is to solve Eq. (2) for $C = 7 - 8S$ and substitute in (1), obtaining $S^2 + (7 - 8S)^2 = 1$, or $65S^2 - 112S + 48 = 0$ again. The solution is then completed as shown above. Readers familiar with trigonometric equations will see that the solution of $8 \sin \theta + \cos \theta = 7$ for $\sin \theta$ and $\cos \theta$, using the identity $\sin^2 \theta + \cos^2 \theta = 1$, can be obtained as done in this problem.

Problem Set 1-12

Solve the following equations for real roots unless otherwise instructed.

Basic Algebraic Theory

1. $\sqrt{x-3} = x - 5$.

2. $1 + \sqrt{3x+3} = \sqrt{6x+7}$.

3. $\sqrt{(x-6)} = \sqrt[3]{x-2}$. (HINT: $x = 10$ is a root. Are there others?)

4. $\sqrt{3-2x} + 5 = 0$.

5. $\dfrac{x-7}{x-2} - \dfrac{3}{x} = \dfrac{3x+4}{x(x-2)}$.

6. $\dfrac{t}{t+2} - \dfrac{5}{(t-1)(t+2)} = \dfrac{1}{t^2+t-2}$.

7. $\sqrt{w+4} + w - 2 = 0$.

8. $\dfrac{3}{z+1} + \dfrac{4}{z-1} = \dfrac{5}{1-z}$.

9. $\sqrt[3]{2x^2 - 8} = 4$.

10. $3.5y^2 - 4.1y + 7.2 = 6.1$.

11. $4\sqrt{2x+3} + 5 = 0$.

12. $z^3 - 8 = 0$. (Obtain all three solutions.)

13. $\begin{cases} x + 2y = 12. \\ x^2 + y^2 = 29. \end{cases}$

14. Without solving, show that $\sqrt{3x^2 + 7} + 2 = 0$ has no solution.

15. It is not difficult to distinguish between extraneous values and values produced through arithmetical error. If the extraneous value is introduced by squaring, it will satisfy an equation obtained by changing the sign of some (not necessarily all) radicals in the original equation. In the above problems having extraneous values, obtain the corresponding equations which are satisfied by the extraneous values.

16. $\dfrac{1}{\sqrt{x} - \sqrt{2x}} = 5$. (HINT: It is unnecessary to determine an auxiliary equation if you consider whether or not the left member can possibly be a positive real number for $x < 0$, for $x = 0$, and for $x > 0$.)

17. $\dfrac{1}{\sqrt{x}} - \dfrac{1}{\sqrt{2x}} = 5$. (HINT: Factor $1/\sqrt{x}$ from each term of the left member and divide by the constant $1 - 1/\sqrt{2}$.)

18. $x\sqrt{5-x} = \sqrt{5-x}$. Some students will see at once that two possible solutions are $x = 1$ and $x = 5$. Are there other solutions?

19. If $g(t) = 4/3t^2$, then $g(t)$ is never zero. For what, if any, values of t will $g(t+2) - g(t) = 0$?

20. $\sqrt{3x-11} + \sqrt{5-x} = 0$.

21. $\sqrt{3x-11} - \sqrt{5-x} = 0$. (Compare with Prob. 20.)

Sec. 1-13. $\Delta f = f(x + \Delta x) - f(x)$ 27

22. $1 + 2z = \sqrt{5z + 4}$.

23. $\sqrt{4z - 3} - \sqrt{8z + 1} = 2$.

24. $x - \sqrt{x} - 6 = 0$.

25. $\dfrac{x - \sqrt{x} - 6}{x^2 - 1} = 0$.

1-13. $\Delta f = f(x + \Delta x) - f(x)$.

The function $f(x + \Delta x) - f(x)$ occurs so often in mathematics that a special symbol is used to represent it. The symbol is Δf. The notation $\Delta f(x, \Delta x)$ is sometimes used since, for a given function $f(x)$

$$\Delta f(x, \Delta x) = f(x + \Delta x) - f(x)$$

is a function of the two variables x and Δx. It represents the change in $f(x)$ between the values at x and at $x + \Delta x$.

You should not try to "multiply out" the function $\Delta f(x, \Delta x)$, any more than you would any function of two variables. By definition

$$\Delta f \equiv \Delta f(x, \Delta x) \equiv f(x + \Delta x) - f(x).$$

Example 1

Using $f(x) = x^2 - 7x + 2$, find Δf. By definition,

$$\Delta f = f(x + \Delta x) - f(x)$$
$$= [(x + \Delta x)^2 - 7(x + \Delta x) + 2] - (x^2 - 7x + 2)$$
$$= 2x\,\Delta x + (\Delta x)^2 - 7\Delta x.$$

The definition of Δf states that Δf is the change in $f(x)$ as x changes from x to $x + \Delta x$.

Example 2

If $f(y) = \dfrac{y}{y - 3}$, determine $f(2)$ and $\Delta f(2, \Delta y)$. Clearly

$f(2) = \dfrac{2}{2 - 3} = -2$. From the definition of $\Delta f(2, \Delta y)$,

$$\Delta f(2, \Delta y) \equiv f(2 + \Delta y) - f(2) = \dfrac{2 + \Delta y}{2 + \Delta y - 3} - \dfrac{2}{2 - 3}$$

$$= \dfrac{2 + \Delta y}{\Delta y - 1} + 2 = \dfrac{2 + \Delta y + 2\Delta y - 2}{\Delta y - 1} = \dfrac{3\Delta y}{\Delta y - 1}.$$

If you had trouble with this example, consult Example 3, Sec. 1-4. Before going further, you should show that if

$$g(z) = 3z^2 - 2z + 5 \quad \text{then} \quad \Delta g(5, \Delta x) = 28\Delta x + 3(\Delta x)^2.$$

Example 3

If $f(x) = 3/x$, find Δf.

$$\Delta f = \frac{3}{x + \Delta x} - \frac{3}{x} = \frac{3}{(x + \Delta x)} \cdot \frac{x}{x} - \frac{3}{x} \cdot \frac{(x + \Delta x)}{(x + \Delta x)}$$

$$= \frac{3x - 3x - 3\Delta x}{x(x + \Delta x)} = \frac{-3\Delta x}{x(x + \Delta x)}.$$

Consult Example 4, Sec. 1-4.

Example 4

If $g(t) = 5/t^2$, find $\dfrac{\Delta g}{\Delta t}$.

$$\frac{\Delta g}{\Delta t} = \frac{\dfrac{5}{(t + \Delta t)^2} - \dfrac{5}{t^2}}{\Delta t} = \frac{5}{(t + \Delta t)^2 \, \Delta t} - \frac{5}{(t)^2 \, \Delta t} = \cdots$$

$$= \frac{5t^2 - 5[t^2 + 2t\,\Delta t + (\Delta t)^2]}{(t + \Delta t)^2 t^2 \, \Delta t}$$

$$= \frac{-5(2t + \Delta t)\,\Delta t}{(t + \Delta t)^2 t^2 \, \Delta t} = \frac{-5(2t + \Delta t)}{(t + \Delta t)^2 t^2}$$

$$\text{if } \Delta t \neq 0 \text{ and } t \neq -t \text{ and } t \neq 0.$$

Consult Example 5, Sec. 1-4.

The division of numerator and denominator by Δt in the last step is valid only if Δt is not zero. If Δt is zero, the entire problem is without meaning since division by zero is not defined. Similar comments apply to the restrictions $\Delta t \neq -t$ and $t \neq 0$.

Since $\Delta g(t, \Delta t)$ represents the change in the function $g(t)$ corresponding to a change Δt in t, it follows that $\dfrac{\Delta g(t, \Delta t)}{\Delta t}$ is the average rate of change of the function $g(t)$ over the interval of length Δt which begins at t_1. (Reread this sentence several times.)

Locate Δt and $\Delta f(t, \Delta t) = f(t + \Delta t) - f(t)$ in Fig. 1-3, page 29. Construct similar sketches for yourself, using other curves.

Problem Set 1-13

1. If $f(x) = x^2 - 7x + 2$, find $\Delta f(x, \Delta x)$.
2. If $g(z) = z/(z - 3)$, find $\Delta g(z, \Delta z)$.
3. If $w(x) = 4/x^2$, find $\dfrac{\Delta w(x, \Delta x)}{\Delta x}$. What restrictions are placed on Δx?

Sec. 1-13. $\Delta f = f(x + \Delta x) - f(x)$ 29

4. If $J(x) = 1/x$, find $\dfrac{\Delta J(x, \Delta x)}{\Delta x}$.

5. If $f(x) = 3x/(2 - 5x^2)$, find $\Delta f(x, \Delta x)$.

6. If $f(x) = 5x/(2 + x)$, find $\Delta f(2y, \Delta y)$ and $\Delta f(2t, \Delta t)$.

7. If $f(t) = 2/t$, find $\Delta f(y, \Delta y)$ and $\Delta f(2t, \Delta t)$.

8. If $f(x) = 7/x^2$, find $\dfrac{\Delta f(z, \Delta z)}{\Delta z}$.

9. If $g(y) = y - 3$, find $\dfrac{\Delta g(t, \Delta t)}{\Delta t}$, and rationalize the numerator.

10. Draw a curve and locate Δx and $\Delta f(x, \Delta x)$ in a manner similar to that shown in this section.

11. Find $\dfrac{\Delta g(x, \Delta x)}{\Delta x}$, given $g(x) = 6x/(3x + 1)$.

12. $f(x) = (3x - 1)/2x$. Find (a) $f(1/x)$; (b) $f(t + \Delta t) - f(t)$; (c) $f(1.2) - f(1)$.

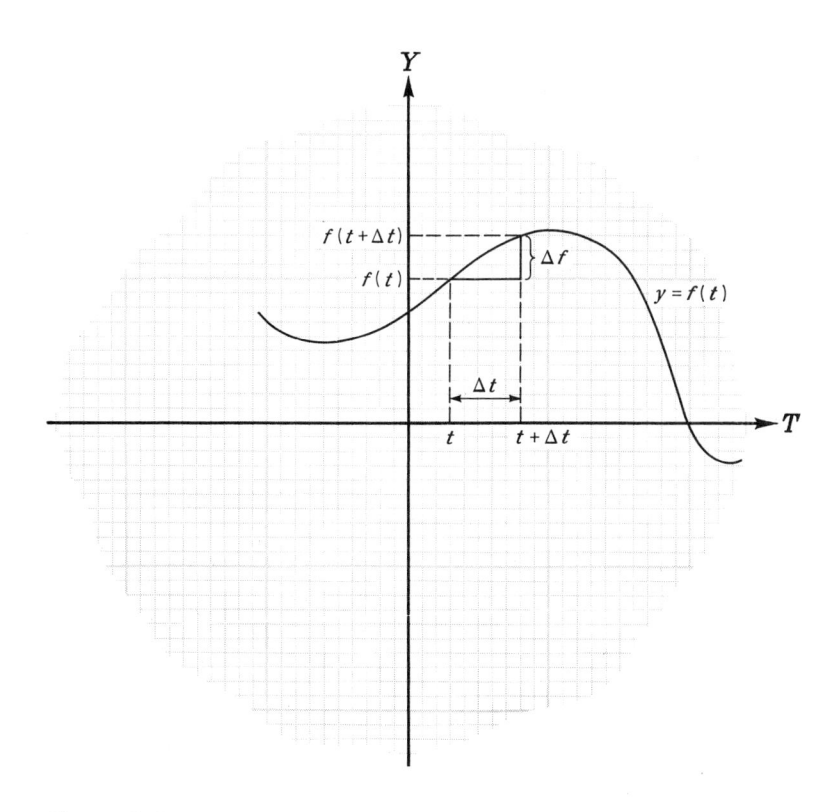

Figure 1-3

1-14. Self-test

self-test

Self-tests have been included to help *you* see whether or not your mastery of the material presented is sufficient. The student should not expect any marked similarity between these self-tests and his classroom examinations, since they are designed for different purposes. It is hoped that, by taking these self-tests under examination conditions, the student may overcome some of his fear of tests and, through better preparation, be able to obtain better grades on classroom tests.

1. Shade the following regions on a number axis:

   (a) $x < -3$.
   (b) $4 < x \leq 7$.
   (c) $x \neq 2$.
   (d) $|x| < \frac{1}{2}$.

2. State some special properties of the numbers zero and one.
3. State the two basic properties upon which the solution of equations is based.
4. If $f(x) = 4x^2 - 7x + 2$, find $\dfrac{\Delta f}{\Delta x}$.
5. Determine g in lowest terms where $g(t) = 7/2t$.
6. Determine the complete solution set of the equation $x^3 - 27 = 0$.
7. Determine the real solution(s) of the equation $\sqrt{x+4} + x = 2$.
8. Solve for x: $\dfrac{x^2 - 7x + 5}{x^4 + 8} = 0$.

2

Basic Geometric Theory

2-1. Analytic (Coordinate) Geometry

It is the coming practice to include rather extensive analytic (coordinate) geometry in the tenth grade course in geometry, partly because analytic geometry is in many cases a more powerful and a more easily used tool than Euclidean synthetic geometry, and partly because analytic geometry closely relates geometry and algebra. Actually, algebra and geometry are merely different facets of the same jewel, and each requires the dispersive effects of the other to achieve maximum brilliance.

axes

origin

One way of describing the position of a point in a plane is to use two intersecting (not necessarily perpendicular) lines, called *axes*, as a frame of reference for the points. The point of intersection O of the reference lines is called the *origin*. It is usual to take one of the reference lines horizontal and to call it OX, or the X axis. The other line is called OY or the Y axis. The direction to the right along a horizontal line is taken as positive. The upward direction is taken as positive along a nonhorizontal line. The four quadrants into which the axes divide the plane are numbered as indicated in Fig. 2-1, page 32.

Cartesian coordinate system

This reference system is known as the *Cartesian coordinate system* in honor of René Descartes (pronounced "Day-cart"). Descartes lived from 1596 to 1650, a contemporary of Shakespeare, Fermat, Galileo, and Milton. He helped introduce the coordinate system, and with it the study of analytic geometry, without which the

31

Basic Geometric Theory

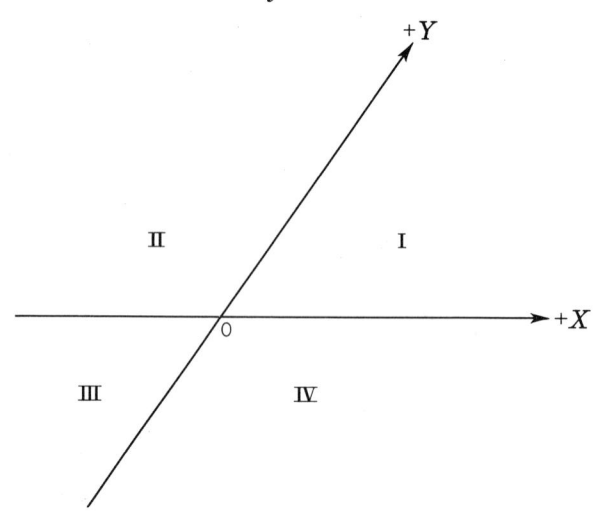

Figure 2-1

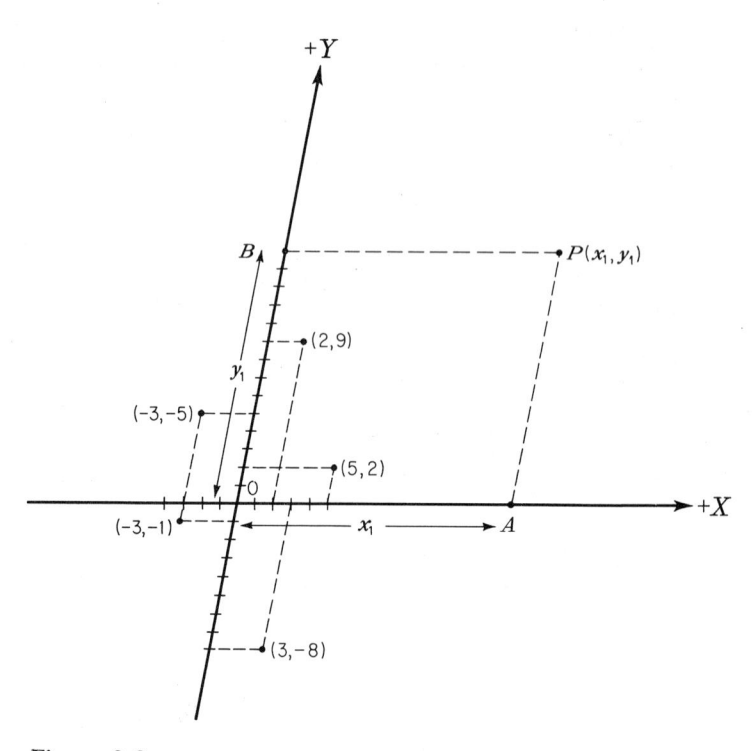

Figure 2-2

Sec. 2-1. Analytic (Coordinate) Geometry 33

calculus and hence modern physics and engineering would be almost impossible. An interesting account of Descartes's life is contained in Chap. 3 of E. T. Bell's *Men of Mathematics*.

A point P may be coordinated with respect to the reference frame by means of directed line segments through P parallel to each of the axes. In Fig. 2-2, two axes are chosen with origin O and positive

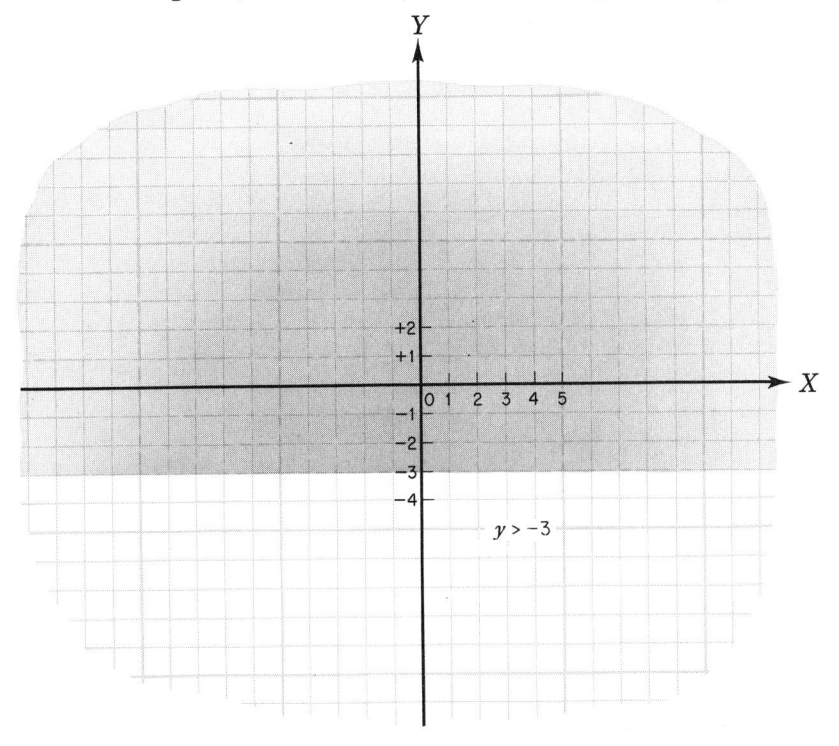

Figure 2-3

direction as indicated by the arrowheads. A unit of measure is chosen for each of the axes. The point P is located by directed line segments BP and AP parallel to the x and y axes. $BP = OA = x_1$ of the x units, while $AP = OB = y_1$ of the y units. The value x_1 is called the *abscissa* and y_1 the *ordinate* of the point P. Together they are called *coordinates* of the point P and are indicated symbolically by the number pair (x_1, y_1). To each point P there corresponds just one number pair (x_1, y_1), and to each pair of real numbers there corresponds just one point P. Several points are located on the diagram. The reader should check his understanding by

abscissa
ordinate

determining the numerical coordinates of each point. The point P_2 having coordinates (x_2, y_2) is often written using the notation $P_2(x_2, y_2)$. In a similar fashion $Q(3, -4)$ indicates that the point Q has x coordinate 3 and y coordinate -4.

All points having x coordinate zero lie on the y axis. (Why?) Furthermore, any point on the y axis has zero as its x coordinate. (What word describes the logical difference in the two statements just made? Does either imply the other? Why not?) Since each point having $x_1 = 0$ lies on the y axis and each point on the y axis has $x_1 = 0$, the equation $x = 0$ is called the equation of the y axis.

What *points* correspond to the equation $y = 0$?

All points for which $x_1 = 7$ are to be found on a line L parallel to the y axis and $+7$ units in the x direction to the right of the y axis. Furthermore, every point on the line L has $x_1 = 7$. Consequently, $x = 7$ is called the *equation of the line L*.

What points correspond to $x = -3$? To $y = 5$?

The point having coordinates $(-3, 5)$ is the point of intersection of the two lines $x = -3$, $y = 5$. The origin $(0,0)$ is the intersection of the lines $x = 0$ and $y = 0$.

In most applied problems the axes are taken perpendicular, but not always. The main advantage of perpendicular axes is the ease with which we may then determine the distance between two points. This is discussed in Sec. 2-2.

Example 1

Where are all points having $y > -3$?

These points are the points *above* (but not on) a line parallel to the x axis and three units in the y direction below the x axis (see Fig. 2-3).

Example 2

Where are the points having $y = -5$?

These points are the points of the line parallel to the x axis and five units below the x axis in the y direction (see Fig. 2-4).

Example 3

Where are the points having $x \neq 2$?

These are all the points in the plane *except* those points on a line

Sec. 2-1. Analytic (Coordinate) Geometry

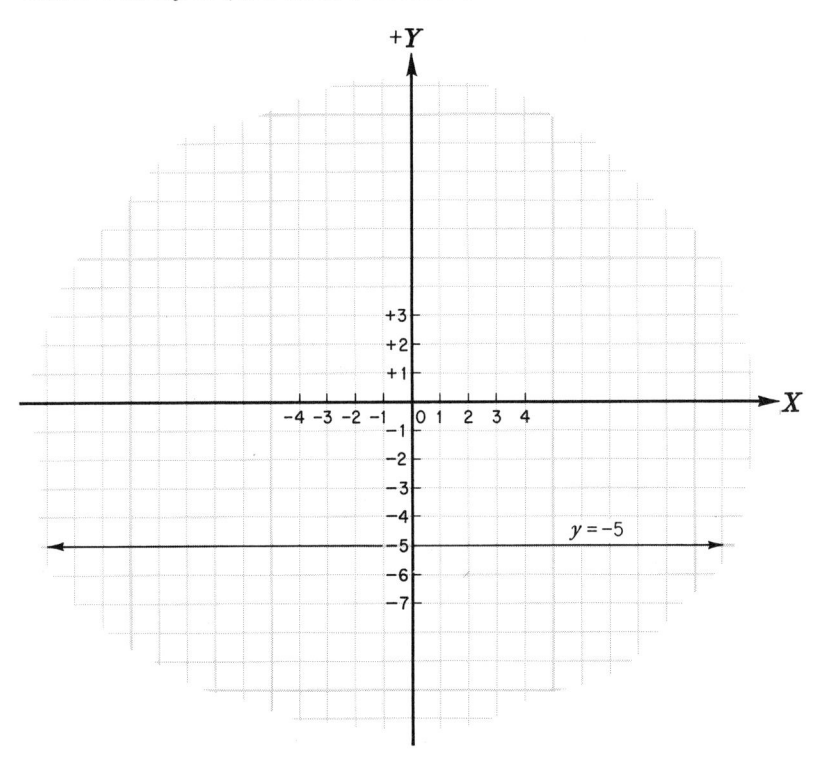

Figure 2-4

parallel to the y axis and two units in the x direction to the right of the y axis (see Fig. 2-5).

In terms of the function-relation concept (Sec. 1-2), a subset of the x axis (or possibly the entire x axis) represents the domain of definition X of the function f, while a subset of the y axis represents the range Y of the function. The graph of $y = f(x)$ (which can be a series of disconnected points or segments, or even an area) then displays the correspondence (which is the function or the relation) since for each permissible $x \in X$, corresponding y values are determined.

Problem Set 2-1

Locate the following points by using a set of coordinate axes in which the angle XOY is approximately 60°. Name the quadrant in which each point lies.

Basic Geometric Theory

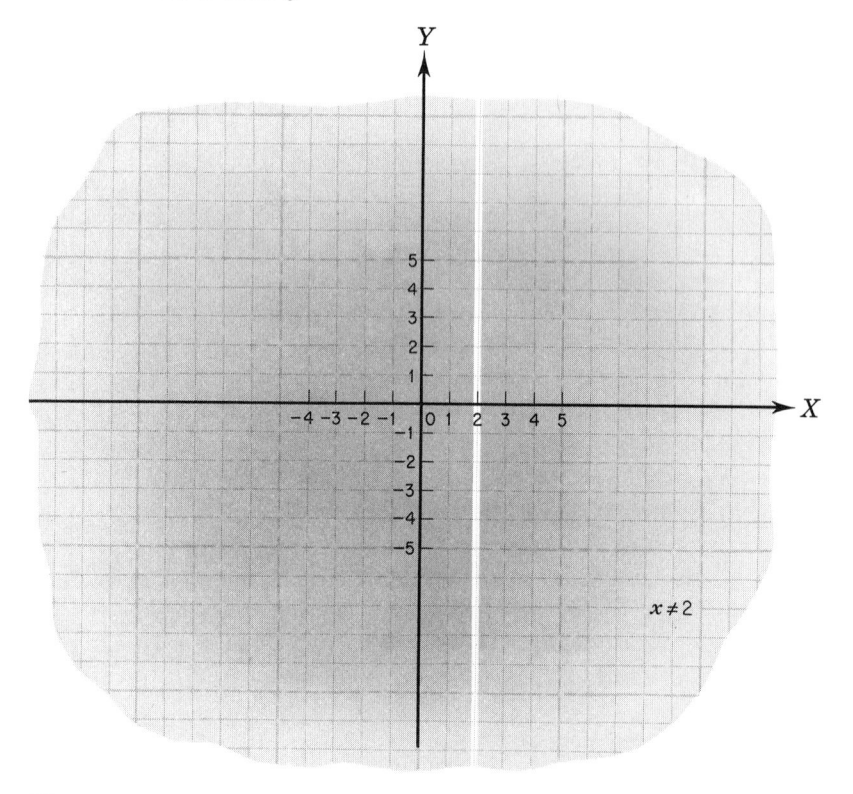

Figure 2-5

1. $(3,1)$, $(2,-5)$, $(-3,2)$.   **2.** $(3,5)$, $(2,-1)$, $(-6,3)$.
3. $(-1,-4)$, $(-5,6)$, $(\pi,2)$.

4. $(0,1)$, $(0,5)$, $(0,-17)$. What, if any, common property do points $(0,y_2)$ having x coordinate zero have?

5. $(-2,7)$, $(-2,5)$, $(-2,-1)$, $(-2,0)$, $(-2,-5)$. What can you state about the position of points having $x_1 = -2$?

6. $(-4,5)$, $(-1,5)$, $(0,5)$, $(3,5)$, $(7,5)$. What can you state about the position of points having $y_1 = 5$?

7. Pretend you are working on the blackboard. Do not attempt to rule off coordinate lines, but draw two intersecting axes and mark a scale on each. Locate each of the following points by eye and mark the point: $(3,1)$, $(2,-5)$, $(-1,3)$, $(-2,4)$.

8. Same as Prob. 7 for points $(1,2)$, $(3,-11)$, $(-6,-2)$, $(-5,1)$.

9. Locate the points $A(3,1)$ and $B(7,4)$ on a rectangular Cartesian $(XOY = 90°)$ coordinate system. Such paper may be purchased

Sec. 2-1. Analytic (Coordinate) Geometry

from your bookstore. Using another piece of paper as a measuring stick, determine the distance between the two points. Draw lines parallel to each coordinate axis in such a manner that line AB is the hypotenuse of a right triangle. How long are the legs of the triangle? From this how long would you expect the hypotenuse to be?

10. Same as Prob. 9, for $A(-1,2)$ and $B(7,17)$.

11. Same as Prob. 9, for $A(4,-1)$ and $B(9,11)$.

On rectangular coordinate paper, shade the regions represented by the following conditions:

12. $x < 3$.   *13.* $y + 4 < 0$.   *14.* $x \neq 8$.

15. $x < 3$ and at the same time $y < 5$.

16. $x > 2$ and at the same time $y > 7$.

17. $2 \leq x < 7$.   *18.* $-3 < y \leq 15$.

19. $-2 < x < 1$ and at the same time $-5 < y < -2$.

20. $x > 5$ and at the same time $x \neq 7$.

21. $x \neq 3, y + 4 < 0$.   *22.* $1 < x < 3, 2 < y < 3$.

23. $x - 5 \neq 0$ and at the same time $x \neq -3$.

24. $(x - 5)(x + 3) \neq 0$. Compare with Prob. 23.

25. $0 < (x - 5) \leq 3$.   *26.* $x + 3 + 4x \neq 7$.

27. $\sqrt{x + 1} < 3$.   *28.* $|x - 5| < 2$.

29. First-class mail costs 4 cents per ounce or fraction thereof. Make a graph showing the cost of first-class mail as a function of weight.

30. Graph $f(x) = \begin{cases} 3 & \text{for } x > 7 \\ 2x & \text{for } x < 7 \\ -10 & \text{for } x = 7 \end{cases}$

31. Graph $y = \begin{cases} 3x + 1 & \text{if } x < 1 \\ 4 & \text{if } 1 < x < 5 \\ 10 & \text{if } x > 5 \\ 10 & \text{if } x = 1 \\ 7 & \text{if } x = 5 \end{cases}$

32. Calvin Butterball dreams he is the swashbuckling hero of a science fiction novel and has just landed on Planet 3XX. What

2-2. The Distance between Two Points

essential features must he explain to the natives before he can tell them where he came from? (Consider the essentials of a coordinate system.)

In the preceding section no restriction was made on the size of the angle XOY. We shall now take the two axes *perpendicular and the same unit of length on each axis.* If the axes are perpendicular, the distance between two points can be determined by constructing right triangles similar to those of Probs. 9, 10, and 11 of Set 2-1. However, it is convenient to solve this problem in general and obtain a formula much as we did in finding the general solution of the quadratic equation. Consider the case where the two points $B(x_2, y_2)$ and $A(x_1, y_1)$ lie in the first and second quadrants, respectively (Fig. 2-6).

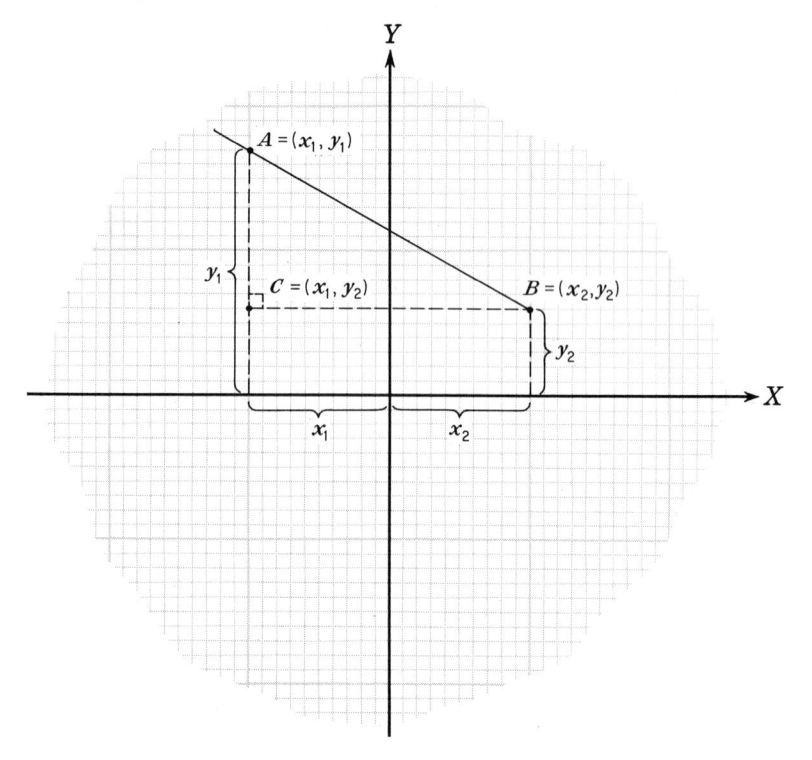

Figure 2-6

Sec. 2-2. The Distance between Two Points

Proceeding as in Prob. 9 of Set 2-1, we construct lines through $A(x_1,y_1)$ and $B(x_2,y_2)$ parallel to the axes so that AB is the hypotenuse of a right triangle ACB. The coordinates of C are (x_1,y_2). (Why?) The length of CA is $y_1 - y_2$. (Why?) The length of CB is $x_2 - x_1$. (Why isn't $x_1 + x_2$ the length of CB?) Since ACB is a right triangle, the theorem of Pythagoras gives:

$$d^2 = (CB)^2 + (CA)^2 = (x_2 - x_1)^2 + (y_1 - y_2)^2,$$

where d is the distance from $A(x_1,y_1)$ to $B(x_2,y_2)$. Then

$$d = \pm \sqrt{(x_2 - x_1)^2 + (y_2 - y_1)^2}$$

since $(y_2 - y_1)^2 = (y_1 - y_2)^2$. In speaking of the distance *from* one point *to* another, we shall select the sign as indicated in the preceding section. Frequently we are interested in the numerical or absolute value of the distance from one point to another. This is spoken of as the distance *between* the two points. The formula $d = \pm \sqrt{(x_2 - x_1)^2 + (y_2 - y_1)^2}$ also holds when A and B are permitted to lie in other quadrants. The formula is valid for every possible choice of A and B as long as the axes are perpendicular (see Probs. 34 and 35 of this section).

Example 1

Find the distance between $(-3,4)$ and $(2,-8)$.

$$d = \overset{+}{\sqrt{(-3-2)^2 + [4-(-8)]^2}} = \overset{+}{\sqrt{25 + 144}} = \overset{+}{\sqrt{169}}$$
$$= +13.$$

Example 2

Give an algebraic condition on points such that their distance from $(5,-3)$ is less than 8.

Clearly these points lie inside (but not on) a circle of radius 8 with center at $(5,-3)$. To obtain an algebraic condition, let the point be (x_1,y_1). Then if and only if $\sqrt{(5-x_1)^2 + (-3-y_1)^2} < 8$ will there be fewer than eight units between (x_1,y_1) and $(5,-3)$.

The restriction that the two axes have equal units, or even that they both represent the same type of measurement, is merely a convenience in beginning a presentation. In actual work it is often an undesirable restriction. In electrical work, for example, units on the x axis may be units of resistance and units on the y axis may be units of inductance (negative inductance is called *capacitance*). In this case the "distance" between two "points" will be measured in units of impedance.

Problem Set 2-2

1. Show that the distance formula holds if A is taken in quadrant II and B in quadrant III.

In Probs. 2 to 10, find the distance between the given points. In each problem begin by making a quick sketch (see Prob. 7, Set 2-1) showing the points and the distance found. Before using the formula to determine the distance in Probs. 2 to 5, write down an estimate of the approximate distance. This ability to estimate the answer in various problems is often the difference between an experienced engineer or scientist and a new graduate. If you always attempt to estimate the answer to a problem before working it, you too will develop this valuable skill. It takes practice.

2. $(2,-1)$, $(13,-7)$.
3. $(3,5)$, $(2,-1)$.
4. $(-3,4)$, $(7,2)$.
5. $(-4,-1)$, $(-1,-4)$.
6. (h,k), $(3,1)$.
7. $(3,r)$, $(-1,s)$.
8. $(0,0)$, (a,b).
9. $(0,7)$, $(5,13)$.
10. $(17,-5r)$, $(12,2r)$.

11. How far is the point $(3,-9)$ from the origin?

12. How far is the point $(2,7)$ from the origin?

13. What are the lengths of the sides of a triangle with vertices $(2,1)$, $(3,5)$, $(4,6)$?

14. Same as Prob. 13 for $(6,4)$, $(2,2)$, $(-1,-5)$.

15. Same as Prob. 13 for $(-1,3)$, $(2,6)$, $(0,-3)$.

16. Show that $(15,4)$, $(-7,8)$, and $(-1,-4)$ are vertices of a right triangle. HINT: The converse of the law of Pythagoras is valid. It states that if a triangle has sides of lengths a, b, c and if we have $a^2 + b^2 = c^2$, then the triangle is a right triangle having its right angle opposite the side of length c.

17. Same as Prob. 16, for the points $(1,4)$, $(2,1)$, $(3,2)$.

18. Show that if $A(0,0)$, $B(9,-2)$, and $C(2,4)$ are vertices of a triangle, then $\angle BAC = \angle BCA$. HINT: Show that the triangle is isosceles.

19. Same as Prob. 18 for $A(2,1)$, $B(7,4)$, $C(2,7)$.

20. Show that the three points $(-2,1)$, $(3,11)$, $(-5,-5)$ are collinear (lie on a straight line).

Sec. 2-2. The Distance between Two Points

21. In Prob. 6, you found the distance between the point (h,k) and the point $(3,1)$. Set this distance equal to 2. Make a sketch showing several possible positions of the point (h,k) so that its distance from the point $(3,1)$ is equal to 2. Where do all such points (h,k) lie? Will there be any points (h,k) satisfying the conditions in quadrant IV? In quadrant II? In quadrant III?

22. If $A(-1,5)$, $B(-1,-2)$, and $C(6,-2)$ are vertices of a triangle, show that $\angle BAC = \angle BCA$. Sketch the figure.

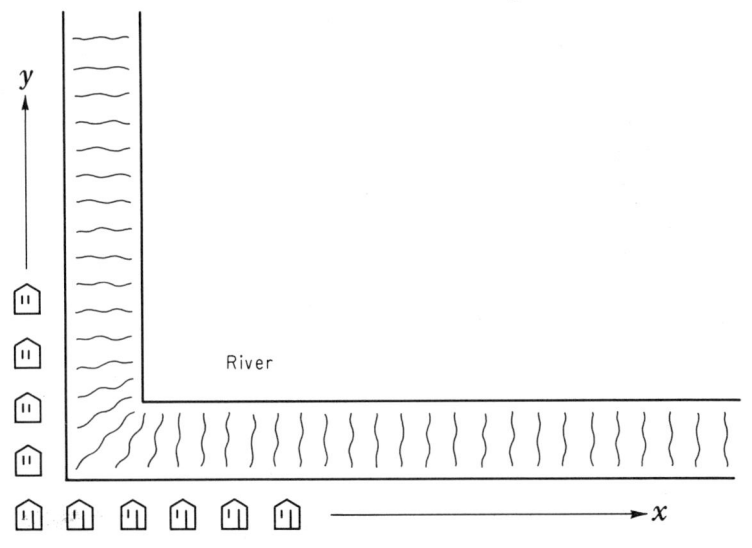

Figure 2-7

23. Same as Prob. 22, for $A(0,1)$, $B(3,7)$, $C(3 + 2\sqrt{5}, 2)$.

24. Make a sketch showing the possible locations of points (a,b) such that $(a - 1)^2 + (b + 3)^2 = 4$.

25. Make a sketch showing the possible locations of points (r,s) such that $(1 + r)^2 + (4 + s)^2 = 9$.

Show in a sketch where the points (x,y) must lie to satisfy the following conditions:

26. $(1 + x)^2 + (4 + y)^2 \leqq 9$.   **27.** $(x - 1)^2 + (y - 5)^2 > 4$.

28. $(x - 2)^2 < 4$.

29. The condition $y > 4$ and at the same time $(x - 3)^2 + y^2 < 25$.

30. $x^2 + y^2 < 9$ and at the same time $(y - 2)^2 + (x + 1)^2 < 1$.

31. Write an algebraic condition such that (x,y) lies inside a circle of radius 5 with center $(3,-2)$.

32. Write an algebraic condition such that (x,y) lies in the ring determined by concentric circles with radii 2 and 7 and center $(-1,2)$.

33. Suppose that a nonaquatic community sprang up at the edge of a river bend as in Fig. 2-7. In order to get from place to place the natives always went around the corner. What sort of distance formula might arise in such an environment?

34. (a) Devise a formula for the distance between two points if XOY is an angle of $60°$ rather than $90°$. HINT: Use the law of cosines.

*(b)† If you have studied vector subtraction in your physics course or elsewhere, compare the results of (a) with vector subtraction.

35. In a certain problem it is convenient to take the physical distance representing y units to be 15 times as long as those representing x units, as shown in Fig. 2-8.

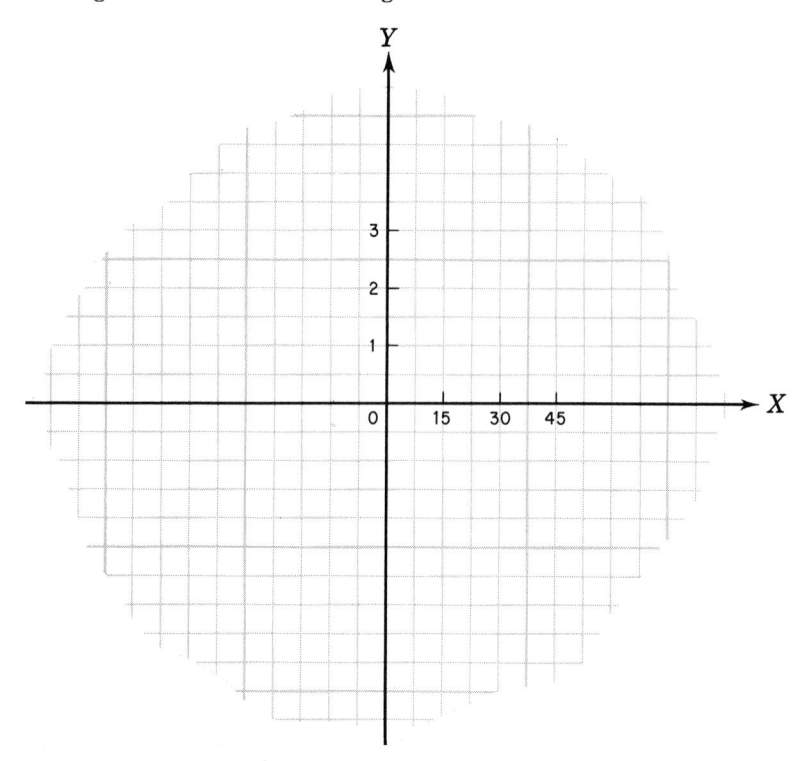

Figure 2-8

† An asterisk on a problem indicates that the author feels that better students will enjoy that particular problem.

Sec. 2-3. Loci

(a) Devise a formula which will determine the actual physical distance p, measured in y length units, between points having coordinates (x_1,y_1) and (x_2,y_2).

(b) Explain why it is impossible to devise a formula which will convert the number d, obtained from the usual distance formula $d = \sqrt{(x_1 - x_2)^2 + (y_1 - y_2)^2}$ to the number p obtained in part (a), unless something about the coordinates of the points is known as well as the number d.

(c) Exactly what is the "something about the coordinates of the points" which is needed in part (b)?

36. Given that a and b are legs of a right triangle of hypotenuse c, we all know that $a^2 + b^2 = c^2$. This problem is a challenge. Determine a geometric length d such that $a + b = c + d$ and show how d is related to the right triangle. [HINT: The first letter of the word diameter is d.]

2-3. Loci

The correspondence between pairs of real numbers (a,b) and points in a plane has been discussed. A correspondence also exists between geometrical figures and equations or inequalities. A number of prior examples and problems (Secs. 2-1 and 2-2) illustrate such correspondences. In general, not all points will satisfy a given equation. For example, the point $(2,5)$ will satisfy the equation $x^2 - 2y + x + 4 = 0$, but the points $(5,2)$ and $(-3,4)$ do not. Verify each of these statements.

The points whose coordinates do satisfy a given equation will form a geometric figure. This figure is called the *locus* of the equation. Actually the equation and the figure are so intimately linked that the usual practice is to refer to both the figure and the equation by the geometric name of the figure. These remarks also apply to inequalities. The point $(1,-2)$ satisfies the inequality $x^2 + y^2 \leqq 9$ but $(-3,-4)$ does not.

locus of an equation The *locus of an equation* (or inequality) may be defined as the totality of all points whose coordinates satisfy the equation (or inequality). This is a two-edged statement. It implies that:

1. The coordinates of every point on the locus satisfy the equation (or inequality).

2. Every point whose coordinates satisfy the equation (or inequality) lies on the locus.

Basic Geometric Theory

Example 1

Show that the locus of the equation $(x - 3)^2 + (y - 1)^2 = 4$ is a circle† with the center $(3,1)$ and radius 2. Let r be the distance between the point $(3,1)$ and the point (x,y). Then by the distance formula

$$r = \sqrt{(x - 3)^2 + (y - 1)^2}.$$

Squaring yields $r^2 = (x - 3)^2 + (y - 1)^2$.

If (x,y) is any point on the circle, then $r = 2$ or $r^2 = 4$, and the given equation is satisfied. (Does this fulfill requirement (1) or (2) of the preceding paragraph?)

Furthermore, if (x_1,y_1) is a point whose coordinates satisfy the equation, we have $(x_1 - 3)^2 + (y_1 - 1)^2 = 4$, so $r^2 = 4$ and $r = 2$. Hence the point (x_1,y_1) lies on the circle. (Which requirement does this fulfill?) What is the locus of all points (x,y) such that $(x - 3)^2 + (y - 1)^2 \leq 4$? Such that $(x - 3)^2 + (y - 1)^2 > 4$?

Example 2

perpendicular bisector

Obtain the equation of the perpendicular bisector of the line segment joining $(-1,3)$ and $(2,5)$. Make a sketch (Fig. 2-9) and recall from geometry that a point is on the perpendicular bisector of a line segment if and only if it is equally distant from the end points of the segment. (Can you still prove this theorem? Try it. Remember that "if and only if" requires *two* proofs.)

Let d_1 be the distance between $(-1,3)$ and (x,y) while d_2 is the distance between $(2,5)$ and (x,y), if (x,y) is a point on the desired locus (the perpendicular bisector). Then the reader should show $d_1 = \sqrt{(x + 1)^2 + (y - 3)^2}$ while $d_2 = \sqrt{(x - 2)^2 + (y - 5)^2}$. Since, by geometry, $d_1 = d_2$, $(d_1)^2 = (d_2)^2$, or

$$(x + 1)^2 + (y - 3)^2 = (x - 2)^2 + (y - 5)^2, \tag{1}$$

which simplifies to

$$6x + 4y - 19 = 0. \tag{2}$$

(Do not take the book's word for it; simplify it yourself. Putting in these missing steps is an important part of *reading* a mathematics

---

† Here, circle refers to the points on the circumference. The interior of the circle is denoted by $(x - 3)^2 + (y - 1)^2 < 4$.

Sec. 2-3. Loci

text. Reading scientific material requires a somewhat different technique from reading literature. This is where you are expected to learn it, if you have not already done so. Always read with pencil and paper at hand, and supply *all* missing steps.)

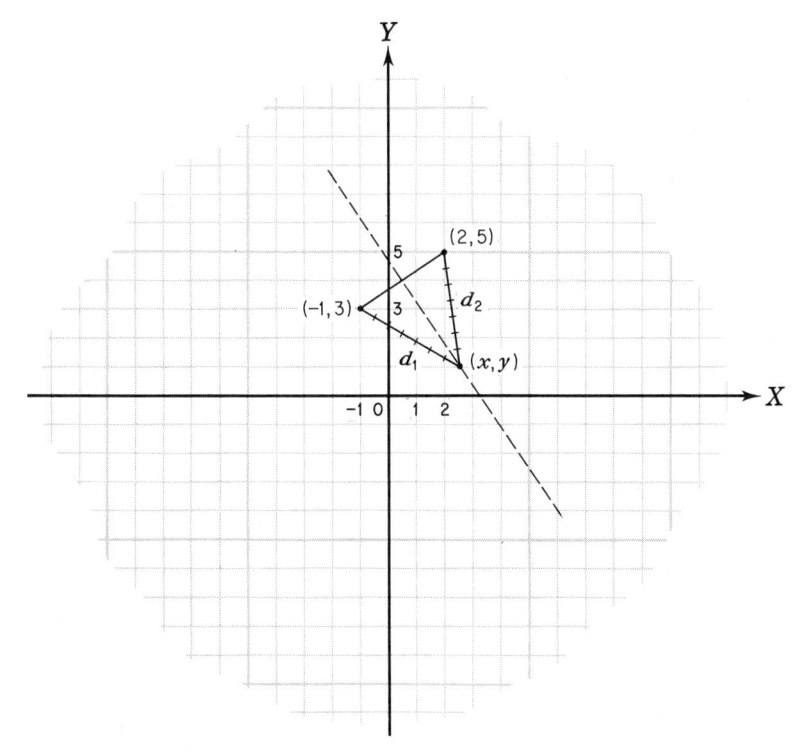

Figure 2-9

Every point on the desired locus satisfies Eq. (2). Furthermore, every point (a,b) which satisfies Eq. (2), or Eq. (1) since they are equivalent equations, must have $d_1 = d_2$, and hence the point lies on the perpendicular bisector. Since both requirements are fulfilled, the desired locus is $6x + 4y - 19 = 0$.

Example 3

Determine the locus of the equation $x = -2$. Every point having x coordinate -2 is on the locus, and every point on the locus has x coordinate -2. The locus is a straight line parallel to the y axis and two units to the left of it. Rework Prob. 5 of Set 2-1 before continuing.

Basic Geometric Theory

Example 4

Sketch the locus of the equation $x^2 - 2x + y - 2 = 0$. Obtain y as a function of x,

$$y = -x^2 + 2x + 2.$$

Substituting various values for x, we glean corresponding values for y. For example, if $x = 3$, $y = -(3)^2 + 2(3) + 2 = -1$. The point $(3, -1)$ therefore lies on the locus. Construct a table by giving values to x and computing the corresponding values of y.

x	-2	-1	0	1	2	3	4	5
y	-6	-1	2	3	2	-1	-6	-13

Plotting these points as coordinates and connecting them by a smooth curve, we obtain the sketch shown in Fig. 2-10. Note that $y > 0$ if $-x^2 + 2x + 2 > 0$.

This happens if $\quad 3 > x^2 - 2x + 1 = (x - 1)^2$.
That is, if $\quad -\sqrt{3} < x - 1 < \sqrt{3}$
or $\quad 1 - \sqrt{3} < x < 1 + \sqrt{3}$
or approximately $0.7 < x < 2.7$.

Also $\quad\quad y < 0$ if $-x^2 + 2x + 2 < 0$
or $\quad\quad 3 < (x - 1)^2$.
That is, if $\quad x - 1 < -\sqrt{3} \quad\quad x - 1 > \sqrt{3}$
or $\quad\quad x < 1 - \sqrt{3} \quad\quad x > 1 + \sqrt{3}$
or approximately $\quad x < -0.7 \quad\quad x > 2.7$.

You should realize that our sketch does *not* contain the entire locus, since very large values can be given to x and the corresponding y determined. The reader will note that we have not verified that every part of our sketch is actually part of the locus. Only eight points have been checked. Later you will learn more about this important and interesting locus. It is closely related to the path of a projectile after it is fired, or of a baseball after it is hit. The curve of water from a garden hose or a waterfall is also closely related to this locus, and one of the strongest possible bracings for a bridge follows the outline of the curve. The curve is called a *parabola*.

parabola

Sec. 2-3. Loci 47

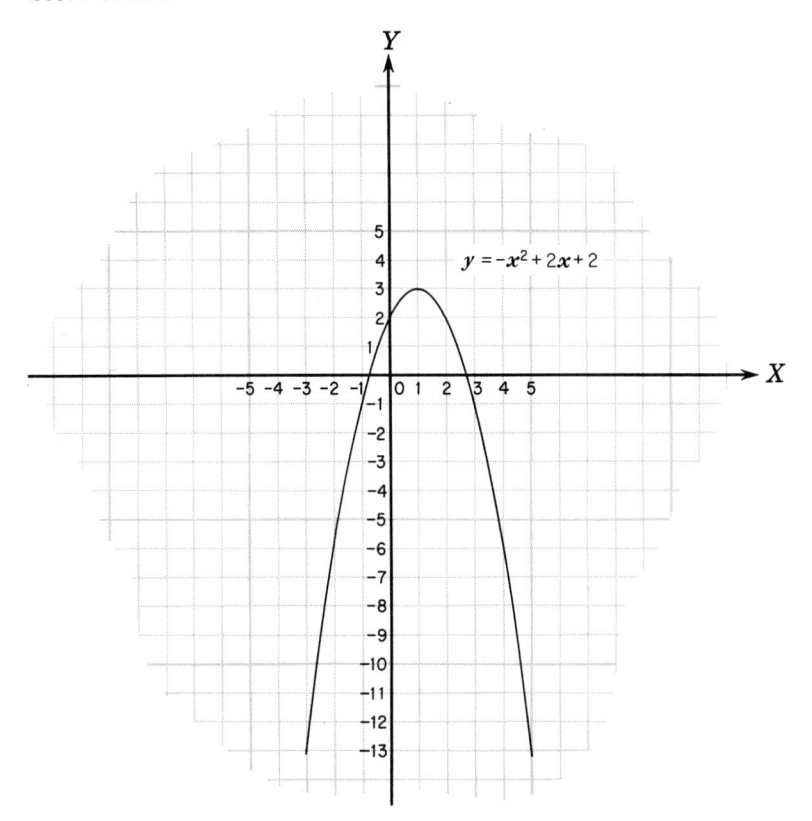

Figure 2-10

Problem Set 2-3

Find the equation of each of the following loci unless otherwise instructed.

1. A circle with radius 3 and center $(-1,2)$.

2. A circle with radius 5 and center $(4,3)$.

3. A circle with radius 8 and center $(-3,-7)$.

4. A circle with radius 1 and center $(0,0)$. This circle is known as a *unit* circle.

5. A circle with radius 2 and center (h,k).

6. A circle with radius r and center (h,k).

7. The perpendicular bisector of the line segment joining $(3,-1)$ and $(5,7)$.

8. The perpendicular bisector of the line segment joining $(2,6)$ and $(3,5)$.

Basic Geometric Theory

9. In Example 2, where are all points such that $d_2 > d_1$? Is this equivalent to $6x + 4y - 19 > 0$?

10. The perpendicular bisector of the segments joining $(0,0)$ and (x_1, y_1).

11. The locus of a point equidistant from the points $(2,3)$ and $(1,7)$.

12. The locus of points equidistant from $(0,3)$ and $(2,-5)$.

13. A straight line parallel to the y axis and five units to the right of it.

14. A straight line parallel to the y axis and passing through the point $(5,-11)$.

15. Shade the locus of all points such that $(x-1)^2 + (y+2)^2 \leq 9$.

16. Shade the locus of all points such that $(x-1)^2 + (y+2)^2 < 9$.

17. A line parallel to the x axis and two units above it.

18. A line parallel to the x axis and passing through $(4,-2)$.

19. A line bisecting the first quadrant.

20. Determine equations of three different lines, each of which passes through the point $(2,-6)$.

21. The x axis.   **22.** The y axis.

*__23.__ The locus of a point which is always as far from the x axis as it is from the point $(1,3)$.

*__24.__ Same as Prob. 23, but using the point $(-2,5)$.

*__25.__ Find the coordinates of all points which lie on both the locus of Prob. 7 and the locus of Prob. 8. Using your knowledge of plane geometry, how many such points would you expect to find?

Identify the locus of each of the following equations or inequalities. Sketch the locus:

26. $(x-4)^2 + (y+1)^2 = 0$.   **27.** $x = 7$.   **28.** $y + 2 = 0$.
29. $x + 4 = 0$.   **30.** $2y - 5 = 0$.   **31.** $3x = \pi$.

32. $(x-4)^2 + (y+1)^2 < 9$.
33. Discuss $(x-4)^2 + (y+1)^2 < 0$.

Sketch each of the following loci:

34. $y - 2x + 3 = 0$.   **35.** $x^2 + y^2 - 25 = 0$.
36. $x^2 + y^2 - 25 = 0$, $y \geq 0$.   **37.** $x^2 + (y+9)^2 + 1 < 5$.

38. $x^2 - 6x - y + 3 = 0$. For what x range is $y < 0$? $y = 0$? $y > 0$?

Sec. 2-4. Solution of Inequalities of First Degree   49

*39. $x + 2y^2 - 3y = 0$. For what y range is $x > 0$? $x = 0$? $x < 0$?

*40. $y < x^2 - 2x + 1$.

2-4. Solution of Inequalities of First Degree

The statement "a is greater than b" ($a > b$) means that $(a - b)$ is positive. The statement "c is less than d" ($c < d$) means that $(c - d)$ is negative. From these properties we see that an inequality remains an inequality of the same order if the same number is added to (or subtracted from) both members.

Illustration

If $a < b$, then $a + 3 < b + 3$ or $a - 17 < b - 17$. Since $5 > 2$, $5 + 3 > 2 + 3$ and $5 - 17 > 2 - 17$.

Furthermore, if both members are multiplied by the same *positive* constant, the order of the inequality is unchanged.

If $a > b$, then $a - b$ is positive; and $ka - kb$ is positive if $k > 0$. Hence $ka > kb$.

Illustration

If $a > b$, then $7a > 7b$. Since $5 > 2$, $7 \cdot 5 > 7 \cdot 2$.

If both members of an inequality are multiplied by the same *negative* constant, the order of the inequality is reversed. If $a > b$, then $-a < -b$.

This rule may be proved by noting that if $a > b$, then $a - b$ is positive. Hence if $-k < 0$, then $(-k)a - (-k)b$ is *negative*. Therefore $(-k)a < (-k)b$.

Illustration

Since $4 < 9$, therefore $(-7)4 > (-7)9$, that is, $-28 > -63$. The latter statement may be written as $-63 < -28$.

unconditional inequality
conditional inequality

An inequality that is true for all values of the letters involved is called an *unconditional inequality*. An inequality that is true only for certain values of the letters involved is called a *conditional inequality*.

Example 1

The inequalities $4 < 9$, $-2 < 1$, $4x + 3 < 4x + 7$, and also $(x + 1)^2 < x^2 + 2x + 9$ are *un*conditional inequalities. The

Basic Geometric Theory

inequality $3x - 4 < 2$ (which is true only if $3x < 6$, that is, if $x < 2$) is a conditional inequality.

The solution of conditional inequalities plays an important role in mathematical analysis. Simple inequalities which do not involve squares or higher powers of the unknown are usually solved using the rules already discussed.

Example 2

Solve the inequality $2x + 7 < 4$. Adding -7 to each member yields $2x < -3$. Multiply each member by $\frac{1}{2}$ to obtain $x < -\frac{3}{2}$.

In terms of a number axis, any x chosen from the heavily shaded portion (Fig. 2-11) will satisfy $2x + 7 < 4$. The number $-\frac{3}{2}$ is *not* included.

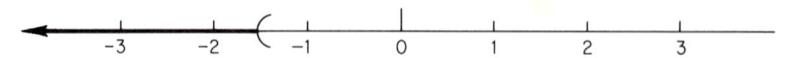

Figure 2-11

Example 3

What values of x satisfy the expression $-1 < 4x - 3 \leq 7$? The addition of 3 to each member, followed by division by 4, gives $2 < 4x \leq 10 \qquad \frac{1}{2} < x \leq \frac{5}{2}$.

Every value of x from the range shaded in Fig. 2-12, including $\frac{5}{2}$ but not $\frac{1}{2}$, will satisfy the original expression.

Figure 2-12

Example 4

What values of x satisfy the condition $\left|\dfrac{x-3}{2}\right| \leq 4$?

$-4 \leq \dfrac{x-3}{2} \leq 4.$

$-8 \leq x - 3 \leq 8.$

$-5 \leq x \leq 11.$

Hence every value of x between -5 and $+11$, including both $x = -5$ and $x = 11$, satisfies the condition (Fig. 2-13).

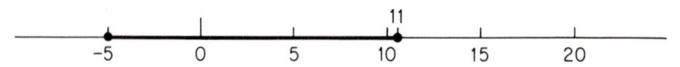

Figure 2-13

Sec. 2-4. Solution of Inequalities of First Degree

Example 5

For what values of x is $|3 - x| < 4$? The given inequality is equivalent to

$$-4 < 3 - x < 4.$$
$$-7 < -x < 1.$$

Since the coefficient of x is negative, multiply by -1, obtaining (Fig. 2-14)

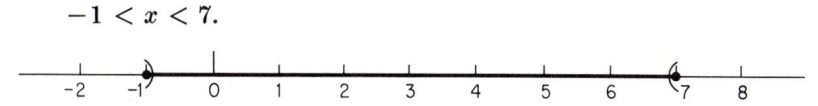

$$-1 < x < 7.$$

Figure 2-14

Example 6

For what values of t is $|t - 3| > 1$? This is equivalent to the two relationships, either $t - 3 > 1$ or $t - 3 < -1$.† Thus $|t - 3| > 1$ is satisfied if either $t > 4$ or if $t < 2$. The shaded portions of the axis in Fig. 2-15 indicate that the values, $t = 2$ and $t = 4$, are excluded.

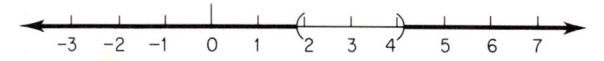

Figure 2-15

Problem Set 2-4

Determine values of the variable for which the following conditions hold. Shade the interval on a number axis.

1. $4x - 3 < 7$.
2. $2x + 5 < 3$.
3. $3x + 1 < 4$.
4. $x - 1 > 11$.
5. $3 - 2x > 13$.
6. $3 - x < 7$.
7. $|x - 2| < 7$.
8. $|2x + 3| < 5$.
9. $3 < x - 41 < 8$.
10. $-1 < 3x + 2 < 9$.
11. $|x - 3| > 7$.
12. $|2x + 1| > 5$.
13. $|3x| > -1$.
14. $|3x| < -1$.
15. $14x - 2 > 71$.
16. $3 - 4x > 7$.
17. $-2 < 13t + 2 < 100$.
18. $-7 < 4t + 2 < -1$.
19. $16 < 4t + 18$.

† The reader should show that the given inequalities are *not* equivalent to $-1 < t - 3 < 1$ or to $1 < t - 3 < -1$. There are no values of t which satisfy the latter. Why?

Basic Geometric Theory

20. Show by shading on a number axis the range of each of the following expressions. Which expressions are equivalent?

(a) $-4 < x < 4$.
(b) $|x| < 4$.
(c) $|x| > 4$.
(d) Either $x > 4$ or $x < -4$.
(e) Either $x < 4$ or $x > -4$.
(f) Both $x < 4$ and $x > -4$.
(g) $|x - 2| < 2$.
(h) $|x - 2| > 4$.

21. Sketch the locus $\sqrt{2x + 3} \leq 8$.

22. Shade the locus represented by $2x - 3y + 5 \geq 0$.

23. Determine an inequality representing all points more than five units from the y axis.

24. Determine the equation of the locus of a point which is always as far from the y axis as from the point (2,3).

25. Consider points (x,y) with $y < x^2 + 5$. For what range of y is $x \geq 0$? For what range of x is $y \geq 0$?

26. Write down the integers (whole numbers) between -5 and $+10$ which satisfy the inequality $-3 < x \leq 7$.

27. If possible, determine one pair of integers x and y which do *not* satisfy the inequality $x^2 + y^2 \leq (x + y)^2$.

28. Solve for t: $3t^2 - 7t - 5 = 0$.

29. Sketch: $y = f(x) = \begin{cases} 4 & \text{if } -3 \leq x \leq 3 \\ 10 - 2x & \text{if } 3 \leq x \leq 5 \\ 10 + 2x & \text{if } -5 \leq x \leq -3 \\ -5 + x & \text{if } 0 \leq x < 5 \\ -5 - x & \text{if } -5 < x < 0 \\ 4 - x & \text{if } 1 < x < 2 \\ 4 + x & \text{if } -2 < x < -1 \\ -3 + |x| & \text{if } -1 < x < 1 \end{cases}$

30. Sketch: $\begin{cases} x = \begin{cases} 1 & \text{if } 1 < y < 3 \\ 2 & \text{if } 1 < y < 3 \\ 3 & \text{if } 1 < y < 3 \end{cases} \\ y = 2 \quad \text{if } 1 \leq x \leq 2 \end{cases}$

31. On what range of values do you think the function determined by the graph of Fig. 2-16 is (a) undefined, (b) single-valued, (c) double-valued, (d) triple-valued, (e) more than triple-valued? The end of each line segment has integral coordinates and is part of the graph.

Sec. 2-5. Slope of a Line   53

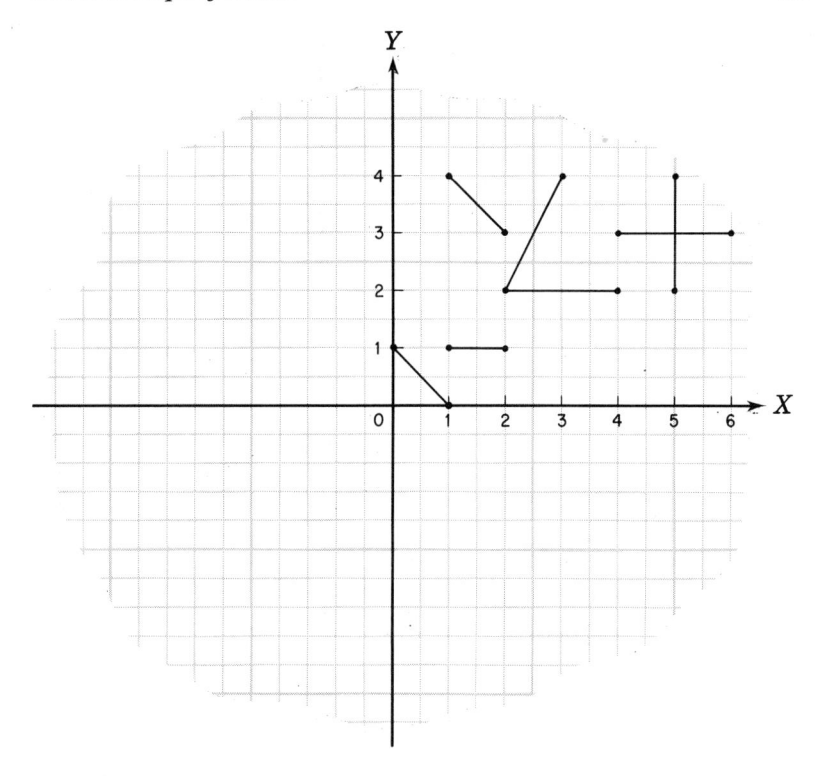

Figure 2-16

2-5. Slope of a Line

slope

Let $C(x_1,y_1)$ and $D(x_2,y_2)$ be two distinct points on a line AB which is not parallel† to the y axis (Fig. 2-17). Construct lines through C and D parallel to the x and the y axes, respectively. Call the intersection Q. The *slope* m of the line AB is the ratio of the directed distances $QD/CQ = m$.

$$m = \frac{QD}{CQ} = \frac{y_2 - y_1}{x_2 - x_1}. \quad \text{(Why?)}$$

The adjective "directed" modifying distances means the distance QD is positive if D lies above Q and negative if D lies below Q. Similarly the distance CQ is positive if Q is to the right of C (see Sec. 2-1) and negative if Q is to the left of C, where it is assumed that the positive x direction is to the right and the positive y direction is upward.

† The slope of a line parallel to the y axis is not defined.

Basic Geometric Theory

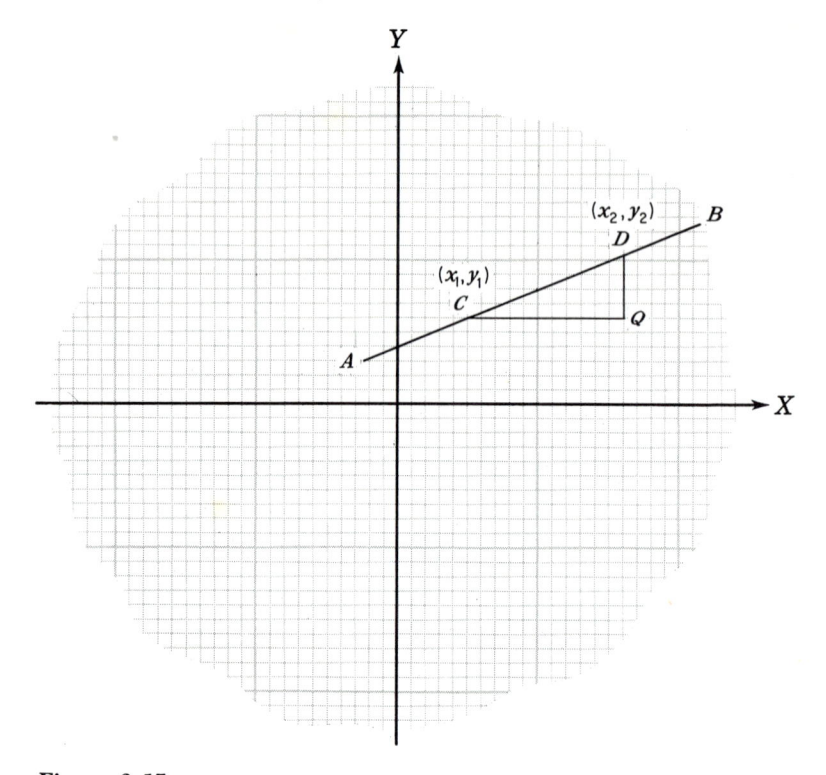

Figure 2-17

Either the ratio QD/CQ or its reciprocal would provide a quantitative measure of the "slant" of a line. The choice $(y_2 - y_1)/(x_2 - x_1)$ has the advantage that a horizontal line (level line) has the numerical slope *zero*. However, the definition of slope makes no provision for vertical lines. Note the distinction between "zero slope" and "the slope is not defined" (no slope) lest horizontal and vertical lines be confused.

If other points C' and D' had been chosen instead of C and D (see Fig. 2-18), then CQD and $C'Q'D'$ would be similar triangles. Consequently

$$m = \frac{QD}{CQ} = \frac{Q'D'}{C'Q'}.$$

Hence (excluding vertical lines) the slope of a straight line is unique. This was taken into consideration when the definition of slope was formulated.

Sec. 2-5. Slope of a Line 55

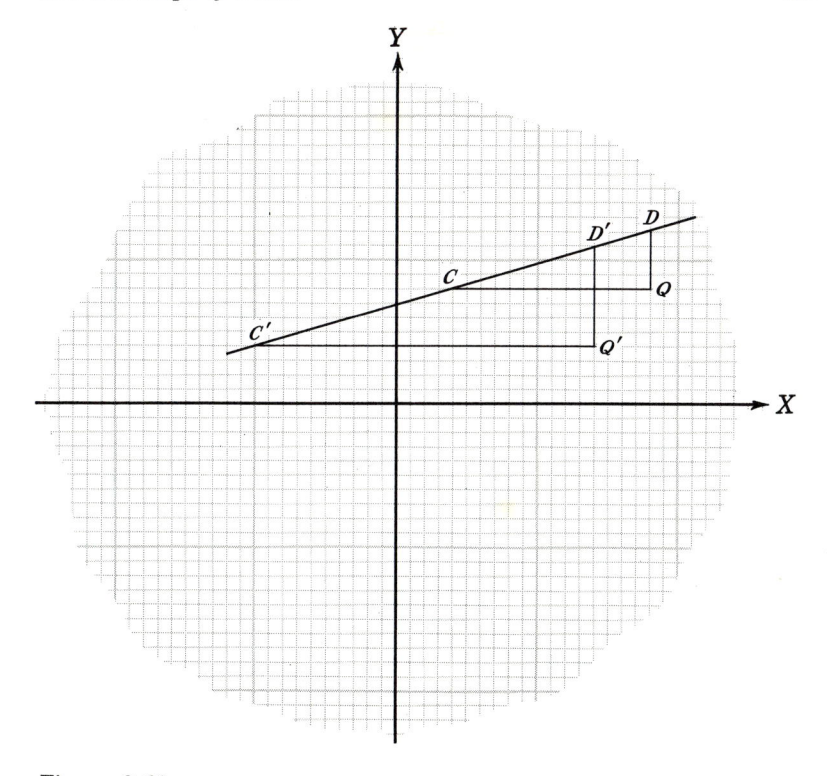

Figure 2-18

Example 1

Find the slope of the line joining $(-2,-3)$ and $(1,2)$.

$$m = \frac{2 - (-3)}{1 - (-2)} = \frac{5}{3}.$$

If the points are considered in the opposite order, then

$$m = \frac{-3 - (2)}{-2 - (1)} = \frac{-5}{-3} = \frac{5}{3}$$

as before.

Example 2

Assume $3x + 2y = 18$ is a line and find its slope. The intercepts provide two convenient points, $(6,0)$ and $(0,9)$. Then

$$m = \frac{9 - 0}{0 - 6} = \frac{9}{-6} = \frac{-3}{2}.$$

Basic Geometric Theory

Problem Set 2-5

In Probs. 1 to 6, determine the slope of a line through the given points. Sketch the line.

1. $(2,-3)$, $(1,5)$.
2. $(3,-1)$, $(4,-6)$.
3. $(2,9)$, $(2,4)$.
4. $(3,2)$, $(1,2)$.
5. $(1,-1)$, $(3,7)$.
6. $(-4,6)$, $(1,6)$.

7. A bridge has trusses as shown in Fig. 2-19. Determine the slope of the 13 trusses if the roadbed is horizontal (zero slope).

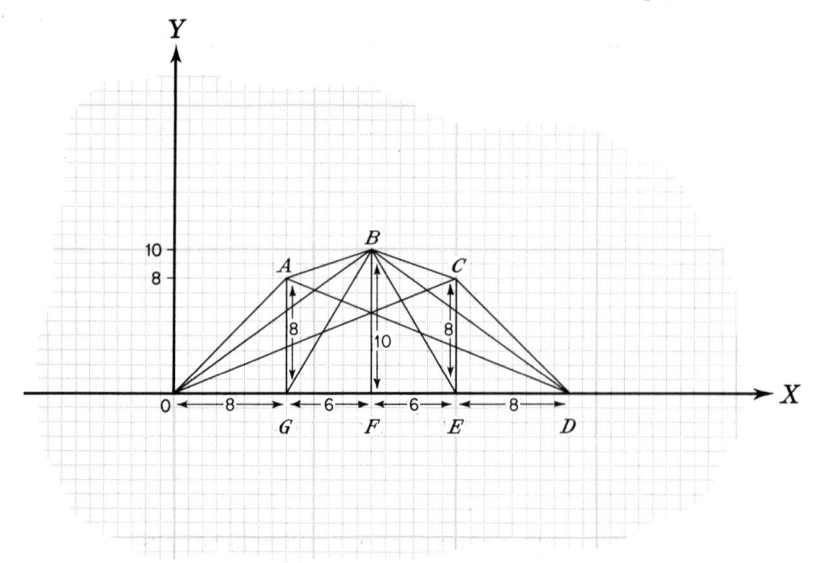

Figure 2-19

In Probs. 8, 9, and 10, find the slope of the given line:

8. $3x + 2y - 5 = 0$.
9. $7x - 3y + 2 = 0$.
10. $4x + 10y = 17$.

11. Sketch the line which passes through $p = (2,-3)$ and has slope $m = 5$.
12. Sketch the line which passes through $p = (1,3)$ and has slope $m = 2$.
13. Sketch the line passing through $(7,-1)$ having slope -3.
14. Sketch the line passing through $(4,-2)$ having slope -1.
15. If $y = L(x)$ is the equation of a line, show that the slope of the line is $\dfrac{\Delta L(x, \Delta x)}{\Delta x}$.

Sec. 2-6. The Equation of a Line

16. Use the result of Prob. 15 to work Probs. 8, 9, and 10.
17. What is the slope of the line $y = 7$?
18. What is the slope of the line $x = 2$?
19. Discuss the slope of the line $6x + 5 = 0$.
20. Discuss the slope of the line $4y - 3 = 15$.
21. Show that a line in the usual coordinate system which rises as we go to the right has positive slope.
22. Show that a line in the usual x, y coordinate system which falls as we go to the right has negative slope.
23. What are the slopes of the two lines through the origin making angles of 45° with the positive half of the x axis?
24. What is the slope of a line parallel to the line of Prob. 23 which lies in quadrants I and III?
25. Determine the slope of a line through the origin which divides the second quadrant into two congruent symmetric regions.
26. Determine the slope of a line parallel to the line of Prob. 25.
27. Determine the slopes of the two lines through (2,1) which make an angle of 30° with the x axis. HINT: Recall from geometry that in a 30°, 60°, or 90° triangle the short side is always half the hypotenuse.
28. Determine the slopes of the two lines through (3,5) which make an angle of 60° with the x axis.
29. Show that the steeper the line the larger the absolute value of the slope becomes.
30. Prove by elementary geometry that two parallel lines have the same slope.
31. Prove that the slope m of a line is the number of units that y increases (decreases if m is negative) for each unit of increase in x. Compare Prob. 15.
*32. If the "slant" of a line were defined as QD/CD in Fig. 2-17 there would be a numerical slant for every line, without exception. In particular, the slant of a horizontal line would be *zero* and of a vertical line *one*. Moreover, parallel lines would have the same slant. Why is "slope" superior to "slant" as a measure of the direction of a line?

2-6. The Equation of a Line

Let $P(x,y)$ be a point on the line through $P_1(x_1,y_1)$ and $P_2(x_2,y_2)$. Construct P_1A and P_2B parallel to the x axis and CP_2 and AP

Basic Geometric Theory

parallel to the y axis (see Fig. 2-20). Then triangles P_1CP_2, P_2BP, and P_1AP are similar right triangles. Therefore

$$\frac{y - y_1}{x - x_1} = \frac{y - y_2}{x - x_2} = \frac{y_2 - y_1}{x_2 - x_1} = m.$$

If P lies on the line joining P_1P_2, then the slope of the segment joining P to P_1 or P to P_2 is the same as the slope of P_1P_2. If a

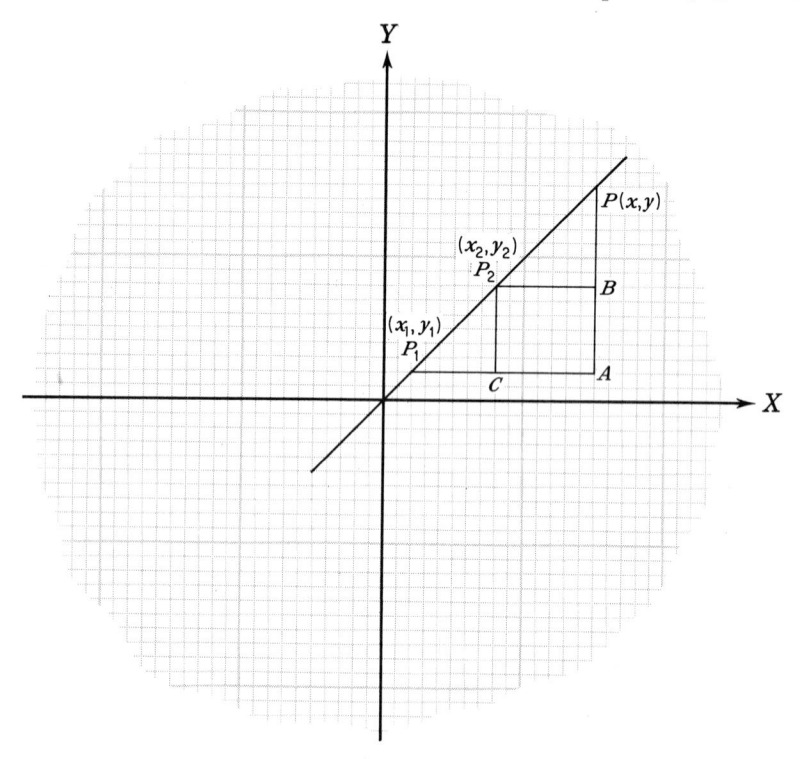

Figure 2-20

point Q, not on the line P_1P_2, is joined to P_1 or P_2, the slopes of QP_1 or QP_2 will not be the same as the slope of P_1P_2. Therefore, those and only those points (x,y) whose coordinates satisfy

$$y - y_1 = \frac{y_2 - y_1}{x_2 - x_1}(x - x_1) \quad \text{or} \quad y - y_1 = m(x - x_1)$$

two-point equation
point-slope equation

lie on the line P_1P_2. The first of these equations is the *two-point equation* of the line. The second is the *point-slope equation* of the line.

Sec. 2-6. The Equation of a Line

Example 1

Write the equation of the line through $(-2,-3)$ and $(1,2)$.

$$y - (-3) = \frac{2 - (-3)}{1 - (-2)} [x - (-2)]$$

$$y + 3 = \frac{5}{3}(x + 2)$$

$$5x - 3y + 1 = 0.$$

We might, instead, first determine the slope

$$m = \frac{2 - (-3)}{1 - (-2)} = \frac{5}{3}$$

and use $(1,2)$. Then, by the point-slope equation,

$$y - 2 = \frac{5}{3}(x - 1).$$

The reader should show that using $m = \frac{5}{3}$ and the point $(-2,-3)$ yields the same equation.

Example 2

Write the equation of the line through $(0,b)$ having slope m (Fig. 2-21).

$$y - b = m(x - 0),$$
$$y = mx + b.$$

slope y-intercept equation   Since b is the y intercept of the line, this is called the *slope y-intercept equation* of the line. Observe that each of the preceding equations of a line may be placed in the form $Ax + By + C = 0$. Such equations are called *linear*.

If $B = 0$, $A \neq 0$, then $x = -C/A$ is the equation of a line parallel to the y axis.

If $B \neq 0$, then $y = -Ax/B - C/B$ is the equation of a line having slope $m = -A/B$, and y-intercept $b = -C/B$.

Example 3

Find slope and y-intercept for the line $3x + 2y = 6$ (Fig. 2-22).

$$y = -\frac{3x}{2} + \frac{6}{2} \qquad m = -\frac{3}{2} \qquad b = 3.$$

Basic Geometric Theory

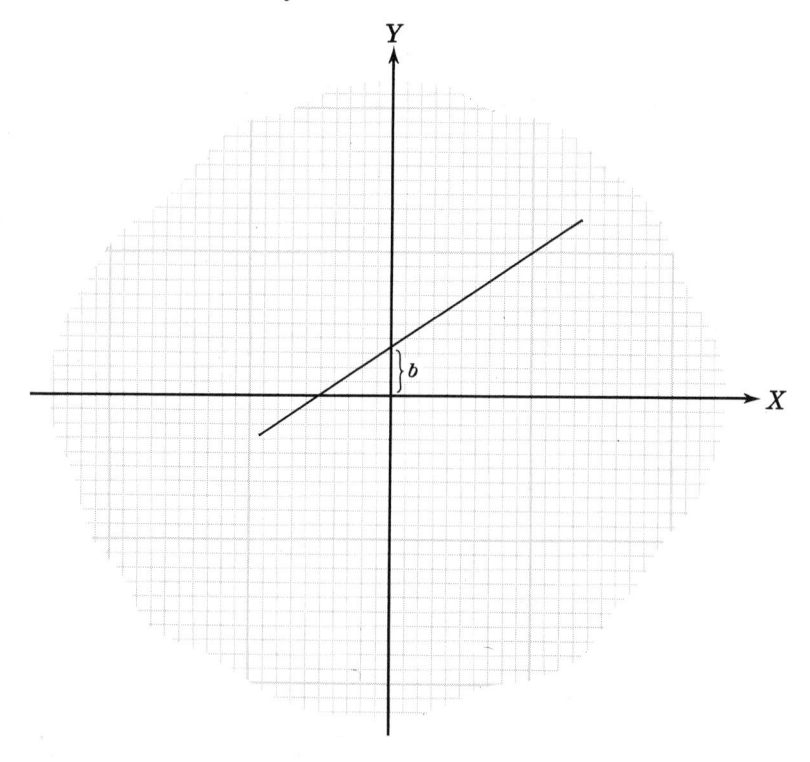

Figure 2-21

Example 4

Show that if two lines have the same slope, the lines are parallel. Consider the two lines:

$$y = mx + K, \quad y = mx + L.$$

Clearly if there is a point (a,b) which satisfies both equations, then $b - ma = K$ and $b - ma = L$. Hence $K = L$. If $K = L$ the two lines have the same equation and hence are the same line. However, if $K \neq L$, the original statement is false and the two lines have no point in common. Hence, by definition, they are parallel.

Example 5

Write the equation of the line through $(-1,-2)$ parallel to the line $3x + 2y = 6$. As above, $m = -\frac{3}{2}$. Then

Sec. 2-6. The Equation of a Line

$$y - (-2) = -\frac{3}{2}[x - (-1)],$$

$$y + 2 = -\frac{3}{2}(x + 1),$$

$$3x + 2y + 7 = 0.$$

family of lines   The phrase *family of lines* is used to describe a set of lines having some common property.

Example 6

Determine an equation of the family of lines having y intercept 7.
The family is represented by the equation

$$y = mx + 7.$$

By assigning various values to m, different lines in the family are obtained.

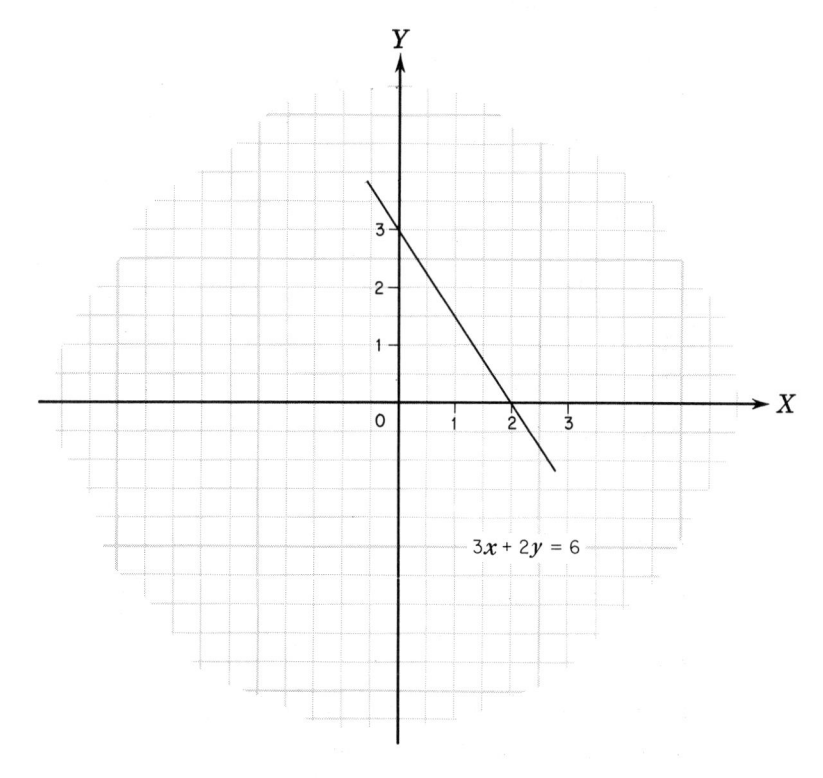

Figure 2-22

Example 7

Determine an equation of the family of lines through $(3,-8)$.

This family is represented by the equation $y + 8 = m(x - 3)$ where the slope m is the parameter. Note that there exists one line, namely $x = 3$, in the family which is not obtained by giving values to the parameter m. This line has no defined slope.

Example 8

Find an equation of the family of lines parallel to $3x + 5y - 7 = 0$.

Since the given line has slope $-\frac{3}{5}$, the desired equation is

$$y = -\frac{3x}{5} + b,$$

where the parameter is the y intercept b.

Problem Set 2-6

In Probs. 1 to 5, determine the equation of a line having the given slope and passing through the given point p.

1. $m = 2$, $p(-1,6)$.
2. $m = 3$, $p(3,4)$.
3. $m = -1$, $p(2,-1)$.
4. $m = \frac{1}{2}$, $p(\frac{1}{2},\frac{1}{3})$.
5. $m = -\frac{1}{3}$, $p(4,-2)$.

In Probs. 6 to 9, determine the equation of a line through the given points. Use two different methods on each problem.

6. $(3,1)$, $(2,4)$.
7. $(4,-3)$, $(-1,-3)$.
8. $(1,6)$, $(-1,-3)$.
9. $(-2,0)$, $(0,7)$.

10. Show that the equation of a line having nonzero x and y intercepts a and b is $x/a + y/b = 1$. Is the restriction necessary?

In Probs. 11 to 15, determine the slope and the y intercept of the given line.

11. $3x + y - 5 = 0$.
12. $7x - 14y + 10 = 0$.
13. $3x + 12y - 4 = 0$.
14. $x = 2y - 3$.
15. $3x + 5 = y$.

In Probs. 16 to 20, find the equation of a line passing through the point p and parallel to the line L.

16. $p(3,-7)$, $L: 4x + y - 3 = 0$.

Sec. 2-6. The Equation of a Line

17. $p(2,1)$, $L: x + 2y - 4 = 0$. **18.** $p(3,4)$, $L: 3x = y + 1$.
19. $p(-1,0)$, $L: 4x + 3y = 6$. **20.** $p(4,2)$, $L: 6x - 2y + 41 = 0$.

In Probs. 21 to 30, find the equation of a line having the designated properties.

21. Through $(3,-7)$ and parallel to the x axis.
22. Through $(2,-4)$ and parallel to the x axis.
23. Through $(-3,1)$ and parallel to the y axis.
24. Through $(4,-7)$ and parallel to the y axis.
25. Through $(4,-6)$ and parallel to the line through the origin which bisects quadrant II.
26. Through $(-7,4)$ and parallel to the line through the points $(3,4)$ and $(2,-7)$.
27. Through the origin with slope m.
28. Through $(3,-4)$ with slope m.
29. Through $(a,0)$ with slope 0.
***30.** Through the intersection of the lines $4x + y = 2$ and $x = 3$, and parallel to the line $3x - 5y + 17 = 0$.
31. In economic theory a "trend line" may be determined by first computing coordinates of two average points and then writing the equation of a line through these two points. Find the equation of a trend line through $(3.1, 2.6)$ and $(6.4, 9.8)$.
32. Find the equation of a trend line through $(6.1, 2.1)$ and $(9.6, 4.1)$.
33. Find the equation of a trend line through $(6.0, 9.3)$ and $(15.1, 1.4)$.
34. Determine the equation of a line which passes through $(2,-3)$ and $(5,7)$.
35. What is the equation of a line having x intercept $\frac{3}{2}$ and y intercept 7?
36. A rectangular box with a square base and an open top is to contain 50 cu ft. Set up a formula for the area of the material used (that is, set up the function you would use to determine the box requiring the least material, but do not bother to determine this minimum).
37. Determine the x and y intercepts of $3x + 7y + 5 = 2$.
38. Determine the equation of a line through $(2,-11)$ parallel to $2x - 5y + 6 = 0$.
39. Write the equation of a line parallel to the x axis and eight units below it.
40. What is the slope of the line $3x = 14$?

41. Write the equation of a line parallel to the y axis and five units to its right.

42. Write the equation of the locus of points equidistant from $(-3,5)$ and $(7,1)$.

43. Write the equation of the line through the origin which bisects that portion of the line $8x - 3y + 24 = 0$ which is included between the coordinate axes.

44. Find the member of the family $y = mx + 2$ which is parallel to $2x + y = 0$.

45. Write the equation of the family of lines parallel to the line through $(7,1)$ and $(-3,5)$ and select the member of the family which passes through $(4,-3)$.

46. Obtain the equation of the line through the intersection of $2x + 3y - 9 = 0$ and $x + 2y - 7 = 0$, and with slope $-\frac{1}{3}$.

47. Find c so that $x - y + 2 = 0$, $2x - 3y + 7 = 0$, and $3x + 2y + c = 0$ shall meet in a point.

48. The point $(5,-3)$ is the mid-point of the segment of a line which is included between the coordinate axes. Write the equation of the line.

49. Sketch the line passing through $(3,-1)$ having a slope -2.

50. Determine the slope of the line $y = p(x) = 5x - 6$ and also determine the coordinates of some one point on this line.

51. Given the triangle with vertices $A(2,-4)$, $B(6,-8)$, $C(2,6)$. Find:

(a) The equation of side AC.
(b) The length of the altitude from B to side AC.
(c) The length of side AC.
(d) The area of the triangle.

52. Write the equation of the locus of points which are six times as far from $x = 1$ as from $x = 8$. Think carefully before you decide whether the locus consists of one, two, or three lines. Under what conditions might an answer other than the one you gave be correct?

53. A rectangular swimming pool is 75 by 20 ft at ground level. The bottom of the pool is a plane which slants uniformly from one end of the pool to the other. If the pool is full, the water is 3 ft deep at the shallow end and 10 ft deep at the deep end. Determine the volume v of water in the pool as a function of the depth h of the water at the deep end. Note that the function has one formula for

Sec. 2-7. Self Test

$0 \leq h \leq 7$ and another for $7 < h \leq 10$. Outside of these ranges the function is not defined. Graph v as a function of h.

54. Determine the exposed surface area of the water in the pool of Prob. 53 as a function of the depth of the water at the deep end. Graph the function.

2-7. Self-test

1. Find the equation of the line joining $(4,7)$ and $(2,-5)$.

2. Determine the solution set of $4x - 3 < 11$.

3. Shade on a number axis the region representing the solution set of $|x - 5| < 3$.

4. Sketch and find the slope of each of the following lines:
(a) $3x + 11y - 5 = 0$.
(b) $2x + 4y = 5x - 2y + 11$.
(c) $x = 4y - 12$.

5. Determine the equation of the locus of all points at a distance of 3 from the point $(2,-7)$.

6. Sketch the locus $y = (x - 3)(x + 2)$.

7. Shade the set of points in the xy plane which satisfy the locus $(x - 2)^2 + y^2 + 3 < 7$.

3

Tangents and Limits

3-1. The Line Tangent to a Curve at a Point

tangent

To give a satisfactory definition of a line tangent to a curve, it is convenient to use the idea of *limit*, which is discussed in detail in later mathematical work. At this point we speak merely of a limiting geometric position, an idea which is already familiar. The tangent PT to a curve at a point P on that curve is the limiting position of the secant line PS as the point S approaches P by moving on the curve, if this limiting position exists *and is unique* (Fig. 3-1). (An example of a curve which has two such limiting positions is given in Prob. 10.)

Note that nothing is said about the secant line when $S = P$. No secant line is determined in this case since one point does not determine a line. If this definition seems difficult to you or if you feel that the inadequate phrase "cuts the curve at only one point" might suffice as a definition for a tangent, you should read the short article, "What Is a Geometric Tangent?" in *The Mathematics Teacher*, November, 1957, p. 498.

Problem Set 3-1

1. You will understand the definition of tangent better if you draw a curve on a piece of paper and locate the point P by inserting a pin or thumbtack. Now rest a ruler on the paper, touching the pin. Think of the edge of the ruler as the secant PS, and move it in such

Sec. 3-1. The Line Tangent to a Curve at a Point

a manner that S approaches P along the curve. Do this several times, and locate what you consider the limiting position or tangent line, taking S on both sides of P.

2. Trace the curve of Fig. 2-10 (page 47). Locate the point $(3,-1)$. Prove that this point lies on the locus by showing that it satisfies

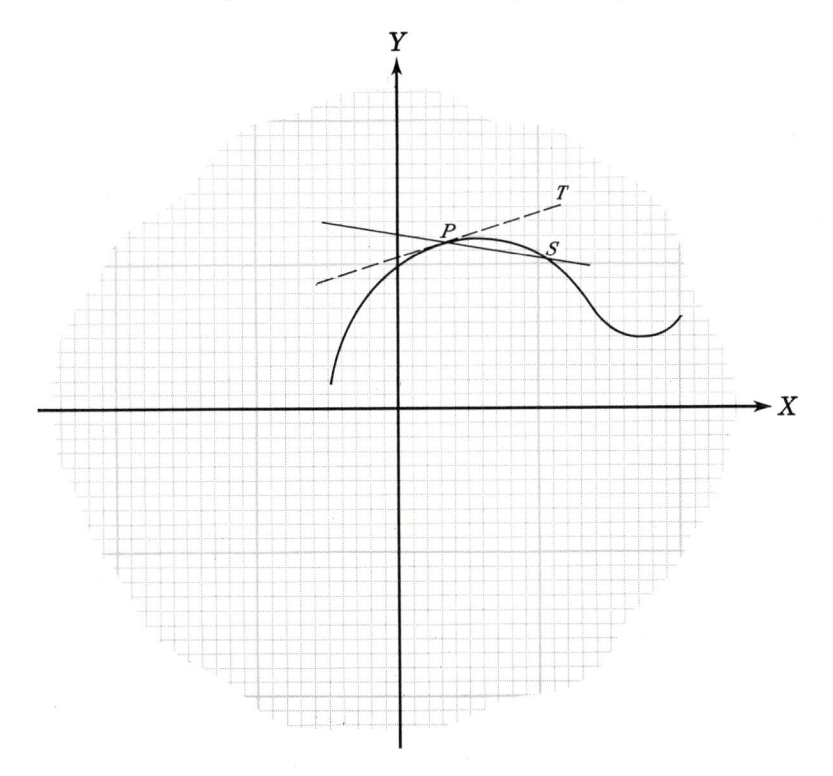

Figure 3-1

$y = -x^2 + 2x + 2$. Using the ruler technique described in Prob. 1, estimate the limiting position of the secant PS and draw a tangent at P. Estimate the slope of this tangent line.

3. Same as Prob. 2 for $P(0,2)$.

4. Show by the ruler technique that the line tangent to the circle $x^2 + y^2 = 25$ at the point $(3,4)$ probably has slope between -0.7 and -0.8, that is, $-0.8 \leq m \leq -0.7$.

5. Show that the slope of the line tangent to $x^2 + y^2 = 169$ at $(-5,12)$ is between 0.4 and 0.5, that is, $0.4 \leq m \leq 0.5$.

Tangents and Limits

6. The locus of $y^3 = x^2$ is shown in Fig. 3-2. Show by the ruler technique that the y axis is probably the tangent line at the point (0,0). What is the equation of this line?

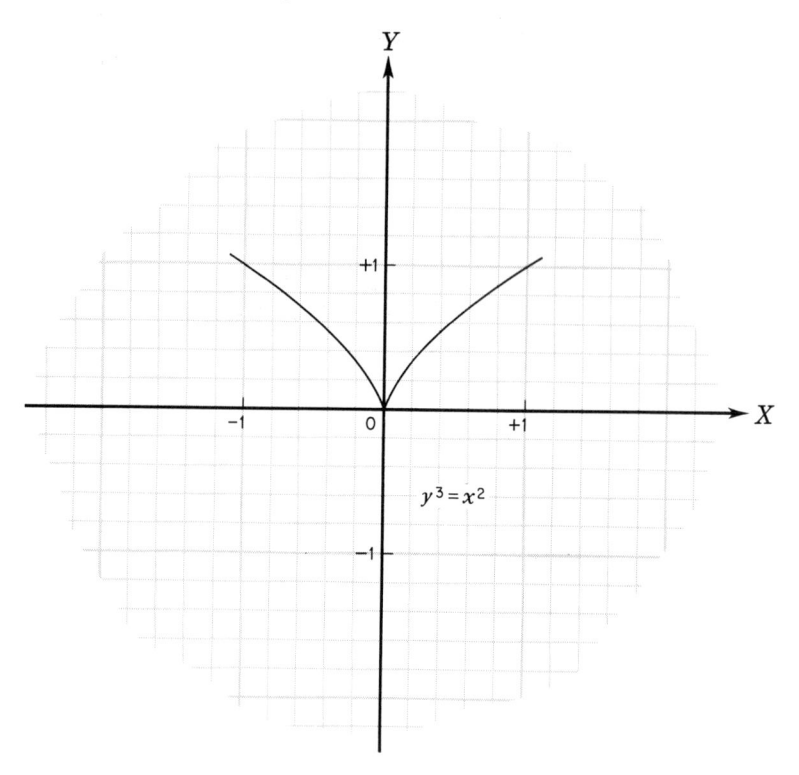

Figure 3-2

7. Show that the tangent to a line AB at any point on a line is the line itself.

8. The locus of $y = 2x + 3$ is a line. What is the slope of the tangent to this line at (4,11)?

9. Sketch the locus $y = 2|x|$. NOTE: $y = 2|x|$ never takes on negative values for y. Determine the slope of the tangent line at the point (3,6) and at the point (−2,4).

10. Attempt to determine the tangent to $y = 2|x|$ at the origin. Let S be to the right of $x = 0$ and then let S be to the left of $x = 0$ and show that these two limiting positions are different; *hence, no tangent line exists at* (0,0). Note the distinction between this and Prob. 6.

Sec. 3-2. Limit of a Function 69

11. Discuss reasons why the statement "a tangent line is perpendicular to the radius at the point of tangency" is not a valid definition of a tangent line unless the locus is a circle.

12. Show that the statement "the tangent line is a line which touches the curve at only one point" is not a valid definition for a tangent line to a general curve. Consider a sine curve.

13. Sketch an example of a curve which is tangent to the line $y = x/2$ three times and crosses it five times.

14. Give an example showing that a line may be tangent to a curve at a point and also cross the curve at the same point. Is $y = x^3$ such a curve? What is the point?

15. Assuming that the slope of the tangent line of Prob. 4 is $-\frac{3}{4}$, find its equation. Solve the equation of the line you obtain and the equation of the circle simultaneously to obtain the points of intersection of the line and the circle. Interpret your results.

3-2. Limit of a Function

In Sec. 3-1, the limiting position of a secant line was considered. We now consider the limiting value of a function. We shall later associate these two ideas to determine the slope of the line tangent to a curve at a given point.

Illustration

If x is near 2, $f(x) = 4x^2 - 3$ is close to 13. The closeness of $f(x)$ to 13 depends upon the nearness of x to 2. If, for example,

$$1.999 < x < 2.001$$
then $$4(1.999)^2 - 3 < f(x) < 4(2.001)^2 - 3$$
$$12.98 < f(x) < 13.02.$$

This could be restated: if

$$-0.001 < x - 2 < 0.001,$$
then $$-0.02 < f(x) - 13 < 0.02,$$

or, using absolute value signs, if

$$|x - 2| < 0.001,$$
then $$|f(x) - 13| < 0.02.$$

Tangents and Limits

In other words, if x is within 0.001 unit of 2, then $f(x)$ will be within 0.02 unit of 13. If x is forced nearer 2, say $|x - 2| < 0.000001$, then $|f(x) - 13| < 0.000016$. This suggests that possibly the difference between $f(x)$ and 13 can be made as small as we wish by taking x sufficiently near 2.

Illustration

Consider another function given by

$$f(x) = \frac{x^2 + 9x - 22}{x - 2}.$$

For this new function it will be found that if x is near 2, *but not equal to* 2, then $f(x)$ is close to 13. Observe that $f(2)$ is meaningless (has no defined value); consequently, x is not permitted to take the value 2. If $x \neq 2$, then we may write

$$f(x) = \frac{x^2 + 9x - 22}{x - 2} = x + 11. \quad \text{(Why?)}$$

From this it is seen that the difference between $f(x)$ and 13 becomes numerically smaller as x becomes closer to but not equal to 2.

In each of these illustrations $f(x)$ is said to approach 13 as a limit as x approaches 2 as a limit. In symbols we write $\lim_{x \to 2} f(x) = 13$ (Fig. 3-3). In the first illustration $f(2) = 13$, but in the second illustration the symbol $f(2)$ is without meaning.

More generally, we say a function f has limit L as x approaches b if the difference between $f(x)$ and L may be made smaller in absolute value than any preassigned positive number ϵ† for each x sufficiently near b but not b. The restriction $x \neq b$ takes care of the difficulties appearing in the second illustration given. In symbols the statement "the limit as x approaches b of the function is L" is written

$$\lim_{x \to b} f(x) = L.$$

We now state a more formal mathematical definition.

---

† The symbol ϵ is the Greek letter epsilon. It is often used in mathematics as a (small) positive number. The symbol δ is the Greek letter delta, the lower-case form of Δ.

limit

Sec. 3-2. Limit of a Function

The function $f(x)$ has the *limit* L as x approaches b [in symbols, $\lim_{x \to b} f(x) = L$] if for every positive number ϵ there is a positive number δ such that for each x satisfying $0 < |x - b| < \delta$, it follows that $|f(x) - L| < \epsilon$.

Note that this definition formalizes the statement given above.

$$f(x) = \frac{x^2 + 9x - 22}{x - 2}$$

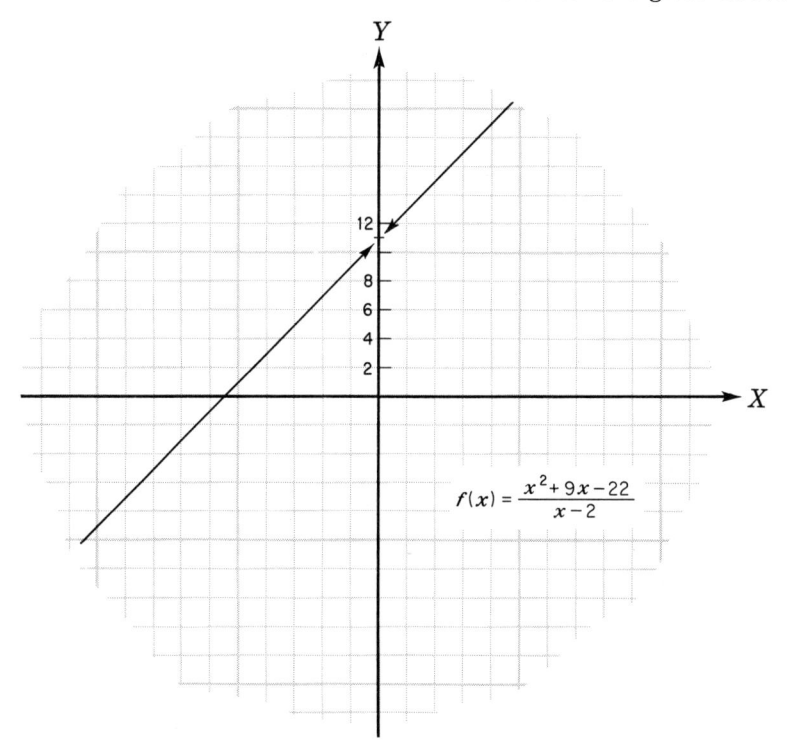

Figure 3-3

Observe that the definition does *not* say that there exists a number δ such that for every $\epsilon > 0$, if $0 < |x - b| < \delta$, then $|f(x) - L| < \epsilon$. The definition states the ϵ must be chosen first; then, no matter what $\epsilon > 0$ is chosen, there exists a δ such that if $x - b$ is smaller in absolute value than δ and $x \neq b$, then $f(x) - L$ is less in absolute value than ϵ. The ϵ is chosen *before* a δ is determined. The order of choice is important. If the ϵ is changed, then it may be necessary to determine a new δ. The reader should note that this definition does *not* say that $f(b) = L$. *The definition says nothing at all about $f(x)$ when $x = b$.* In fact, $f(b)$ need not even be defined. The second illustration of this section is an example of this.

3-3. Continuity

In Sec. 3-2 we pointed out that $\lim_{x \to 2} f(x)$ need not equal $f(2)$. If it does, the function $f(x)$ is said to be continuous at $x = 2$.

continuity   The function $f(x)$ is *continuous* at $x = b$ if the following three conditions are *all* satisfied:

1. $f(x)$ has a definite value $f(b)$ at $x = b$; that is, $f(b)$ is defined.
2. $f(x)$ has a limit L as x approaches b; that is, $\lim_{x \to b} f(x) = L$ exists.
3. The limit L is equal to $f(b)$; that is, $\lim_{x \to b} f(x) = f(b)$.

In symbols, we may say the function f is continuous at b if $\lim_{x \to b} f(x) = f(b)$. The reader may find it helpful to think of a continuous curve as one which can be sketched without removing the pen from the paper.

Example 1

Show that $f(x) = 4x^2 - 3$ is continuous at $x = 2$.

1. $f(2) = 13$
2. $\lim_{x \to 2} f(x) = 13$
3. $13 = 13$

thus $f(x) = 4x^2 - 3$ is continuous at $x = 2$.

Example 2

Is $f(x) = \dfrac{x^2 + 9x - 22}{x - 2}$ continuous at $x = 2$?

The $\lim_{x \to 2} f(x) = \lim_{x \to 2} \dfrac{x^2 + 9x - 22}{x - 2} = \lim_{x \to 2} (x + 11) = 13$† exists as was shown in the previous section. However, $f(2)$ is meaningless and hence, $f(x) = \dfrac{x^2 + 9x - 22}{x - 2}$ is *not* continuous at $x = 2$.

---

† Note that $\dfrac{x^2 + 9x - 22}{x - 2} = \dfrac{(x - 2)(x + 11)}{x - 2} = x + 11$ if $x \neq 2$. Since $\lim_{x \to 2} f(x)$ is not concerned with $f(x)$ at $x = 2$, $(0 < |x - 2|)$, we may write $\lim_{x \to 2} \dfrac{x^2 + 9x - 22}{x - 2}$ $= \lim_{x \to 2} (x + 11) = 13$.

Sec. 3-3. Continuity

Example 3

Consider the functions

$$f_1(x) = \frac{1}{x} \quad f_2(x) = \frac{1}{x^2} \quad f_3(x) = \frac{x}{|x|}.$$

Each of these functions is undefined for $x = 0$ and consequently not continuous at $x = 0$. (Why?) For each of these functions we ask the question, "Does $\lim_{x \to 0} f(x)$ exist?" The sketches of these functions are helpful in understanding the reasoning involved.

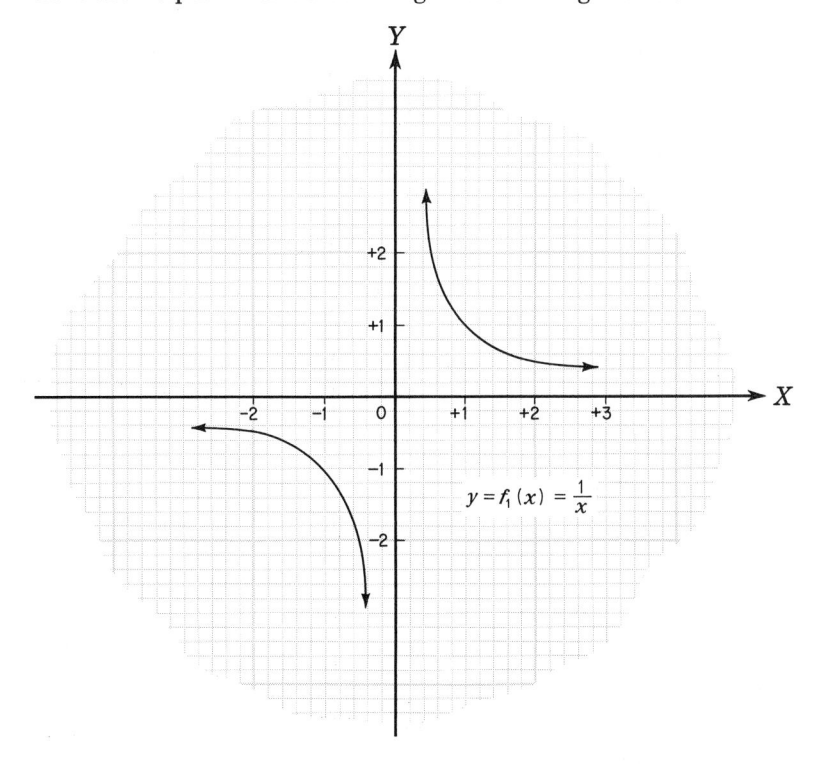

Figure 3-4

In each case the function is defined everywhere except at $x = 0$. The question "Does $\lim_{x \to 0} f(x)$ exist?" asks, "Does the locus $y = f(x)$ become and remain close to some *one* point for x near, but not equal to, zero?"

In the first case, for x near zero, $f_1(x) = 1/x$ may be either large and positive or large and negative; hence the locus does not approach a definite point as x approaches zero (but x is not equal to zero). The $\lim_{x \to 0} 1/x$ does not exist (Fig. 3-4).

The locus of $y = 1/x^2$ is always above the x axis. If x is close to zero, then y is very large. In fact, given any y value, it is always possible to find values of x close to zero so that $1/x^2$ is much larger than the y value in question. Again, the locus does not approach a finite point as x ($x \neq 0$) approaches zero. Hence, $\lim_{x \to 0} 1/x^2$ does not exist (Fig. 3-5).

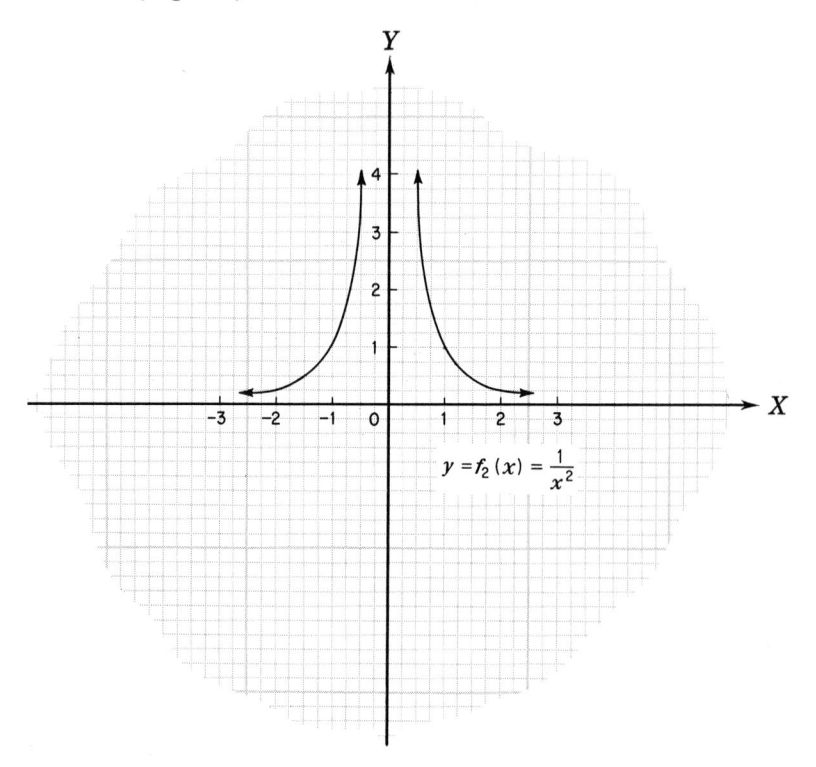

Figure 3-5

The locus $y = x/|x|$ is the line $y = -1$ for $x < 0$, is undefined at $x = 0$, and is the line $y = +1$ for $x > 0$. If $x < 0$ and x approaches

Sec. 3-3. Continuity

zero, then $y = x/|x|$ approaches $(0, -1)$. If $x > 0$ and x approaches zero, then $y = x/|x|$ approaches $(0, 1)$. There is no *one* point that $y = x/|x|$ approaches as x approaches zero. Thus $\lim_{x \to 0} x/|x|$ does not exist (Fig. 3-6).

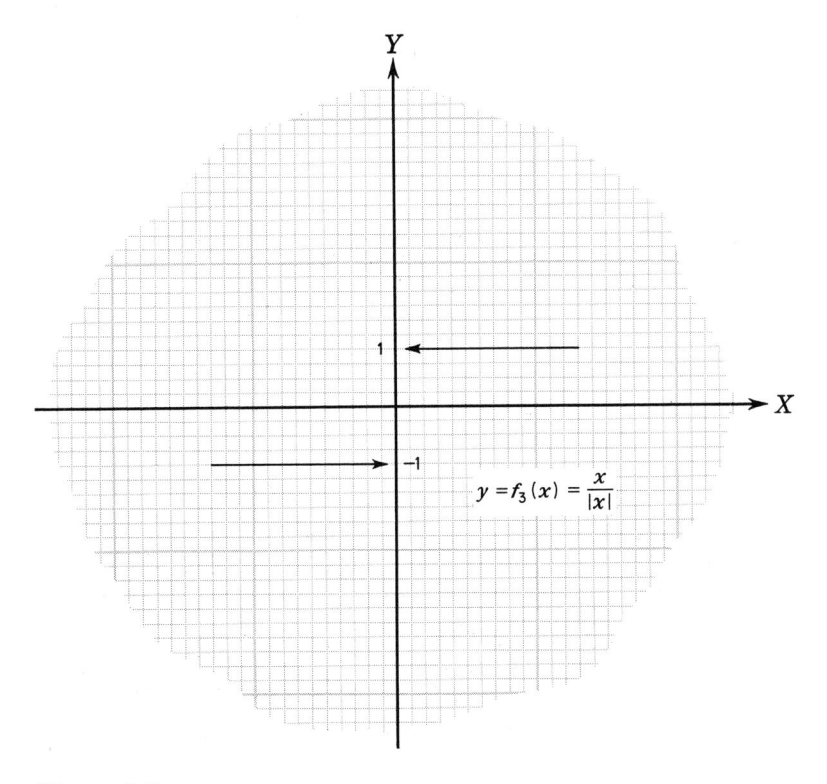

Figure 3-6

Example 4

Consider the functions

$$f_4(x) = \frac{x^2}{|x|} \qquad f_5(x) = \frac{x^2 + x}{x} \qquad f_6(x) = |x|.$$

For each of these functions we ask the question "Does $\lim_{x \to 0} f(x)$

76   *Tangents and Limits*

exist?" Again, we shall examine the sketches to aid our understanding of the reasoning involved. The question "Does $\lim_{x \to 0} f(x)$ exist?" asks, "Does the locus of $y = f(x)$ approach (that is, become and remain close to) some one point for x near, *but not equal to*, zero?" Note that it makes no difference what, if any, value is given to $f(x)$ at $x = 0$. The $\lim_{x \to 0} f(x)$ is concerned with values of $f(x)$ for x near to, but *not* equal to, zero.

The locus of $y = x^2/|x|$ approaches the point (0,0) as x approaches

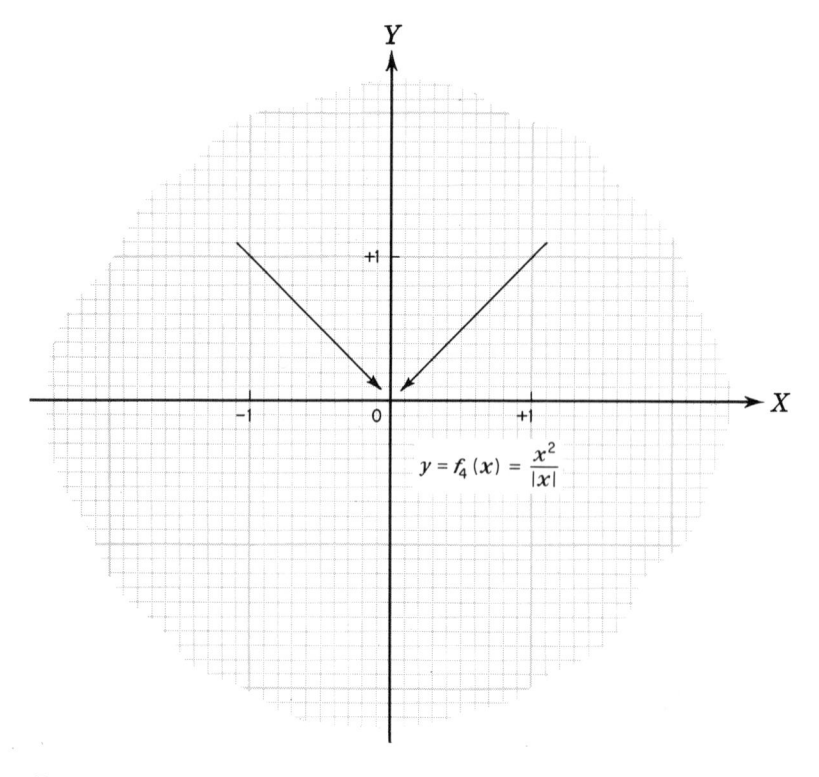

$$y = f_4(x) = \frac{x^2}{|x|}$$

Figure 3-7

Sec. 3-3. Continuity

(but is not equal to) zero (Fig. 3-7). Hence, $\lim_{x \to 0} x^2/|x| = 0$, even though $f_4(0)$ is meaningless.

The locus of $y = (x^2 + x)/x$ is the line $y = x + 1$, with the point (0,1) removed (Fig. 3-8). As x approaches zero, the locus of $y = (x^2 + x)/x$ approaches (0,1) and

$$\lim_{x \to 0} f_5(x) = \lim_{x \to 0} \frac{x^2 + x}{x} = 1,$$

regardless of what value, if any, is given to $f_5(0)$.

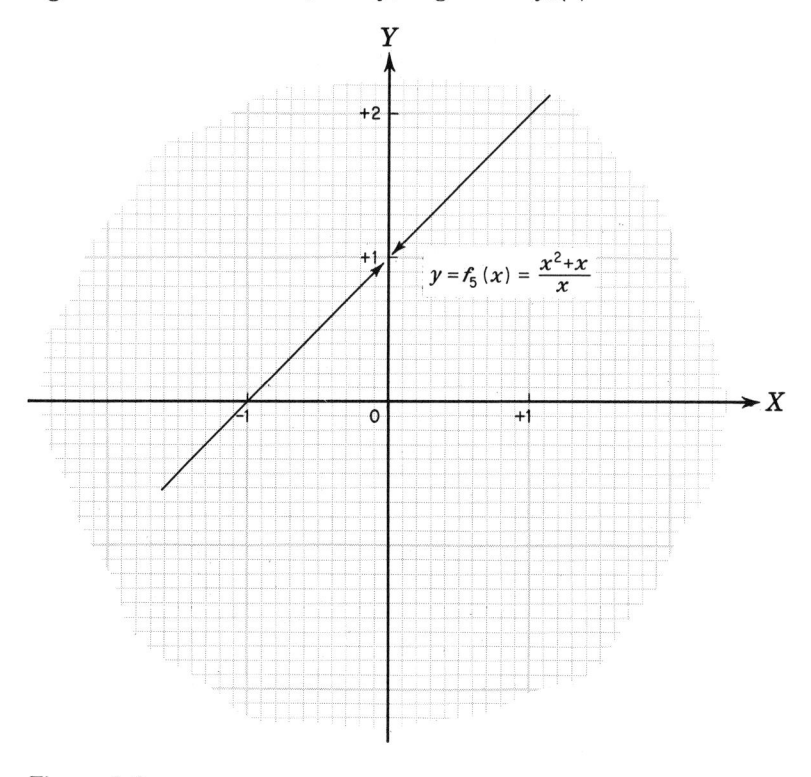

Figure 3-8

The locus of $y = |x|$ becomes and remains arbitrarily close to $(0,0)$ for x near to, but not equal to, zero, and $\lim_{x \to 0} |x| = 0$. In this case it is also true that $f_6(0) = 0$, but this has nothing to do with $\lim_{x \to 0} f_6(x)$ (Fig. 3-9). We have defined a function f to be continuous at b if both $\lim_{x \to b} f(x)$ and $f(b)$ exist, and $\lim_{x \to b} f(x) = f(b)$. The function f_6 just discussed has these properties at zero. Hence we conclude that f_6 is *continuous* at zero.

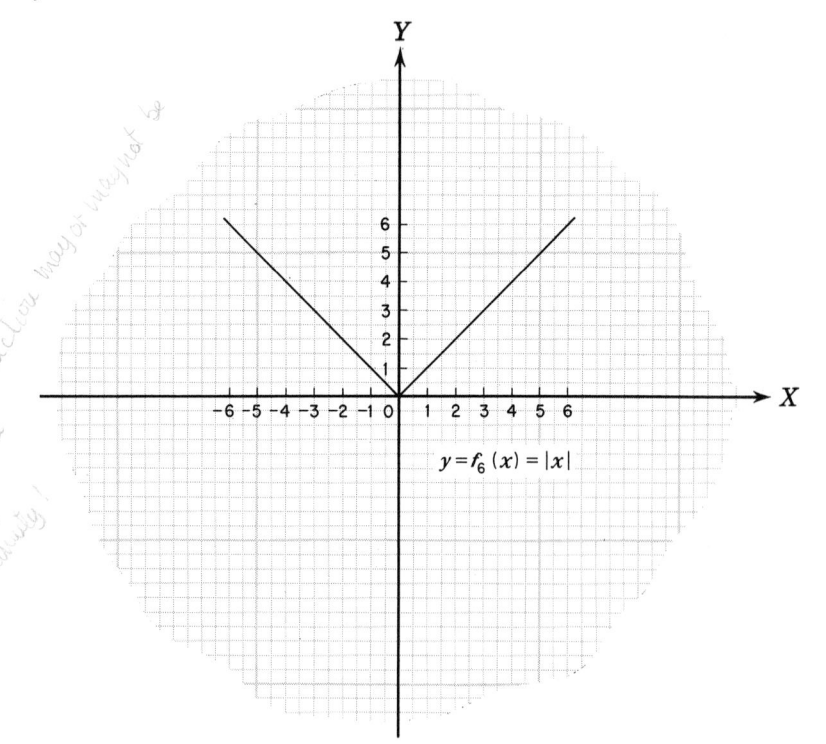

Figure 3-9

The functions f_1, f_2, f_3 described in Example 3 each fail to have $\lim_{x \to 0} f(x)$ defined; hence these functions certainly are *not* continuous at zero.

In the cases of f_4 and f_5, where

$$f_4(x) = \frac{x^2}{|x|} \quad \text{and} \quad f_5(x) = \frac{x^2 + x}{x}$$

Sec. 3-3. Continuity

$\lim_{x \to 0} f(x)$ exists but $f(0)$ was not defined. Whether or not these $f(x)$ are continuous at $x = 0$ depends upon how $f(x)$ is defined at $x = 0$. If we define $f_4(x)$ as follows:

$$f_4(x) = \begin{cases} \dfrac{x^2}{|x|} & \text{if } x \neq 0 \\ 0 & \text{if } x = 0 \end{cases}$$

then the function $f_4(x)$ is continuous at $x = 0$. (Why?) If $f_4(0)$ is undefined or is defined as some value other than zero, then $f_4(x)$ is discontinuous at $x = 0$.

The function $f_5(x) = (x^2 + x)/x$ must have $f_5(0) = 1$ in order to be continuous at $x = 0$. (Why?)

The function $f_6(x) = |x|$ already has $f_6(0) = 0$ and $\lim_{x \to 0} f_6(x) = 0$. Hence, $f_6(x)$ is continuous at $x = 0$, as previously stated.

Problem Set 3-3

1. $\lim_{x \to 4} (3x^2 - 2x + 5) = ?$

2. $\lim_{x \to 3} \dfrac{x^3 - 27}{x - 3} = ?$

3. Is $f(x) = \dfrac{x^3 - 27}{x - 3}$ continuous at $x = 3$? Is it continuous at $x = 5$?

4. Show that $\lim_{x \to 4} \dfrac{3x - 11}{x - 4}$ does not exist, but that $\lim_{x \to 4} \dfrac{3x - 12}{x - 4}$ does exist.

5. Are either of the functions of Prob. 4 continuous at $x = 4$? At $x = 6$?

6. Find $\lim_{t \to 5} (3t^2 + 2t - 8)$.

7. If $\epsilon = 2$ is chosen, find a possible corresponding value of δ such that if $0 < |x - 3| < \delta$, then $|f(x) - 9| < \epsilon = 2$ where $f(x) = 4x - 3$.

8. If ϵ is taken as .01 in Prob. 7, determine a corresponding δ.

9. Will the δ of Prob. 8 also work for Prob. 7? For a given value of ϵ, is the corresponding value of δ uniquely determined?

*10. Does $\lim_{\Delta x \to 0} f(b + \Delta x) = L$ imply that as $x \to b$, $f(x) \to L$?

*11. Does $\lim_{\Delta x \to 0} |f(b + \Delta x) - f(b)| = 0$ imply that $f(x)$ is continuous at b?

*12. Restate the definition of a continuous function, using delta notation.

Tangents and Limits

13. $\lim\limits_{\Delta x \to 0} \dfrac{4\Delta x + (\Delta x)^2}{\Delta x} = \lim\limits_{\Delta x \to 0} (4 + \Delta x) = ?$

14. $\lim\limits_{x \to 1} \dfrac{2(x+1)(x-1)}{x-1} = ?$ 15. $\lim\limits_{\Delta x \to 1} \dfrac{-3\Delta x}{(1+\Delta x)\,\Delta x} = ?$

16. $\lim\limits_{x \to 2} \dfrac{\sqrt{2x+5}-3}{x-2} = ?$ (Hint: Rationalize the numerator.)

17. (a) Does $\lim\limits_{x \to 3} \dfrac{x^2-9}{x-3}$ exist?

   (b) Is $f(x) = \begin{cases} \dfrac{x^2-9}{x-3} & \text{if } x \neq 3 \\ 6 & \text{if } x = 3 \end{cases}$ continuous at 3?

   (c) Is it continuous at $x = -3$?

18. (a) Does $\lim\limits_{x \to -2} \dfrac{x^2-4}{x+2}$ exist?

   (b) Is $f(x) = \begin{cases} \dfrac{x^2-4}{x+2} & \text{if } x \neq -2 \\ 5 & \text{if } x = -2 \end{cases}$ continuous at -2?

   (c) If not, what definition of $f(-2)$ will make $f(x)$ continuous at -2?

19. Give an example of a function which is not continuous at 4.

20. Is $f(x) = \begin{cases} x^2 & \text{if } x \neq 10 \\ 4 & \text{if } x = 10 \end{cases}$ continuous at 10?

21. Use the definition of $\lim\limits_{x \to b} f(x) = L$ to show that if f_1 and f_2 are two functions such that $f_1(x) = f_2(x)$ for all values of x except possibly at $x = b$, then if $\lim\limits_{x \to b} f_1(x)$ exists, so does $\lim f_2(x)$ and furthermore these two limits are equal. Examples of such functions are $f_1(x) = (x^2 + x)/x$ and $f_2(x) = x + 1$, with $b = 0$.

3-4. Slope of a Tangent Line

In Sec. 3-1 the tangent line to a curve $y = f(x)$ at the point $P(a,b)$ on the curve was defined as the limiting position of the secant line PS, as the point S approaches P by moving along the curve if a unique limiting position exists. A method of determining the slope (and hence the equation) of the tangent line to the curve $y = f(x)$ at the point $P(a,b)$ on the curve will be developed next. To do this a theorem is needed which will not be proved here.

Sec. 3-4. Slope of a Tangent Line

Theorem

Given a curve which has a tangent line at point P on the curve. Then

Slope of [the limiting position of the secant line PS]
$\quad\quad\quad\quad\quad {\scriptstyle S \to P \text{ along the curve}}$
$\quad\quad\quad\quad\quad\quad\quad\quad = $ limit of [the slope of the secant line PS].
$\quad\quad\quad\quad\quad\quad\quad\quad\quad {\scriptstyle S \to P \text{ along the curve}}$

We use this theorem to determine the slope of the tangent line to a curve $y = f(x)$ at the point P on the curve.

Example 1

Find the slope of the line tangent to $y = 2x^2$ at $(1,2)$ (Fig. 3-10).

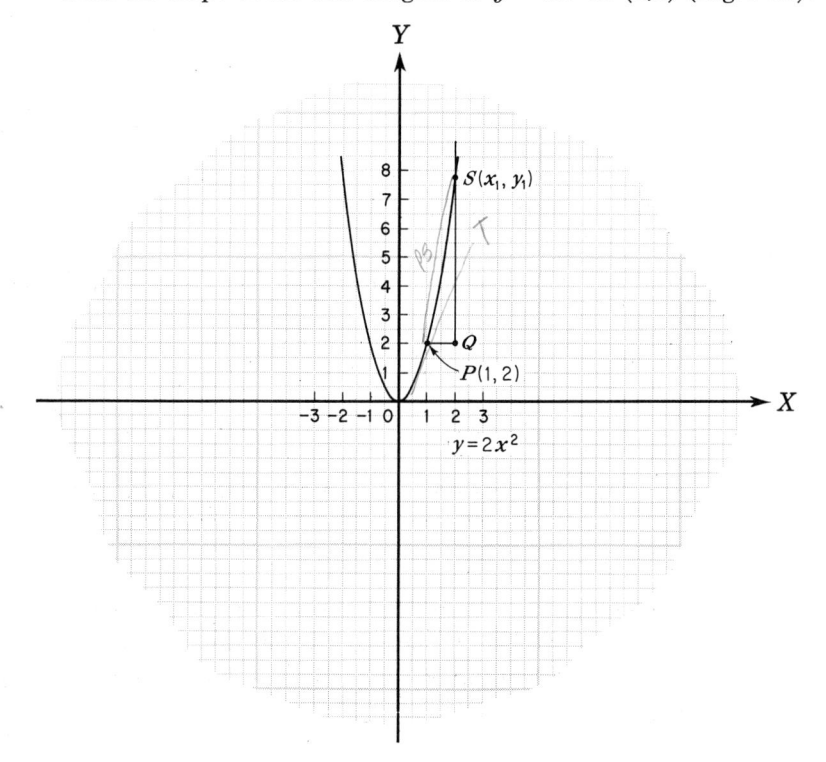

Figure 3-10

Let $P(1,2)$ and $S(x_1,y_1)$ be two points on $y = 2x^2$. The slope of the secant PS will be determined as in Sec. 2-5. The point Q has coordinates $Q(x_1,2)$. (Why?) The directed distance $PQ = x_1 - 1$ while $QS = y_1 - 2$.

Tangents and Limits

The secant line PS has slope $= QS/PQ = (y_1 - 2)/(x_1 - 1)$. Since $S(x_1,y_1)$ is on the locus $y = 2x^2$, its coordinates satisfy the equation of the locus; that is, $y_1 = 2x_1^2$. Hence

$$\text{Slope of secant } PS = \frac{y_1 - 2}{x_1 - 1} = \frac{2x_1^2 - 2}{x_1 - 1} = \frac{2(x_1 + 1)(x_1 - 1)}{x_1 - 1}.$$

From the definition of tangent line and using the theorem of this section, we obtain

$$\text{Slope of tangent line} = \text{slope of (limiting position of secant } PS)$$
$$\text{as } S \to P \text{ along } y = 2x^2$$

$$= \text{limit of (slope of secant line } PS)$$
$$\text{as } S \to P \text{ along } y = 2x^2$$

$$= \text{limit of } \frac{2(x_1 + 1)(x_1 - 1)}{x_1 - 1}.$$
$$\text{as } S \to P \text{ along } y = 2x^2$$

Since $S(x_1,y_1)$ is on the curve, $S(x_1,y_1) \to P(1,2)$ may be replaced by $x_1 \to 1$. Hence

$$\text{Slope of tangent line} = \lim_{x_1 \to 1} \frac{2(x_1 + 1)(x_1 - 1)}{x_1 - 1}.$$

Since $\lim_{x_1 \to 1} f(x)$ is not concerned with $f(x_1)$ at $x_1 = 1$, this becomes

$$\text{Slope of tangent line} = \lim_{x_1 \to 1} \frac{2(x_1 + 1)(x_1 - 1)}{x_1 - 1}.$$

$$= \lim_{x_1 \to 1} 2(x_1 + 1) = 2(2) = 4.$$

Hence the line tangent to $y = 2x^2$ at $(1,2)$ has slope 4.

Example 2

The problem of Example 1 may also be worked in the following manner, which will lead to more general methods in later sections. Let $P(1,2)$ and $S(1 + \Delta x, 2 + \Delta y)$† be two points on the locus $y = 2x^2$ (Fig. 3-11). Then

$$\text{Slope of secant } PS = \frac{\Delta y}{\Delta x}.$$

The theorem of this section gives

$$m = \text{slope of tangent line at } P = \lim_{\Delta x \to 0} \frac{\Delta y}{\Delta x}.$$

† Recall that Δx and Δy are single elements, not products.

Sec. 3-4. Slope of a Tangent Line

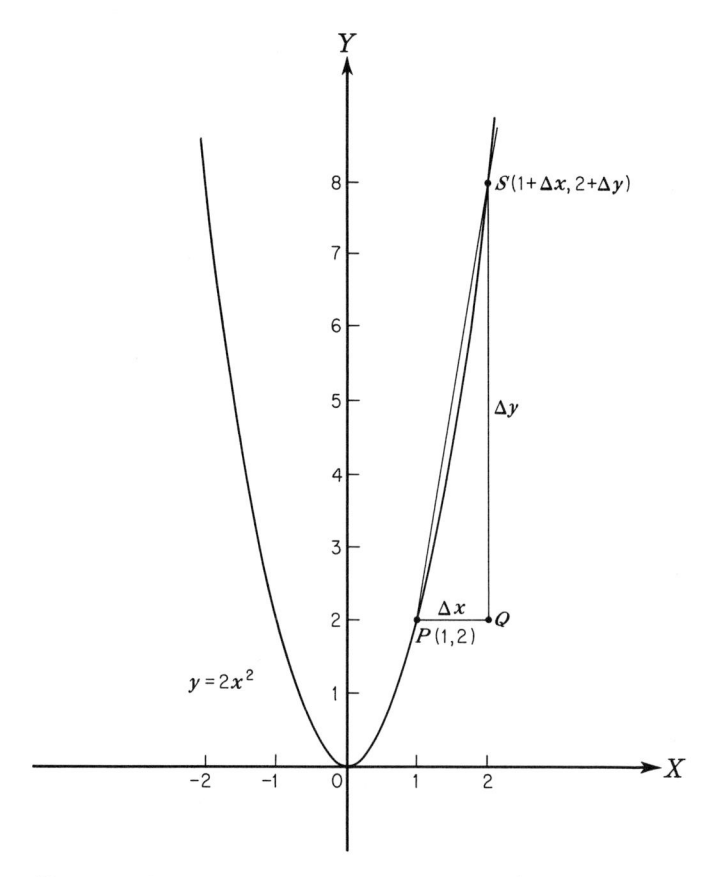

Figure 3-11

Since $S(1 + \Delta x, 2 + \Delta y)$ is on the locus $y = 2x^2$, the coordinates of S must satisfy the equation $y = 2x^2$.

$$2 + \Delta y = 2(1 + \Delta x)^2$$

and
$$\begin{aligned}\Delta y &= 2(1 + \Delta x)^2 - 2 \\ &= 2 + 4\Delta x + 2\Delta x^2 - 2\dagger \\ &= 4\Delta x + 2\Delta x^2.\end{aligned}$$

Therefore $m = \lim\limits_{\Delta x \to 0} \dfrac{\Delta y}{\Delta x} = \lim\limits_{\Delta x \to 0} \dfrac{4\Delta x + 2\Delta x^2}{\Delta x}$.

Since $\lim\limits_{\Delta x \to 0}$ specifically states $\Delta x \neq 0$, we have

$$m = \lim_{\Delta x \to 0} \frac{(4 + 2\Delta x)\,\Delta x}{\Delta x} = \lim_{\Delta x \to 0}\,(4 + 2\Delta x) = 4 + 0 = 4.$$

† By convention, the notation Δx^2 means $(\Delta x)^2$, not $\Delta(x^2)$. If the latter is intended, parentheses will be used as indicated.

84   *Tangents and Limits*

If the equation of the tangent line is desired, it can be written using the point-slope form of Sec. 2-6, with $m = 4$ and $P(1,2)$, to obtain

$$y - 2 = 4(x - 1)$$
or $\quad y = 4x - 2.$

Example 3

Determine the slope and the equation of the line tangent to $y = 3/x$ at the point $(1,3)$ (Fig. 3-12). Consider the points $P(1,3)$ and

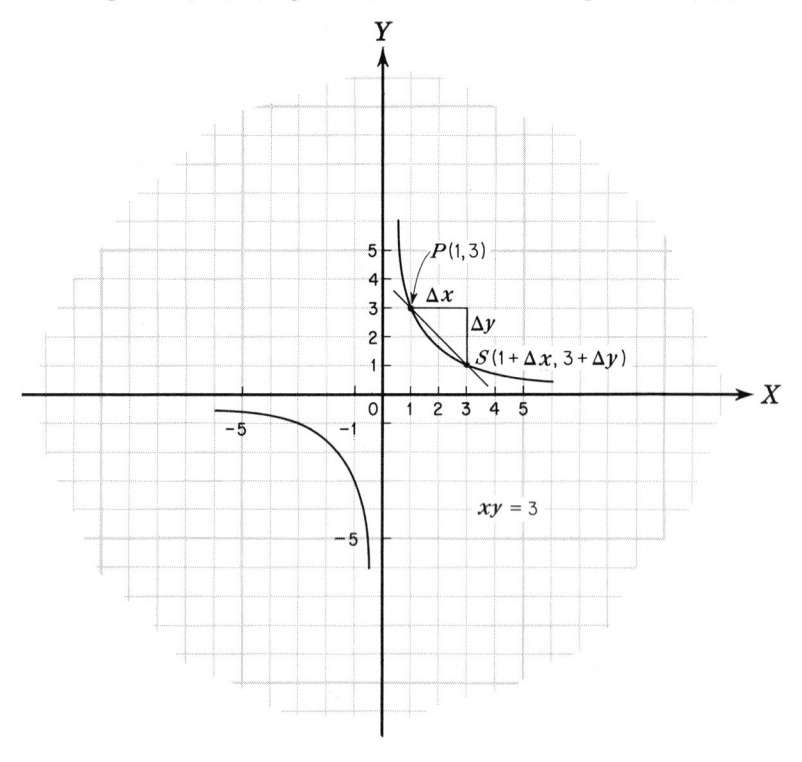

Figure 3-12

$S(1 + \Delta x, 3 + \Delta y)$ on the graph of $y = 3/x$. Since S is on the graph, its coordinates satisfy the equation of the locus; therefore

$$3 + \Delta y = \frac{3}{1 + \Delta x}$$

$$\Delta y = \frac{3}{1 + \Delta x} - 3 = \frac{-3\Delta x}{1 + \Delta x}.$$

Sec. 3-4. Slope of a Tangent Line

$$m = \text{slope of tangent line} = \lim_{\Delta x \to 0} \frac{\Delta y}{\Delta x} = \lim_{\Delta x \to 0} \frac{\frac{-3\Delta x}{1 + \Delta x}}{\Delta x}$$

$$= \lim_{\Delta x \to 0} -\frac{3\Delta x}{(1 + \Delta x)\, \Delta x}.$$

Since $\lim_{\Delta x \to 0} \frac{\Delta y}{\Delta x}$ implies $\Delta x \neq 0$, we have

$$m = \lim_{\Delta x \to 0} \frac{-3\Delta x}{(1 + \Delta x)\, \Delta x} = \lim_{\Delta x \to 0} \frac{-3}{1 + \Delta x} = -3.$$

Note that the slope of the tangent line is negative, as you would expect from the sketch.

The equation of the tangent line at $P(1,3)$ is

$$y - 3 = -3(x - 1) \quad \text{or} \quad y = -3x + 6.$$

Problem Set 3-4

1. Find the slope of the line tangent to $y = 2x^2$ at $(-1,2)$ by the method of Example 1.
2. Find the slope of the line tangent to $y = 2x^2$ at $(3,18)$ by the method of Example 1.
3. Work Prob. 1 by the method of Example 2.
4. Work Prob. 2 by the method of Example 2.
5. Determine the slope and equation of the line tangent to $y = 6/x$ at $(2,3)$.
6. Carefully sketch the locus $y = 2x^2$ and the line which passes through $(1,2)$ with slope 4. Does your sketch look as Example 1 indicates it should?

In Probs. 7 to 19, determine the slope and the equation of the tangent line to the given curve at the given point.

7. $y = 5x^2$ at $(1,5)$.
8. $y = 2x^2 - 3x + 5$ at $(2,7)$.
9. $y = 2/x$ at $(2,1)$.
10. $y = 3x^3$ at $(2,24)$.
11. $y = 5x^2 - 4$ at $(-1,1)$.
12. $y = 2x^3 - 3x$ at $(-2,-10)$.
13. $y = -3x + 2$ at $(5,-13)$.
14. $y = x^2 - 7x$ at $(0,0)$.
15. $y = 4x^2 + 1$ at $(1,5)$.

16. $y = 5x^2 + 3$ at $(1,8)$. Compare Prob. 7.
17. $y = 2x^2 - 3x + 1$ at $(2,3)$. Compare Prob. 8.
18. $y = 2/x + 4$ at $(2,5)$. Compare Prob. 9.
19. $y = 3x^3 - 10$ at $(2,14)$. Compare Prob. 10.

86   *Tangents and Limits*

*20. Show that lines tangent to $y = x^2$ and $y = x^2 + k$ have the same slope for corresponding values of x, no matter what value k has.

3-5. Increments

It is convenient to use the notation $P(x_1, y_1)$ and $S(x_1 + \Delta x, y_1 + \Delta y)$, where (x_1, y_1) is a fixed point on the curve, to represent two points on the locus. The slope of the secant line PS is $\Delta y / \Delta x$. If P and S lie on the locus $y = f(x)$, their coordinates satisfy the equation $y = f(x)$. Thus

$$y_1 = f(x_1) \tag{1}$$

and

$$y_1 + \Delta y = f(x_1 + \Delta x). \tag{2}$$

Subtracting Eq. (1) from Eq. (2) yields

$$\Delta y = f(x_1 + \Delta x) - f(x_1). \tag{3}$$

Then

$$\text{Slope of secant line } PS = \frac{\Delta y}{\Delta x} = \frac{f(x_1 + \Delta x) - f(x_1)}{\Delta x}. \tag{4}$$

Thus

$$\text{Slope of tangent line at } P = \lim_{\Delta x \to 0} \frac{\Delta y}{\Delta x}$$

$$= \lim_{\Delta x \to 0} \frac{f(x_1 + \Delta x) - f(x_1)}{\Delta x}. \tag{5}$$

Thus Δy and the Δf of Sec. 1-13 are identical.

Example 1

Find the slope of the tangent line to $y = 2x^2$ at the point (x_1, y_1) on the locus.

$$y_1 = 2x_1^2 \tag{1}$$
$$y_1 + \Delta y = 2(x_1 + \Delta x)^2$$
$$= 2x_1^2 + 4x_1 \Delta x + 2(\Delta x)^2. \tag{2}$$

Subtract Eq. (1) from Eq. (2) to obtain

$$\Delta y = 4x_1 \Delta x + 2(\Delta x)^2. \tag{3}$$

Then

$$\frac{\Delta y}{\Delta x} = \frac{4x_1 \Delta x + 2\Delta x^2}{\Delta x} = \frac{(4x_1 + 2\Delta x)\Delta x}{\Delta x}. \tag{4}$$

Sec. 3-5. Increments

If $\Delta x \neq 0$, we have $\dfrac{\Delta y}{\Delta x} = 4x_1 + 2\Delta x$. Since $\lim\limits_{\Delta x \to 0}$ includes the restriction $\Delta x \neq 0$ in its definition, we have

$$m = \lim_{\Delta x \to 0} \frac{\Delta y}{\Delta x} = \lim_{\Delta x \to 0} (4x_1 + 2\Delta x) = 4x_1. \tag{5}$$

Thus the slope of the tangent line to $y = 2x^2$ at any point on the locus is four times the x coordinate of that point. In particular, when $x_1 = 1$, we find $m = 4(1) = 4$, which is the problem of Examples 1 and 2 of Sec. 3-4.

One advantage of this method over that of Sec. 3-4 is that, given the x coordinates of several points on the locus, we may find the slopes of the tangent lines at these points by direct substitution. Even more important, it permits us to determine all points at which a curve has a given slope. This converse property is useful in applied problems.

Example 2

Determine the slope of the tangent line to $y = x^3$ at the points $(1,1)$, $(-2,-8)$, and $(3,27)$. First determine the slope at the point $P(x_1, y_1)$ on $y = x^3$.

$$y_1 = x_1^3 \tag{1}$$

$$\begin{aligned} y_1 + \Delta y &= (x_1 + \Delta x)^3 \\ &= x_1^3 + 3x_1^2\, \Delta x + 3x_1\, (\Delta x)^2 + (\Delta x)^3. \end{aligned} \tag{2}$$

Subtracting Eq. (1) from Eq. (2),

$$\Delta y = 3x_1^2\, \Delta x + 3x_1\, (\Delta x)^2 + (\Delta x)^3 \tag{3}$$

$$\begin{aligned} \frac{\Delta y}{\Delta x} &= \frac{3x_1^2\, \Delta x + 3x_1\, (\Delta x)^2 + (\Delta x)^3}{\Delta x} \\ &= 3x_1^2 + 3x_1\, \Delta x + (\Delta x)^2, \quad \text{if } \Delta x \neq 0. \end{aligned} \tag{4}$$

$$\begin{aligned} m = \text{slope of tangent line} &= \lim_{\Delta x \to 0} \frac{\Delta y}{\Delta x} \\ &= \lim_{\Delta x \to 0} (3x_1^2 + 3x_1\, \Delta x + (\Delta x)^2) = 3x_1^2. \end{aligned} \tag{5}$$

Hence at $(1,1)$ the tangent line has slope $3(1)^2 = 3$; at $(-2,-8)$ the tangent line has slope $3(-2)^2 = 12$; and at $(3,27)$ the tangent line has slope $3(3)^2 = 27$.

Example 3

Find all points of $y = x^3$ at which the tangent line is parallel to the line $y = 48x - 15$. Example 4 of Sec. 2-6 shows that parallel lines have the same slope. Hence the problem is reduced to finding those points at which $y = x^3$ has slope $m = 48$. By Example 2, the tangent line has slope $m = 3x_1^2$ where x_1 is the abscissa of the point. Hence we solve

$$m = 3x_1^2 = 48$$
$$x_1^2 = 16$$
$$x_1 = \pm 4.$$

Hence $(4,64)$ and $(-4,-64)$ are the required points.

Example 4

Determine the point or points at which the tangent line to the curve $y = 3x^2 - 7x + 5$ has slope 5.

Let $P(x_1,y_1)$ and $S(x_1 + \Delta x, y_1 + \Delta y)$ be two points on the curve $y = 3x^2 - 7x + 5$.

$$y_1 = 3x_1^2 - 7x_1 + 5 \qquad (1)$$
$$y_1 + \Delta y = 3(x_1 + \Delta x)^2 - 7(x_1 + \Delta x) + 5$$
$$= 3x_1^2 + 6x_1\,\Delta x + 3(\Delta x)^2 - 7x_1 - 7\Delta x + 5. \qquad (2)$$

Subtracting Eq. (1) from Eq. (2),

$$\Delta y = 6x_1\,\Delta x + 3(\Delta x)^2 - 7\Delta x \qquad (3)$$
$$\frac{\Delta y}{\Delta x} = \frac{6x_1\,\Delta x + 3(\Delta x)^2 - 7\Delta x}{\Delta x}$$
$$= 6x_1 + 3\Delta x - 7, \qquad \text{if } \Delta x \ne 0. \qquad (4)$$

Since $\lim_{\Delta x \to 0} \frac{\Delta y}{\Delta x}$ is not concerned with $\frac{\Delta y}{\Delta x}$ at $\Delta x = 0$, we obtain

$$m = \text{slope of tangent} = \lim_{\Delta x \to 0} \frac{\Delta y}{\Delta x}$$
$$= \lim_{\Delta x \to 0} (6x_1 + 3\Delta x - 7) = 6x_1 - 7. \qquad (5)$$

To determine where $m = 5$, set

$$m = 6x_1 - 7 = 5$$
$$x_1 = 2$$

and the desired point is $(2,3)$.

Sec. 3-5. Increments

Problem Set 3-5

1. Find the slope of the line tangent to $y = 5x^2$ at the point (x_1, y_1) by the method of this section. Assume (x_1, y_1) lies on the locus $y = 5x^2$.

2. Determine the slopes of the lines tangent to $y = 3x^2$ at $(-1, 3)$, $(1, 3)$, $(2, 12)$, $(5, 75)$, and $(10, 300)$. Use the method of this section.

3. Determine the slopes of the lines tangent to $y = 12/x$ at $(-1, -12)$, $(2, 6)$, $(3, 4)$, $(6, 2)$, and $(24, \frac{1}{2})$ by using the method of Example 2.

4. Determine the coordinates of all points at which $y = 3x^2$ has slope 12.

5. Determine the coordinates of all points at which $y = 12/x$ has slope -3.

6. Just before Example 1 of this section, the statement is made: "Thus Δy and the Δf of Sec. 1-13 are identical." Explain this statement in more detail.

7. Sketch the locus $y = x^3$ and its tangent lines near $(2, 8)$ and near $(0, 0)$.

8. Show that, if the slope of the tangent line to a curve is positive at a point P, then the curve is rising near P (that is, as x increases, so does y, near P).

9. Show that, if the slope of the tangent line to a curve is negative at a point P, then the curve is falling near P (that is, near P, as x increases, y decreases).

10. Find the equation of a line tangent to $y = 2x^2 + 5x - 3$ at $(2, 15)$.

11. Find the equations of all lines tangent to $y = x^3$ and having slope 12.

12. Find the equation of the line tangent to $y = 3/x$ at $(6, \frac{1}{2})$. (Note Example 3, Sec. 3-4.)

It is common practice to drop the subscripts on $P(x_1, y_1)$ and write $P(x, y)$, $S(x + \Delta x, y + \Delta y)$. Work the remaining problems using this modification. If an *equation* (in x and y) of the tangent line at (x_1, y_1) is requested, then subscripts are used to distinguish the x of the linear equation from the x_1 occurring in its slope.

13. Find the equations of all lines tangent to $y = 7x^2 - 6x$ and parallel to $2x - y + 5 = 0$.

14. Find the equations of all lines tangent to $y = x^3 + 2$ and parallel to $y = 3x + 5$.

Tangents and Limits

15. Determine the slope of the line which is tangent to the curve $y = 200x - 60x^2 + 4x^3$ at the point (a,b).

16. Determine the points of $y = 200x - 60x^2 + 4x^3$ at which the tangent is horizontal. On what intervals is the function increasing as x increases?

17. What is the slope of the line tangent to $6x^2 + 3y - 12 = 0$ at the point where it crosses the line $x = 2$?

18. Determine the points on $y = x^3 - 2x^2 - 7x + 5$ where the tangent line is horizontal. On what intervals is the function increasing as x increases?

19. At what points does $y = (x - 5)^2 + 2$ have horizontal tangents? Is this point a maximum or a minimum point of the curve?

***20.** Sketch $y = 2x^3$. Find all points at which the line tangent to $y = 2x^3$ at $(2,16)$ touches the curve. HINT: Solve the equation of the tangent line simultaneously with the equation of the curve. The fact that $(2,16)$ is such a point may help you solve the resulting equation.

3-6. Self-test

1. Write the equation of the line through
  (a) $(-4,1)$ and $(2,-3)$.
  (b) $(-2,3)$ with slope $-\frac{2}{3}$.
  (c) $(0,0)$ and parallel to $2x - 3y + 5 = 0$.

2. (a) Write $3x + 4y - 5 = 0$ in slope y-intercept form.
  (b) Write $4x - 3y = 12$ in intercept form.

3. (a) Find the slope of the tangent line to $y = x^2 - 7x$ at $(0,0)$.
  (b) Write the equation of this tangent line.

4. Find the slope of $y = 2/x$ at $(2,1)$.

5. Find the coordinates of all points at which the tangent line to $y = x^3$ at $(2,8)$ intersects the curve.

6. Find $\lim_{x \to 2} f(x)$, given $f(x) = \begin{cases} 10 & \text{when } x = 2 \\ \dfrac{x^3 - 8}{x - 2} & \text{when } x \neq 2. \end{cases}$

7. (a) Find $\lim_{x \to 1} \dfrac{x^2 + 2x - 3}{x - 1}$.

8. Is $f(x) = \dfrac{x^2 + 2x - 3}{x - 1}$ continuous at $x = 1$? Explain.

4
Differential Calculus

4-1. The Derivative

The geometric meaning of a line tangent to a curve at a point P on the curve (that is, the limiting position of the secant line PS as S approaches P along the curve, providing this limit exists and is unique) and the meaning of $\lim_{x \to a} f(x) = L$ have been discussed. A method of combining these two concepts to determine the slope of a line tangent (if one exists) to a given curve, $y = f(x)$, at a point $[x_1, f(x_1)]$ on the curve was obtained in Chap. 3. This method is essentially to compute

$$\lim_{\Delta x \to 0} \frac{f(x_1 + \Delta x) - f(x_1)}{\Delta x}$$

This limit is of such great mathematical importance that it is given a special name, *derivative*. To represent $\lim_{\Delta x \to 0} \frac{f(x + \Delta x) - f(x)}{\Delta x}$, it is common practice to use the symbols $f'(x)$, y', $\frac{dy}{dx}$, $\frac{df(x)}{dx}$, or $D_x F$. The symbol $f'(x)$ is read "f prime of x" while y' is read "y prime." The symbols $\frac{dy}{dx}$ and $\frac{df(x)}{dx}$ should be read "the derivative of y with respect to x" and "the derivative of $f(x)$ with respect to x." The symbols $\frac{dy}{dx}$ and $\frac{df(x)}{dx}$ are *not* fractions. Each represents a single concept,

derivative

Differential Calculus

not a quotient, just as the symbol Δx represents a single concept, not a product. The first two notations are preferable since they emphasize this point, but all four symbols are common in current usage. The symbol $\dot{y}$ was used to represent $\frac{dy}{dt}$ by Isaac Newton (English, 1642–1727), one of the inventors of the calculus. The symbol $\frac{dy}{dt}$ is due to Gottfried Wilhelm Leibnitz (German, 1646–1716) who also invented calculus. Current historians credit each with the independent invention of this important branch of mathematics. However, in the early 1700s a heated feud developed, with supporters of each man claiming the other had plagiarized the calculus.

4-2. The Delta Process

differen-tiation

The process of finding the derivative of y with respect to x is called *differentiation*. A function which has a derivative at a point P is said to be differentiable at P. Not all functions have derivatives at a given point P. The so-called "delta process" used in the preceding chapter provides a method of computing $\frac{\Delta y}{\Delta x}$ needed to find $y' = \lim_{x \to 0} \frac{\Delta y}{\Delta x}$. It is summarized here and two examples computed.

To find $y' = f'(x)$, where $y = f(x)$:

1. Form $\Delta y = f(x + \Delta x) - f(x)$.
2. Divide by Δx to obtain

delta process

$$\frac{\Delta y}{\Delta x} = \frac{f(x + \Delta x) - f(x)}{\Delta x}.$$

3. If the limit as $\Delta x \to 0$ exists, then this limit is the desired derivative:

$$y' = \lim_{\Delta x \to 0} \frac{\Delta y}{\Delta x} = \lim_{\Delta x \to 0} \frac{f(x + \Delta x) - f(x)}{\Delta x}.$$

Example 1

Compute y' if $y = f(x) = 5x^2 - 3$.

1. $\Delta y(x, \Delta x) = f(x + \Delta x) - f(x)$
$= 5(x + \Delta x)^2 - 3 - (5x^2 - 3)$
$= 5x^2 + 10x(\Delta x) + 5\, \Delta x^2 - 3 - (5x^2 - 3)$
$= 10x(\Delta x) + 5\, \Delta x^2.$

Sec. 4-2. The Delta Process

2. Divide by Δx.

$$\frac{\Delta y}{\Delta x} = \frac{10x(\Delta x) + 5\,\Delta x^2}{\Delta x} = 10x + 5\,\Delta x \qquad \text{if } \Delta x \neq 0.$$

3. Take the limit of $\frac{\Delta f}{\Delta x}$ as $\Delta x \to 0$ to determine

$$y' = \lim_{\Delta x \to 0} \frac{\Delta y}{\Delta x} = \lim_{\Delta x \to 0} (10x + 5\,\Delta x) = 10x + 0 = 10x.$$

Hence $y' = 10x$, or $f'(x) = 10x$.

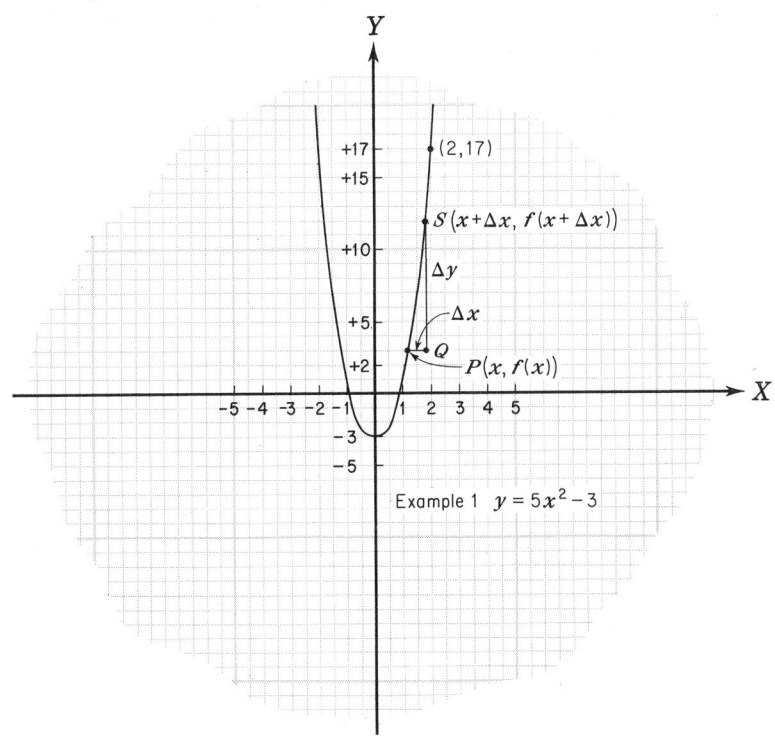

Figure 4-1

Example 2

Compute $f'(x)$ if $f(x) = \sqrt{x+3}$.
 1. $\Delta y = f(x + \Delta x) - f(x) = \sqrt{x + \Delta x + 3} - \sqrt{x+3}$.
 2. Divide by Δx.

$$\frac{\Delta y}{\Delta x} = \frac{\sqrt{x + \Delta x + 3} - \sqrt{x+3}}{\Delta x}.$$

Differential Calculus

In the present form we cannot divide Δx from numerator and denominator. It will be helpful to rationalize the numerator (a process similar to rationalizing the denominator). Think of the fraction as $\dfrac{A - B}{\Delta x}$. Multiply numerator and denominator by $A + B$ to obtain

$$\frac{(A - B)(A + B)}{\Delta x \,(A + B)} = \frac{A^2 - B^2}{\Delta x \,(A + B)}.$$

In symbols this does not appear to be a simplification, but in the actual problem

$$\begin{aligned}
\frac{\Delta y}{\Delta x} &= \frac{(\sqrt{x + \Delta x + 3} - \sqrt{x + 3})(\sqrt{x + \Delta x + 3} + \sqrt{x + 3})}{\Delta x \,(\sqrt{x + \Delta x + 3} + \sqrt{x + 3})} \\
&= \frac{(\sqrt{x + \Delta x + 3})^2 - (\sqrt{x + 3})^2}{\Delta x \,(\sqrt{x + \Delta x + 3} + \sqrt{x + 3})} \\
&= \frac{x + \Delta x + 3 - (x + 3)}{\Delta x \,(\sqrt{x + \Delta x + 3} + \sqrt{x + 3})} \\
&= \frac{\Delta x}{\Delta x \,(\sqrt{x + \Delta x + 3} + \sqrt{x + 3})} \\
&= \frac{1}{\sqrt{x + \Delta x + 3} + \sqrt{x + 3}} \qquad \text{if } \Delta x \neq 0.
\end{aligned}$$

It is now a simple task to find $\lim\limits_{\Delta x \to 0} \dfrac{\Delta y}{\Delta x}$. Using the previous form the task was rather ominous. This technique is often used if the numerator of $\dfrac{\Delta y}{\Delta x}$ contains radicals.

3. Taking the limit as $\Delta x \to 0$,

$$\begin{aligned}
f'(x) &= \lim_{\Delta x \to 0} \frac{\Delta y}{\Delta x} = \lim_{\Delta x \to 0} \frac{1}{\sqrt{x + \Delta x + 3} + \sqrt{x + 3}} \\
&= \frac{1}{\sqrt{x + 3} + \sqrt{x + 3}} = \frac{1}{2\sqrt{x + 3}}.
\end{aligned}$$

The student should consider the meaning of each step in the above examples by actually sketching the steps involved geometrically on the graphs in Figs. 4-1 and 4-2.

Sec. 4-2. The Delta Process

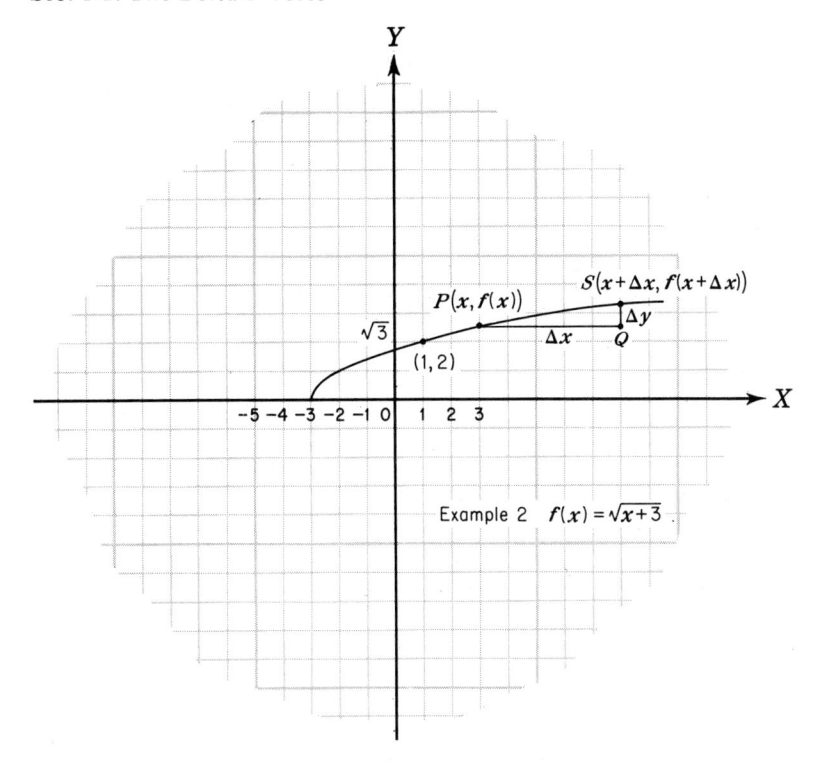

[Figure: Graph showing $f(x) = \sqrt{x+3}$ with points $P(x, f(x))$ at $(1,2)$ and $S(x+\Delta x, f(x+\Delta x))$, with Δx and Δy marked, and $\sqrt{3}$ indicated on the y-axis.]

Example 2 $f(x) = \sqrt{x+3}$

Figure 4-2

Problem Set 4-2

1. Explain Example 1 using the sketch given in Fig. 4-1. Be sure you understand each step and tell what you have when you arrive at $f'(x) = 10x$.

2. Explain Example 2 using the sketch given in Fig. 4-2. Why does the sketch show only positive values for y? Does the given curve ever have a tangent with negative slope? Reconcile your answer with the fact that $f'(x) = \dfrac{1}{2\sqrt{x+3}}$. What happens to the tangent line at $[x, f(x)]$ as x becomes large? Does the given curve have a horizontal tangent line? Does it have a vertical tangent line? Justify your answer algebraically.

In Probs. 3 to 8, compute the derivative of $f(x)$ with respect to x.

3. $f(x) = 17 - x^2$.   4. $f(x) = 1/x$.   5. $f(x) = 1/x^2$.
6. $y = f(x) = 3x^2 + x$.   7. $y = f(x) = 7x^3$.

Differential Calculus

8. $y = f(x) = 2x^5$.   HINT: Use the binomial expansion to obtain
$$(x + \Delta x)^5 = x^5 + 5x^4 \, \Delta x + 10x^3(\Delta x)^2 + 10x^2(\Delta x)^3 + 5x(\Delta x)^4 + (\Delta x)^5.$$

9. Let $f(x) = 7$ for all x.   Determine $f'(x)$.

10. Compute y' if $y(x) = 3$.   Interpret this result graphically.

11. Prove that if $y(x) = C$ (a constant) then $y' = 0$, no matter what constant value C is given.

12. If $y(t) = -3/t$ for $t > 1$, find the slope of the curve at the point $(6, -\frac{1}{2})$.

13. If $y(x) = \sqrt{x-5}$, compute y'.   HINT: Rationalize the numerator as in Example 2.

14. If $f(x) = \sqrt{7-2x}$, compute $f'(x)$.

15. If $f(t) = \sqrt{7-2t}$, compute $f'(t) = \lim_{\Delta t \to 0} \dfrac{\Delta f}{\Delta t}$.

16. If $F(x)$ is a function such that $F'(x) = 3x^2 + 2$, what is the derivative of $y = F(x) + 4$?

17. If $g(x) = 3x^2 - 6x - 5$, find $\dfrac{dg(x)}{dx}$.

18. If $y = 2x^3$, find $\dfrac{dy}{dx}$.

19. If $f(x) = g(x) + k$, show that $f'(x) = g'(x)$.   HINT:
$$\Delta f = [g(x + \Delta x) + k] - [g(x) + k] = \cdots = \Delta g.$$

20. If $y = 2x^3 + 27$, find $\dfrac{dy}{dx}$.   Use the results of Probs. 18 and 19 to ease your computation.

21. If $y^2 = 2x + 3$, find $\dfrac{dy}{dx}$.   **22.** If $y = 3x + \pi^2$, find y'.

23. If $g(x) = 1/\sqrt[3]{4x^2-1}$, find $g'(x)$.

24. Show that $\lim_{\Delta x \to 0} \Delta f = 0$ if $\dfrac{df(x)}{dx}$ exists.   HINT: Assume $\lim_{\Delta x \to 0} \Delta f \neq 0$ and show that this would imply $f'(x)$ does not exist. Does this prove the theorem requested?

25. In Sec. 3-3 the function f was defined to be *continuous* at b if $\lim_{x \to b} f(x) = f(b)$. Show that if $\dfrac{df(x)}{dx}$ exists at b, then $f(x)$ is necessarily continuous at b.

4-3. Generalization

Difficulty in solving certain quadratic equations by factoring led us to solve the problem *in general*, obtaining the quadratic formula. Rather than construct triangles each time to find the distance between two points, we again solved the problem in general, obtaining a distance formula. A general method of determining the derivative of a polynomial would be most useful. Three preliminary theorems (or lemmas) are derived first. Then, the theorem on the derivative of a polynomial can be deduced from these lemmas.

4-4. Preliminary Theorems on Differentiation

Theorem 1

The derivative of a sum of two functions is equal to the sum of the derivatives of the functions; that is, if $f(x) = g(x) + h(x)$, then $f'(x) = g'(x) + h'(x)$.

Use the delta process of Sec. 4-2.

If $f(x) = g(x) + h(x)$, then $f(x + \Delta x) = g(x + \Delta x) + h(x + \Delta x)$,

$$\Delta f = f(x + \Delta x) - f(x)$$
$$= [g(x + \Delta x) + h(x + \Delta x)] - [g(x) - h(x)]$$
$$= [g(x + \Delta x) - g(x)] + [h(x + \Delta x) - h(x)]. \tag{1}$$

$$\frac{\Delta f}{\Delta x} = \frac{g(x + \Delta x) - g(x)}{\Delta x} + \frac{h(x + \Delta x) - h(x)}{\Delta x}. \tag{2}$$

$$f'(x) = \lim_{\Delta x \to 0} \frac{\Delta f}{\Delta x}$$
$$= \lim_{\Delta x \to 0} \left[\frac{g(x + \Delta x) - g(x)}{\Delta x} + \frac{h(x + \Delta x) - h(x)}{\Delta x} \right] \tag{3}$$

But, $f'(x) = \lim_{\Delta x \to 0} \left[\frac{g(x + \Delta x) - g(x)}{\Delta x} \right] + \lim_{\Delta x \to 0} \left[\frac{h(x + \Delta x) - h(x)}{\Delta x} \right]$[†]

$$= g'(x) + h'(x).$$

Theorem 2

The derivative of a constant is zero; that is, if $f(x) = c$, a constant, then $f'(x) = 0$ for all x.

[†] This step assumes that if the individual limits exist, then the limit of a sum is equal to the sum of the limits.

98 Differential Calculus

This was essentially proved in Prob. 11, Set 4-2. The proof is repeated here for the sake of completeness. Let
$$f(x) = c,$$
then $f(x + \Delta x) = c$
and $\Delta y = f(x + \Delta x) - f(x) = c - c = 0.$ (1)
$$\frac{\Delta y}{\Delta x} = \frac{0}{\Delta x} = 0.$$ (2)
$$f'(x) = \lim_{\Delta x \to 0} \frac{\Delta y}{\Delta x} = \lim_{\Delta x \to 0} (0) = 0.$$ (3)

Theorem 3

If $f(x) = kx^n$, where k is a constant and n a positive integer, then $f'(x) = knx^{n-1}$.

The proof of this important theorem uses the identity†
$$h^n - x^n = (h - x)(h^{n-1} + h^{n-2}x + h^{n-3}x^2 + \cdots + hx^{n-2} + x^{n-1})$$
with $h = x + \Delta x$.

The delta process of Sec. 4-2 will be employed. If
$$f(x) = kx^n$$
$$f(x + \Delta x) = k(x + \Delta x)^n$$
and $\Delta f = f(x + \Delta x) - f(x)$ (1)
$$= k(x + \Delta x)^n - kx^n$$
$$= k[(x + \Delta x)^n - x^n]$$
$$= k[(x + \Delta x) - x] \cdot [(x + \Delta x)^{n-1} + (x + \Delta x)^{n-2}x + \cdots + (x + \Delta x)^{n-2} + x^{n-1}]$$
$$= k(\Delta x)[(x + \Delta x)^{n-1} + (x + \Delta x)^{n-2}x + \cdots + (x + \Delta x)x^{n-2} + x^{n-1}]$$
$$\frac{\Delta f}{\Delta x} = k[(x + \Delta x)^{n-1} + (x + \Delta x)^{n-2}x + \cdots + (x + \Delta x)x^{n-2} + x^{n-1}] \quad \text{if } \Delta x \neq 0. \quad (2)$$

Since the brackets contain n terms, each of which has limit x^{n-1} as $\Delta x \geqq 0$, we obtain
$$f'(x) = \lim_{\Delta x \to 0} k[\text{the sum of } n \text{ terms each with limit } x^{n-1} \text{ as } \Delta x \to 0]$$ (3)
$$= knx^{n-1}.$$

† To prove that the given equation is an identity, perform the multiplication indicated in the right member and reduce it to the left member.

Sec. 4-4. Preliminary Theorems on Differentiation

By employing the theorem

$$\lim_{\Delta x \to 0} k\left[\text{the sum of } n \text{ terms each with limit } x^{n-1} \text{ as } \Delta x \to 0\right]$$

$$= k \lim_{\Delta x \to 0} \left[\text{the sum of } n \text{ terms each with limit } x^{n-1} \text{ as } \Delta x \to 0\right] = knx^{n-1}.$$

$$\boxed{\dfrac{d(kx^n)}{dx} = knx^{n-1} \qquad \text{where } k \text{ is a constant and } n \text{ a positive integer.}}$$

This theorem will be generalized in Sec. 5-3 to the case where n is any rational number, but the given proof is valid only for n, a positive integer.

Example

If $y = f(x) = 7x^5$, find $f'(x)$. In this case $k = 7$ and $n = 5$. Hence $f'(x) = 7 \cdot 5x^4 = 35x^4$.

The more mathematically mature student will be able to prove the above theorem in a simple fashion by using the binomial theorem to obtain $(x + \Delta x)^n$ in expanded form. The same student may wish to consider whether or not mathematical induction is (or should be) used in the given proof of Theorem 3.

Problem Set 4-4

In Probs. 1 to 10, use Theorems 1, 2, or 3 to determine y' or $f'(x)$ or $p'(x)$ as appropriate.

1. $y = 4x^9$.
2. $y = 27x^{15}$.
3. $y = 3x^2$.
4. $p(x) = 2x^3$.
5. $f(x) = x^7$.
6. $p(x) = x$.
7. $f(x) = 41$.
8. $y = x\sqrt{31}$.
9. $f(x) = \sqrt{5x^2}$.

10. $f(x) = 2x^{15}$. Also form Δf to see how much work you have really saved by using these theorems.
11. Use the results of this section to determine the slope of $y = x^3$ at the point $(2,8)$.
12. Find the equation of the line tangent to $y = 4x^3$ at $(2,32)$.
13. Determine the slope of the line tangent to $y = 3x^5$ where the curve crosses the line $x = 3$.
14. Determine the slope of the line tangent to $y = 5x^3$ where the curve crosses the line $y = 40$.

15. Determine the coordinates of a point at which $y = 7x^3$ has a slope of $\frac{21}{25}$. *The slope of a curve at the point (a,b) is defined as the slope of the tangent at the point (a,b).*

16. If the line $y = 9$ and the curve $y = x^2$ intersect, find the equation of the line tangent to the curve at the point of intersection. If they do not intersect, prove this.

17. Carry out the steps of the proof of Theorem 3, using $11x^5$ for $f(x)$.

18. Find the equations of all lines tangent to $y = 7x^2 - 6x$ and parallel to $2x - y + 5 = 0$. What theorems are used?

19. Find the equations of all lines tangent to $y = x^3 + 2$ and parallel to $y = 3x + 5$. What theorems are used?

20. Determine the slope of the tangent line at the point (a,b) to the curve $y = 200x - 60x^2 + 4x^3$. What theorems are used?

21. Determine the points of $y = 200x - 60x^2 + 4x^3$ at which the tangent is horizontal. On what intervals is the function increasing as x increases?

22. What is the slope of the line tangent to $6x^2 + 3y - 12 = 0$ at the point where it crosses the line $x = 2$? Show that Theorems 1, 2, and 3 are each used.

23. Determine the points on $y = x^3 - 2x^2 - 7x + 5$ where the tangent line is horizontal. On what intervals is the function increasing as x increases?

24. At what points does $y = (x - 5)^2 + 2$ have horizontal tangents? Is this point a maximum or a minimum point of the curve?

25. Sketch $y = 2x^3$. Find all points at which the line tangent to $y = 2x^3$ at $(2,16)$ touches the curve. HINT: Solve the equation of the tangent line simultaneously with the equation of the curve. The fact that $(2,16)$ is such a point may help you solve the resulting equation.

26. Find the derivative of $y = 3x^4 - 2x + 7$. Show how all three theorems of this section are used to determine this derivative.

4-5. The Derivative of a Polynomial

A polynomial function is represented by a formula of the form

$$P(x) = ax^n + bx^{n-1} + \cdots + rx^2 + sx + t.$$

Sec. 4-5. The Derivative of a Polynomial

The following expressions are polynomials (with real coefficients).

$3x^2 - 7x + 5$

$14x^{17} - 21x^{16} + \sqrt{3}\, x^5 - (14 + 2\sqrt{7})x^4 - 3x^2 - \dfrac{371}{15}$

$\frac{1}{4}$

$5t^{237} - 4t^{17} + 5t - \sqrt{3} + 2\sqrt{5} - 671$

x

However, the expression $2/x$ is *not* a polynomial, nor is the expression $4x^2 - 3x + 1/(2x - 3) + 1$. Why not?

polynomial   A *polynomial* is a sum of terms of the form kx^n where n's are non-negative integers (possibly zero) and the k's are real numbers (or even complex numbers or elements of any integral domain). We do *not* say that a polynomial must have more than one term. The functions represented by $f(x) = x^2$ and $g(x) = 17$ are perfectly valid polynomial functions. Many such polynomial functions arise in applications. By using the three theorems of Sec. 4-4, it is possible to express the derivative of any polynomial function as a sum of derivatives of expressions of the form kx^n and constants (Theorem 1). These, in turn, may be differentiated by using Theorems 3 and 2.

Theorem 4

The derivative of a polynomial is equal to the sum of the derivatives of its individual terms.

Example 1

Write the equation of the line tangent to the curve $y = 3x^4 - 7x + 5$ at the point (2,39).

$$y(x) = 3x^4 - 7x + 5$$
$$y'(x) = 3(4)x^3 - 7$$
$$= 12x^3 - 7.$$

Hence $y'(2) = 12(2)^3 - 7 = 89$.

The equation of the desired tangent line is

$$y - 39 = 89(x - 2).$$

Differential Calculus

Example 2

Determine the coordinates of a point on the curve $y = 7x^2 - 5x + 2$ at which the tangent line has slope 6.

$$y = 7x^2 - 5x + 2.$$
$$y' = 14x - 5.$$

The problem requests a point (x,y) where $y' = 6$.
Solving, $y' = 14x - 5 = 6$

$$14x = 11$$
$$x = \frac{11}{14}.$$

The corresponding y value is determined by substituting $x = \frac{11}{14}$ into the original equation to obtain $y(\frac{11}{14})$. The reader is expected to perform the computation to obtain the desired point $(\frac{11}{14}, ?)$.

Problem Set 4-5

1. Find the derivative of $y = 5x^4 - 3x^2 - 2x + 11$.
2. If $f(x) = 7x^3 - 4x^2 + 3x - 11$, find $f'(x)$ and from this determine the slope of the line tangent to $y = f(x)$ at the point of intersection of the curve $y = f(x)$ and the line $x = -1$.
3. Find the equation of a line tangent to $y = x^3 - 2x^2 + 3x - 5$ and parallel to $y = 2x + 11$.
4. Is there a second line satisfying the conditions of Prob. 3? If so, find its equation.
5. Find the equation of a line tangent to the curve $y = 5x^7 + 3x$ where the curve crosses $x = 3$.
6. Find the slope of the tangent line to the curve $y = f(x) = x^3 - 7x^2 + 12$ at $(2, -8)$.
7. Find the slopes of the lines tangent to the curve $y = x^2 - 3x + 2$ at points where the curve crosses the x axis.
8. Find the slope of the line tangent to the curve $y = x^5 - 7x + 4$ at the point where the curve crosses the y axis.
9. If the line $y = 4$ and the curve $y = x^2 - 5x + 10$ intersect, determine the slopes of the lines tangent to the curve at the points of intersection. If they do not intersect, prove this.
10. Find a point at which the line tangent to $y = 3x^2 + 5x + 9$ has slope -1.

Sec. 4-5. The Derivative of a Polynomial

11. Show that the slope of the line tangent to $y = x^3 - 10$ is never negative. What is the graphical interpretation of this observation?

12. Determine the equation of a line having slope 4 and tangent to $y = 3x^2 + 4x - 7$.

13. Find the coordinates of a point at which $y = 3x^2 - 7x + 2$ has slope -3.

14. Is the slope of $y = 5x^5 + 13$ ever negative? Is it ever zero?

15. If $p(x) = x^3 - 12x$, find the slope of the tangent line at points where $y = p(x)$ crosses the x axis.

16. If $y = g(x) = x^3 + 3x^2 + 9x + 11$ crosses the line $x = 2$, find the equation of the tangent to $y = g(x)$ at this point. If they do not intersect, prove this.

17. Write the equations of the lines tangent to $y = 3x^2 + 4x + 5$ at the points of intersection of the curve and $y = 12$.

18. Find the point of intersection of the two tangents of Prob. 17 if they intersect. If they do not intersect, prove this.

In Probs. 19 to 25, determine the equation of a line tangent to the given curve at each intersection of the given line L and given curve C.

19. $C: y = 3x^2 + 6x + 5$; $L: x = 4$.
20. $C: y = 4x^3 - 7x^2 + 2x - 11$; $L: x = -3$.
21. $C: y = x^5 - 4x^2 + 9$; $L: x = 0$.
22. $C: y = x^3 + 6x + 2$; $L: y = -3$.
23. $C: y = x^3 - 14$; $L: y = 216$.
24. $C: y = 4x^2 + 3x - 5$; $L: y = 2x$.
25. $C: y = 5x^2 - 9$; $L: y = 7x - 3$.

26. Write the equation of a curve whose slope is (a) never positive; (b) never negative; (c) positive for $x > 3$ and negative when $x < 3$.

27. Find a point on the arc of $y = x^2 - 2x - 3$ connecting $(-1,0)$ and $(-2,5)$ at which the tangent is parallel to the secant connecting these points.

28. Determine the points at which $y = 2x^3 - x^2 - 18x + 5$ has slope 2.

29. (a) Determine the points at which $y = 6x^3 - 3x^2 - 60x + 7$ has slope 0.
(b) On what range of x values does y increase as x increases?
(c) On what range of x values does y decrease as x increases?

30. On what x range does y (a) increase; (b) decrease as x increases if $y = 3x^2 - 7x + 5$?

31. Determine a point in quadrant II at which $y = -3/x$ has slope

12. (Since $y = -3/x$ is *not* a polynomial, it will be necessary to use the delta process in computing y'.)

32. Determine a point in quadrant IV at which $y = 3/x$ has slope 12 (see Prob. 31).

33. Sketch $y = -3/x$. Are there any points other than those found in Probs. 31 and 32 at which $y = -3/x$ has slope 12? Demonstrate your reply both by considering the sketch and also by considering the algebraic solution of $y' = 12$.

34. If $y = -x^3$ and $y = 3x^2 + 1$ intersect, determine the equation of the tangent lines to each curve at their intersection. If they do not intersect, prove this.

35. Problem 25, Set 4-2, states that if $h'(x)$ exists at a point b, then $y = h(x)$ is continuous at b. Use this result to guarantee continuity, and the interpretation of $h'(x)$ as the slope of a tangent line to show that if $h'(x) = 0$ *for every* x, then $h(x)$ is constant. Is this result different from that of Problem 11, Set 4-2?

36. Use derivatives to prove that a linear function has a constant slope.

37. Prove, using derivatives, that a linear function is the only polynomial which has constant slope. Be sure you work this problem, not its converse, which is Prob. 36.

38. The point $(2,-3)$ lies on the graph of $p(x) = 3x^2 - 7x - 1$. Determine the slope of the tangent line to $p(x)$ at the point $(2,-3)$ and make a careful sketch of the tangent line.

39. Find the equation of the tangent line to $y = 2x^3 - 5x^2 + 6$ at the point where $x = 3$.

40. Find the equation of the tangent line to $y = 2x^3 - 5x^2 + 6$ at the point where $x = 1$.

41. Find the equation of the line which passes through the point $(5,7)$, and the intersection of the lines $7x + 11y - 12 = 0$ and $2x - 9y + 13 = 0$.

42. Determine the coordinates of each point at which the curve $y = x^3 + 2x^2 - 4x + 2$ has slope -5. How many such points exist?

43. For a certain function $p(x)$, $p(3) = 6$, $p'(3) = 1$, and also $p''(3) = -2$. Sketch the tangent line at $(3,6)$ and state whether the graph is above the tangent line, below the tangent line, both above and below the tangent line, or neither for points of the graph near the point of tangency $(3,6)$.

Sec. 4-6. Maximum and Minimum Values of a Function

44. Determine the slope of the curve $x^2y - yx + 10 = 0$ at the point $(2, -5)$.

45. Determine the equation of each line which has slope 2 and is tangent to the curve $y = x^3 + x^2 - 19x - 40$.

4-6. Maximum and Minimum Values of a Function

Some of the most perplexing problems in life today are those involving superlatives. We seek the *biggest* profit, the *most efficient* method, the *smallest* friction, the *shortest* time, the *least* wear. In short, we often seek the biggest or the smallest. Calculus may assist in solving such problems since many of them reduce to finding the highest (maximum) and the lowest (minimum) points on the graph of a function.

Let us consider the function f, determined by $f(x) = x^4 + 5$. Clearly there is no largest value of the function since $x^4 + 5$ can be made as large as desired by taking x sufficiently large. On the other hand, since x^4 is never negative, the function $x^4 + 5$ will always be at least 5. The value of $f(0)$ is 5. Hence, 5 is the minimum value that the function $f(x) = x^4 + 5$ obtains, and it obtains this minimum when $x = 0$.

In many cases it is not as simple as this to determine the maximum and minimum values of a function. Consider the function $y = x^4 - 32x + 6$. It is fairly obvious that y may be made as large as you wish by taking x sufficiently large; hence the function has no biggest value. However, it is not immediately obvious that the value -42, which occurs when $x = 2$, is the smallest or minimum value of the function determined by $x^4 - 32x + 6$.

Before continuing, it is desirable to define the meaning of two phrases: *absolute maximum value* and *relative maximum value*.

absolute maximum A function f has an *absolute maximum* value at b if $f(b) \geq f(x)$ for every member x in the domain of definition of f; that is, if the value of the function at b is as big as or bigger than the value of the function at every other point. Clearly the absolute maximum value is the largest value the function takes anywhere in its domain of definition.

relative maximum A function f is said to have a *relative maximum* value at a number b if $f(b) \geq f(b + \Delta x)$ for Δx sufficiently near zero. Since Δx may be either positive or negative, this means that $f(b)$ is the largest value of $f(x)$ for x near b.

106   *Differential Calculus*

The absolute maximum value is the largest value of the function in its entire domain of definition. A relative maximum value is merely the largest value of the function in a small neighborhood of the point at which this relative maximum occurs (see Fig. 4-3).

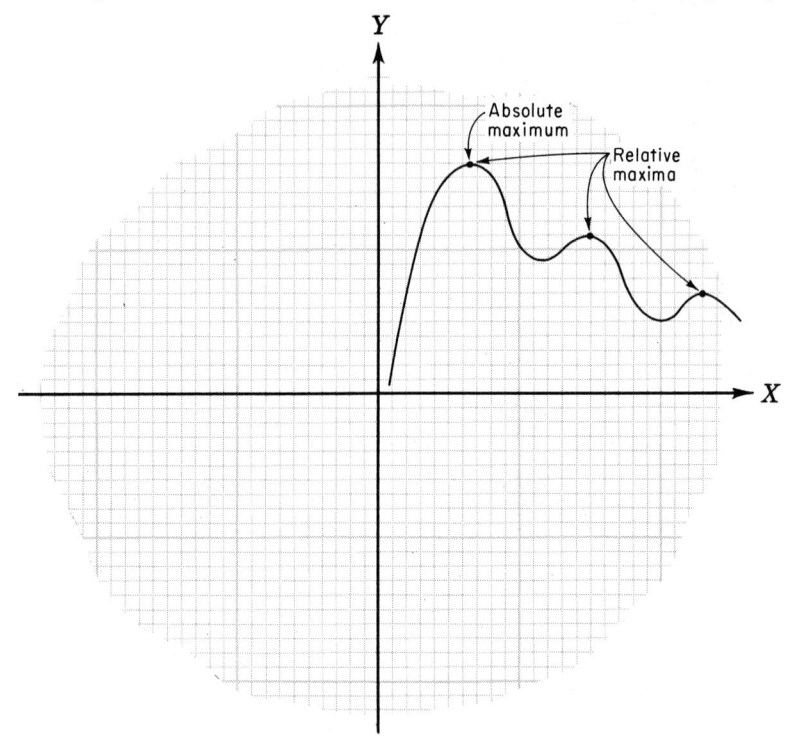

Figure 4-3

The *absolute minimum* and *relative minimum* values of a function are defined similarly to those of the absolute and relative maximum values.

absolute minimum $f(b)$ is an *absolute minimum* of the function f if $f(b) \leq f(x)$ for each x in the domain of definition of f.

relative minimum $f(b)$ is a *relative minimum* of the function f if $f(b) \leq f(b + \Delta x)$ for Δx sufficiently near zero.

The absolute minimum is the smallest value of the function. Stating that a relative minimum $f(b)$ is assumed at b merely implies that $f(b)$ is the smallest value of $f(x)$ for x near b.

In actual practice, it is usually the absolute maximum or the absolute minimum value which is sought. However, like most

Sec. 4-6. Maximum and Minimum Values of a Function 107

absolutes, these are rather elusive. The easiest way to determine the absolute maximum is (usually) to determine the largest relative maximum. The largest relative maximum may not be the absolute maximum. A function need not have either an absolute or a relative maximum. Figure 4-4 gives a function having several relative maximum points but no absolute maximum. Does this function have a relative minimum? An absolute minimum? (Why?)

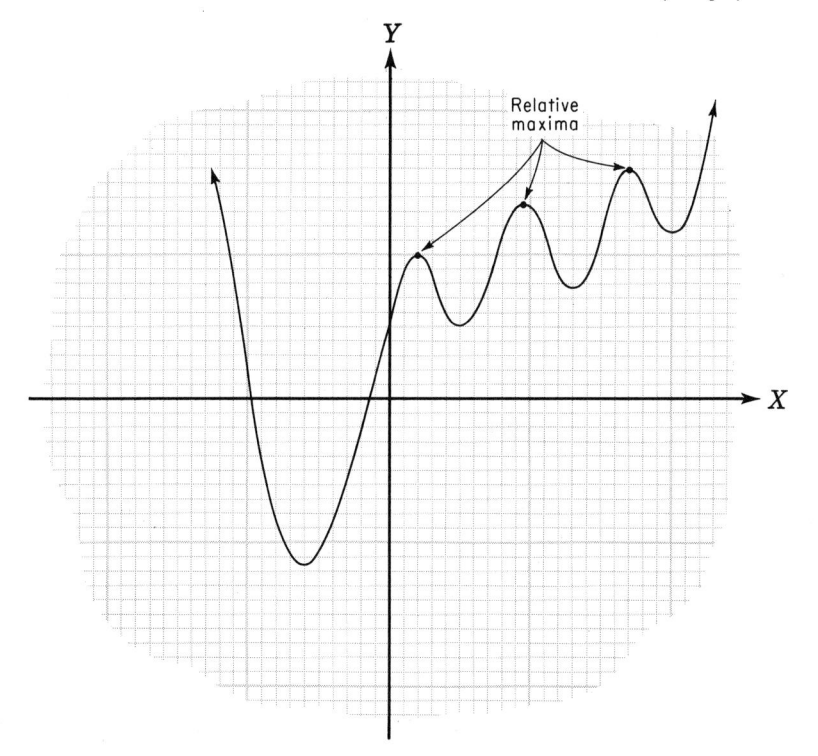

Figure 4-4

Since mathematicians are somewhat lazy, they have adopted the policy of using the words "maximum" and "minimum" in place of "relative maximum" and "relative minimum." In fact, the abbreviations "max" and "min" are often used. However, when *absolute* maximum or minimum is intended, the full term is generally used.

The reader should be careful that he understands these points before continuing. As always in mathematics, what is to come will depend upon what has passed.

108   *Differential Calculus*

The function sketched in Fig. 4-5 has relative maxima at A, D and F, and relative minima at B, E and G. The function value corresponding to C is neither a maximum nor a minimum value of $f(x)$. The value of the function $f(b)$ at a relative maximum is not necessarily the greatest value of $f(x)$. It is true that $f(x) \leq f(b)$ for x close to b, but not necessarily for *all* x. In Fig. 4-5, we have a maximum at A which is smaller than the minimum at G.

It would be very helpful if there were some simple (or even some difficult) way of determining all (relative) maximum and (relative)

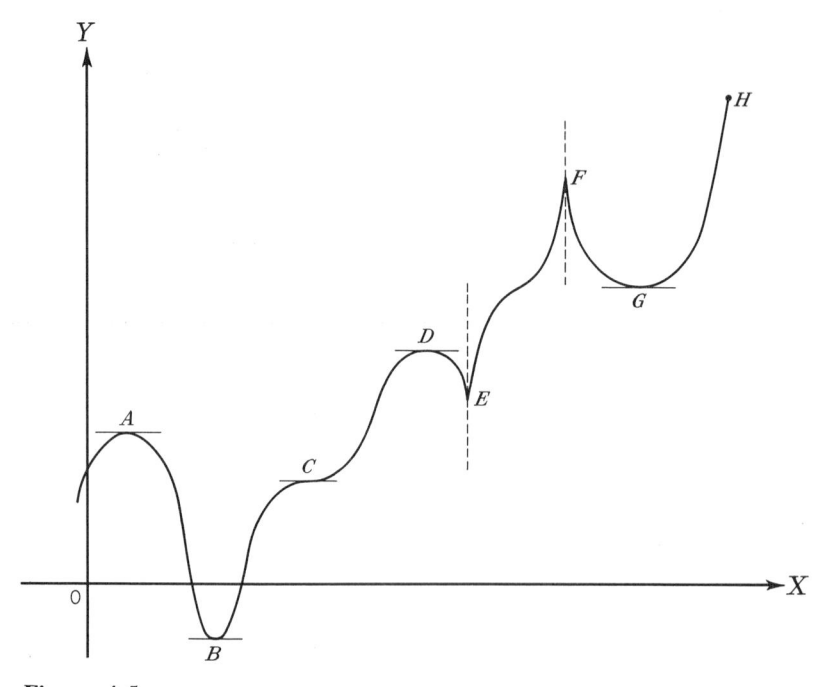

Figure 4-5

minimum values of a given function. Whether or not such a method exists depends upon the complexity of the function.

In general, relative maximum and relative minimum values of a function f will occur only at points where one of the following conditions holds:

1. The derivative $f'(x_1) = 0$ (points A, B, D, and G of Fig. 4-5).
2. The derivative $f'(x_1)$ does not exist (points E and F of Fig. 4-5).
3. x_1 occurs at the end point of the domain of definition (point H of Fig. 4-5).

It is not too difficult to prove this theorem, but the proof would detract from the emphasis on the theorem itself, so it is omitted here.

Sec. 4-6. Maximum and Minimum Values of a Function 109

Interested students should consult Brixey and Andree, *Fundamentals of College Mathematics*, Sec. 6-9, or Begle, *Introductory Calculus*, Sec. 5-1. The converse of the theorem is *not* valid. (What does that mean?)

In Fig. 4-5, points A, B, D, and G represent maximum and minimum points at which the tangent line is horizontal (has slope zero).

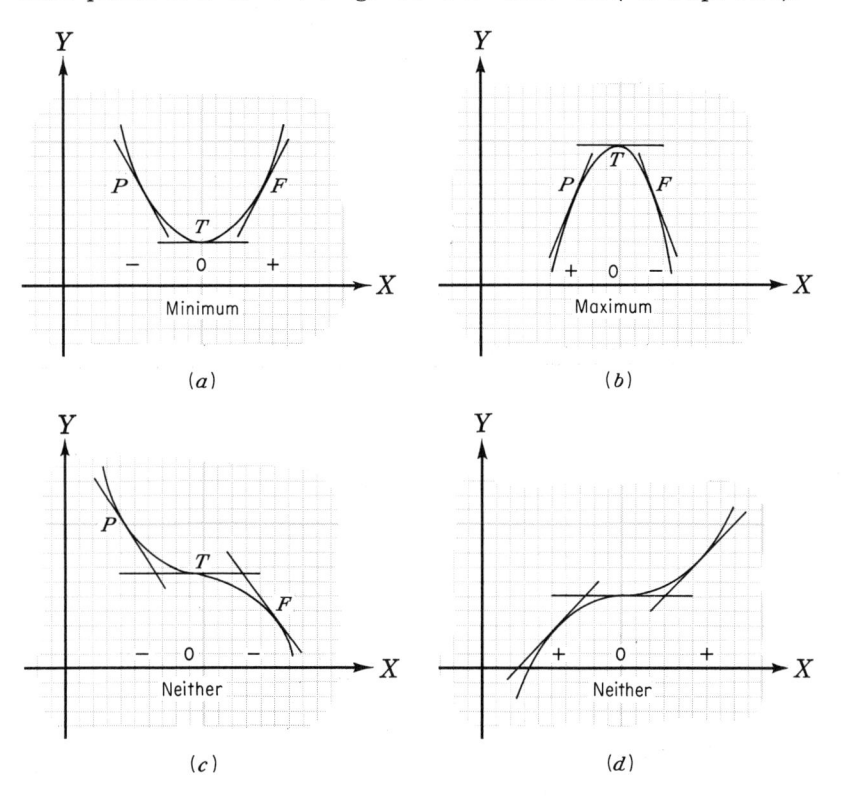

Figure 4-6

However, there may exist points such as E and F which are maximum and minimum points but which do *not* have horizontal tangent lines.

At such points, the tangent line may be vertical (no slope), or there may be no tangent line at all (no derivative). Difficulties of this type do not arise if the function is a *polynomial*. It is also possible to have points at which the tangent line is horizontal (slope zero) but which are neither maximum nor minimum points of the locus. Locate such a point in Fig. 4-5. In spite of these shortcomings, the determination of points at which the tangent line has zero slope is extremely helpful in many applications, as you will soon see.

Tests for Maximum and Minimum Points

If $y = P(x)$, a polynomial in x, there exist four possible cases of a tangent line with slope zero. The sketches in Fig. 4-6 illustrate each case.

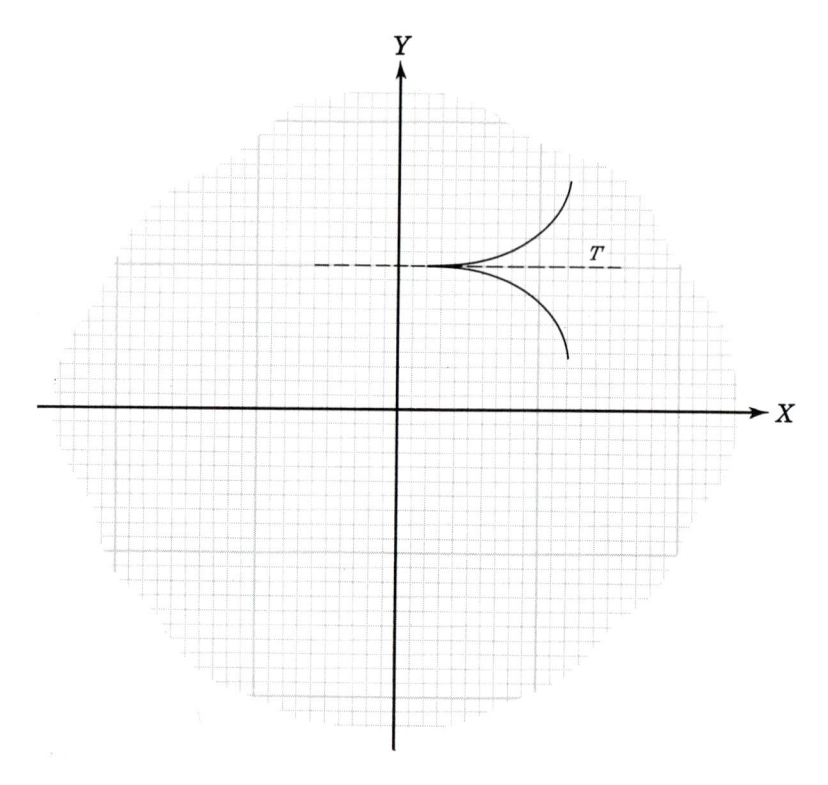

Figure 4-7

Let T be a point at which the locus has a horizontal tangent. To distinguish these four cases, consider the slope of the tangent line at a point P, preceding T, and at a point F following the point T, where the tangent line is horizontal at the point T. In the case of a minimum, the slope changes from negative to zero to positive as x increases. In the case of a maximum, the slope changes from positive to zero to negative as x increases. In the two cases which are neither maximum or minimum points, the slope has the same sign on both sides of the point where the tangent line is horizontal.

Sec. 4-6. Maximum and Minimum Values of a Function 111

For single-valued functions this provides an excellent test for distinguishing these four cases. It is possible for a locus to have a horizontal tangent line not included in these cases. See Fig. 4-7 where this is shown.

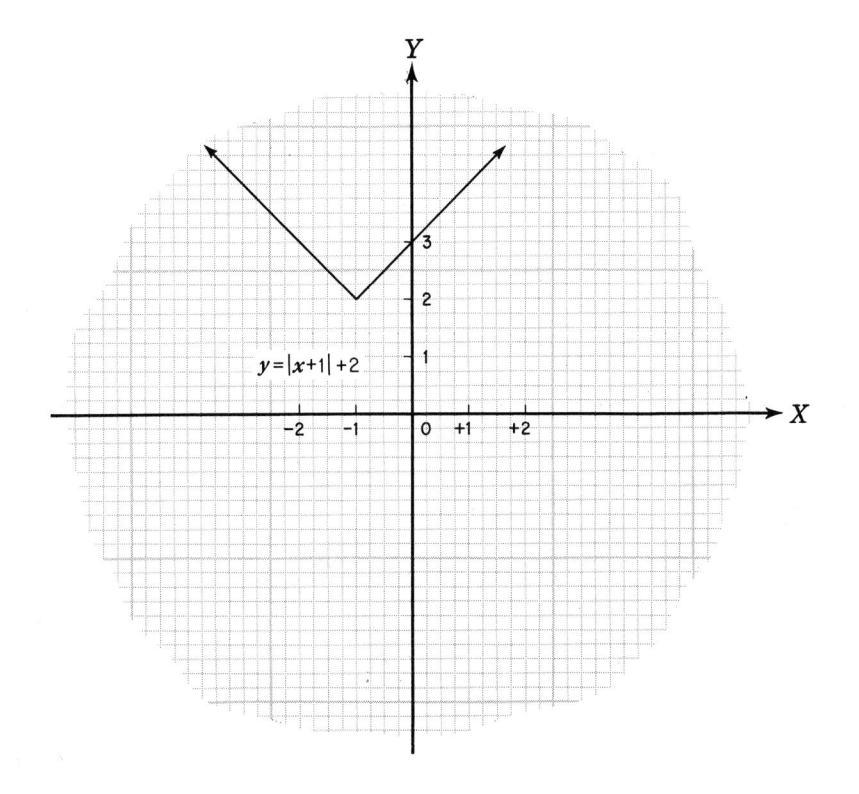

Figure 4-8

This difficulty does not occur with polynomials in x, since polynomials are single-valued. Figure 4-8 shows the graph of the function $y = |x + 1| + 2$ which has a minimum of 2 corresponding to $x = -1$ but fails to have a derivative at $y = -1$. (Why?)

Example 1

Determine, using the derivative, the maximum and minimum points of the locus $y = x^2 - 4x + 3$ and sketch the locus.

112   Differential Calculus

For a polynomial, the slope of the tangent line is zero at maximum and minimum (and possibly at other) points. Since $x^2 - 4x + 3$ is a polynomial, we compute y' and set it equal to zero to determine values of x at which the tangent lines are horizontal.

$$y' = 2x - 4$$
If $y' = 2(x - 2) = 0$,
then $x = 2$.

Thus $(2, -1)$ is the only point at which $y = x^2 - 4x + 3$ has a horizontal tangent line. Consider the slope of the tangent line for $x < 2$ and for $x > 2$.

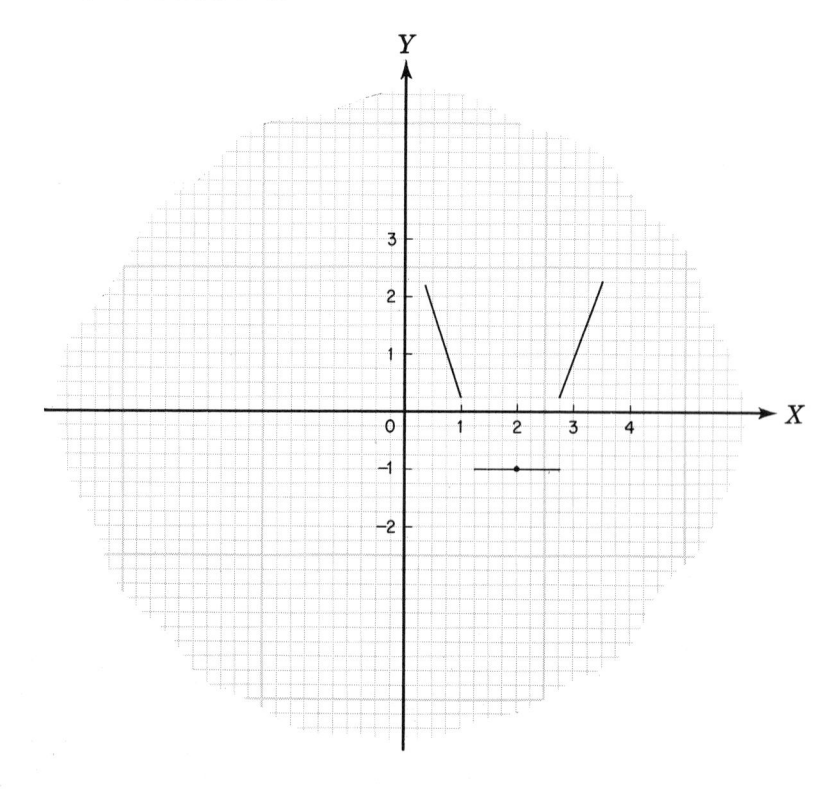

Figure 4-9

If $x < 2$, then $x - 2 < 0$ and $y' = -$.
If $x > 2$, then $x - 2 > 0$ and $y' = +$. Figure 4-9 shows such tangent lines. Clearly $(2, -1)$ is a minimum point and the locus has the sketch given in Fig. 4-10.

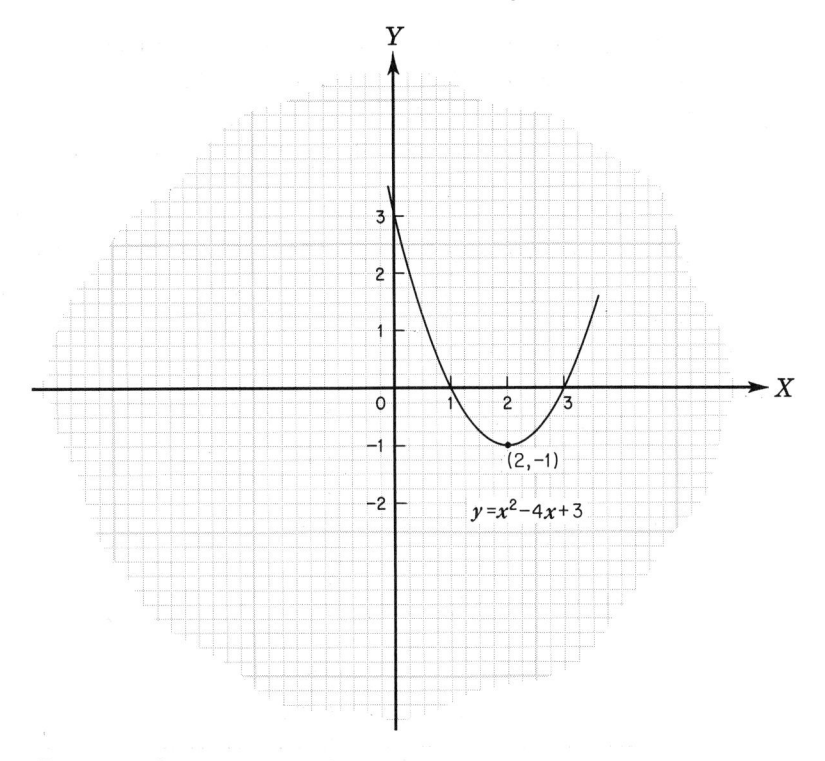

Sec. 4-6. Maximum and Minimum Values of a Function 113

Figure 4-10

Example 2

Determine all maximum and minimum points of the locus of $y = x^3 + 3x^2 - 9x + 4$ and sketch the locus.

$$y' = 3x^2 + 6x - 9$$
$$= 3(x + 3)(x - 1).$$

To determine points of the locus at which the tangent line is horizontal, set $y' = 0$, obtaining

$$3(x + 3)(x - 1) = 0$$
$$x = -3, \quad x = 1.$$

By substitution into $y = x^3 + 3x^2 - 9x + 4$, we find that the locus has horizontal tangents at $(-3, 31)$ and $(1, -1)$. The locus crosses the y axis at $(0, 4)$.

114   *Differential Calculus*

Using the test of this section, we shall first consider the slope $y' = 3(x + 3)(x - 1)$ of the locus for $x < -3$, for $-3 < x < 1$, and for $1 < x$.

If $x < -3$, then $x + 3 < 0$, $x - 1 < 0$ and $y' = +(-)(-) = +$.

If $-3 < x < 1$, then $x + 3 > 0$, $x - 1 < 0$ and we have $y' = +(+)(-) = -$.

If $1 < x$, then $x + 3 > 0$, $x + 1 > 0$ and $y' = +(+)(+) = +$.

Our knowledge of the tangent lines at certain points is presented in Fig. 4-11.

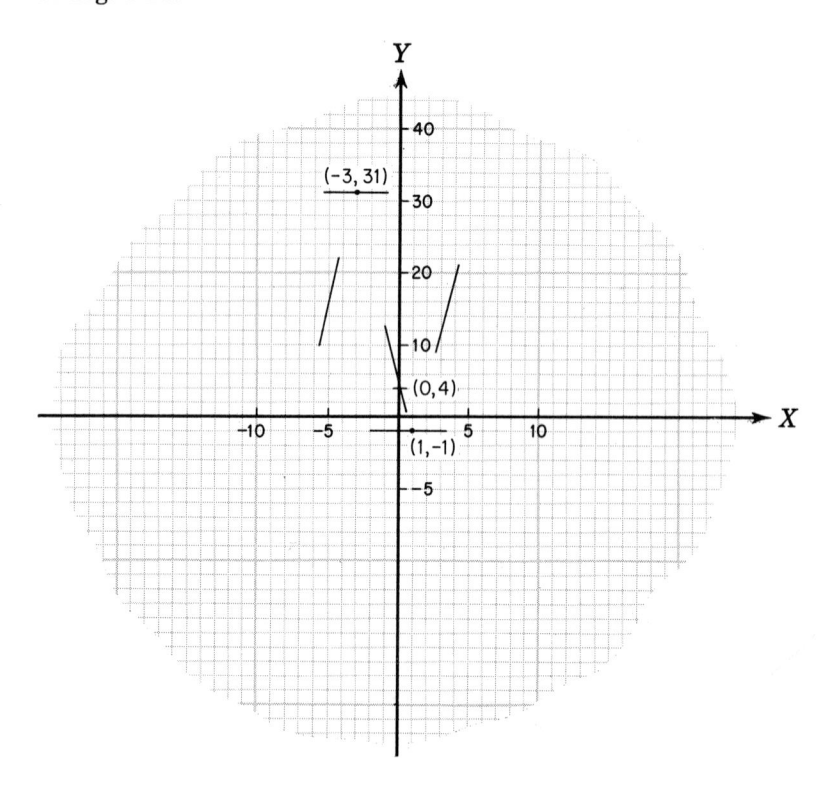

Figure 4-11

This suggests the sketch of Fig. 4-12. The function $y(x)$, where $y(x) = x^3 + 3x^2 - 9x + 4$ has a relative maximum at $(-3, 31)$, since $y(-3) = (-3)^3 + 3(-3)^2 - 9(-3) + 4 = 31$, and a relative minimum at $(1, -1)$.

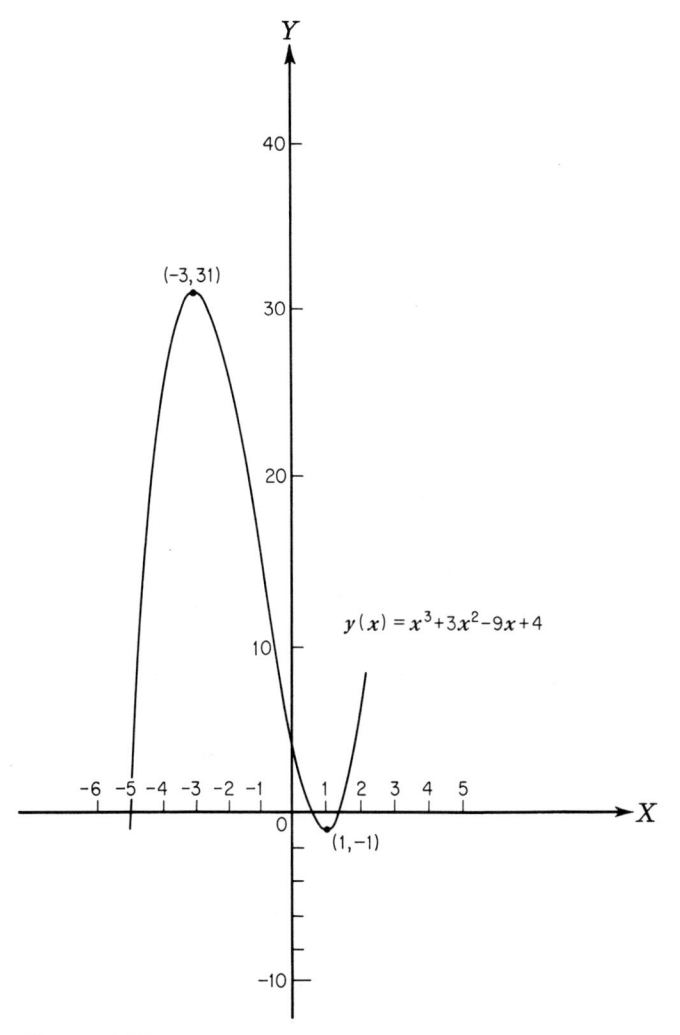

Sec. 4-6. Maximum and Minimum Values of a Function   115

[Graph showing $y(x) = x^3 + 3x^2 - 9x + 4$ with local maximum at $(-3, 31)$ and local minimum at $(1, -1)$]

Figure 4-12

Example 3

Find all points at which $y = x^3 + 7$ has horizontal tangent lines and sketch the locus.

$y' = 3x^2$.

Horizontal tangent lines occur at points where $y' = 0$. This implies

$3x^2 = 0 \quad x = 0.$

Hence, the only horizontal tangent line occurs at (0,7). Consider the slope $y' = 3x^2$ for $x < 0$ and for $0 < x$. In either case $y' = 3x^2$ is $+$. Tangent lines have positive slope except at (0,7). The

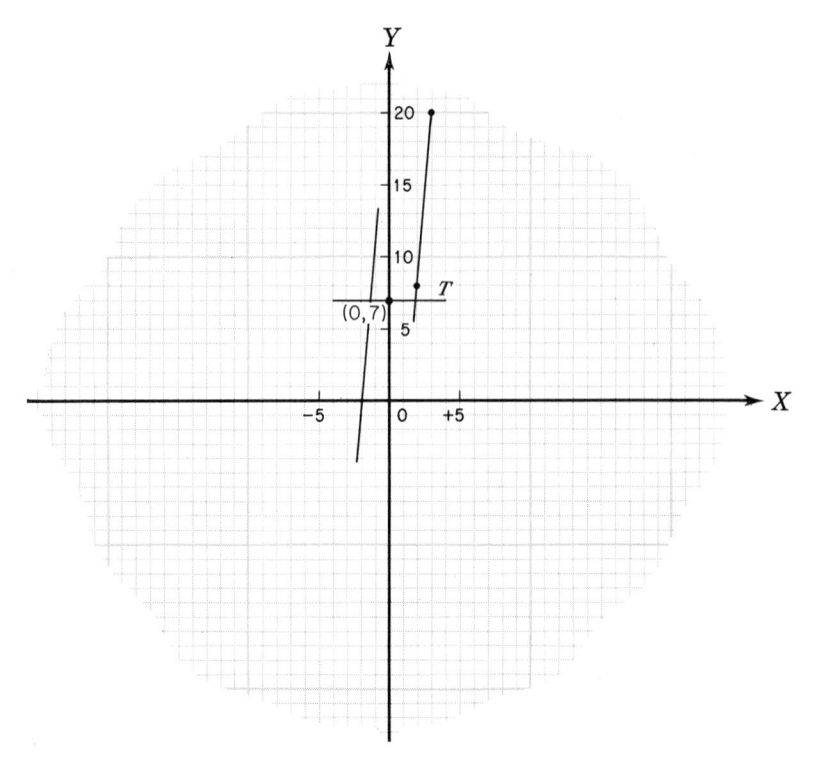

Figure 4-13

point (0,7) is neither a maximum nor a minimum point of the curve (see Fig. 4-13). The locus is sketched in Fig. 4-14.

Problem Set 4-6

Determine all points at which the following loci have horizontal tangent lines and determine whether each is a maximum, a minimum, or neither. Sketch the locus. Do *not* plot more than a few points.

1. $y = 3x^2 - 6x + 5$.
2. $y = x^4 - 32x + 3$.
3. $y = 31x^5 - 17$.
4. $y = x^5 - 20x^2 + 1$.
5. $h = 600 + 120t - 16t^2$.
6. $f(x) = 4x^3 + 6x^2 - 72x - 11$.

Sec. 4-6. Maximum and Minimum Values of a Function

7. $y = x^4 - 32x + 6$.
8. $y = 1/(x - 2) + 4$.†
9. $y = 2 - 7x + 4x^2$.
10. $h = 4t^2 - 7t + 2$.
11. $h = 400 + 120t - t^2$.
12. $h = x^2 - 5x + 6$.
13. $3y = 6x^2 + 5x - 9$.
14. $6y = x^4 - 12x^2 + 12$.
15. $2y = x^3 - 6x^2 + 12x$.
16. $y = 4/x$.†
17. $y = x^4 - 12x^2 + 12$.
18. $y = (x - 5)$.

19. Draw the graphs of three functions which have minimum values corresponding respectively to each of the conditions at which minima may occur. (Reread Sec. 4-6 if in doubt.)

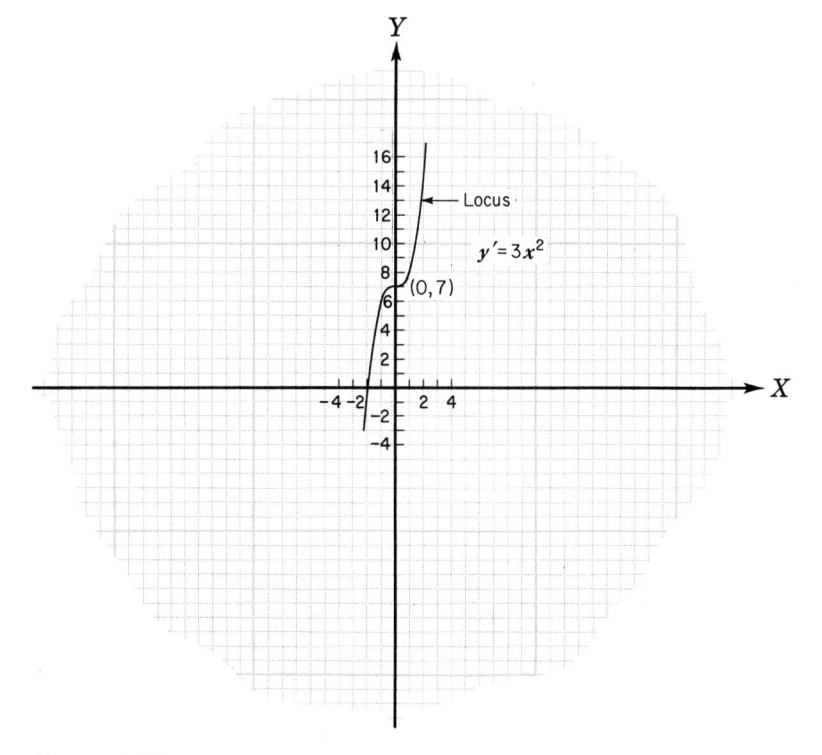

Figure 4-14

20. (a) Draw the graphs of three functions which satisfy respectively three conditions under which a relative minimum may occur, but which fail to have relative minima at these points.
(b) Discuss the "necessity and sufficiency" of the three conditions given for relative maximum and minimum of polynomials.

† Apparently Theorem 3 of Sec. 4-4 does not apply. Use the delta process of Sec. 4-2.

4-7. Applications Involving Maxima and Minima

Example 1

A stone is thrown vertically upward from the top edge of a tower 400 ft high. The height, h ft, of the stone from the ground at any time t sec after it is thrown, until it strikes the ground, is

$$h = 400 + 120t - 16t^2.$$

What is the maximum height reached by the stone (Fig. 4-15)?

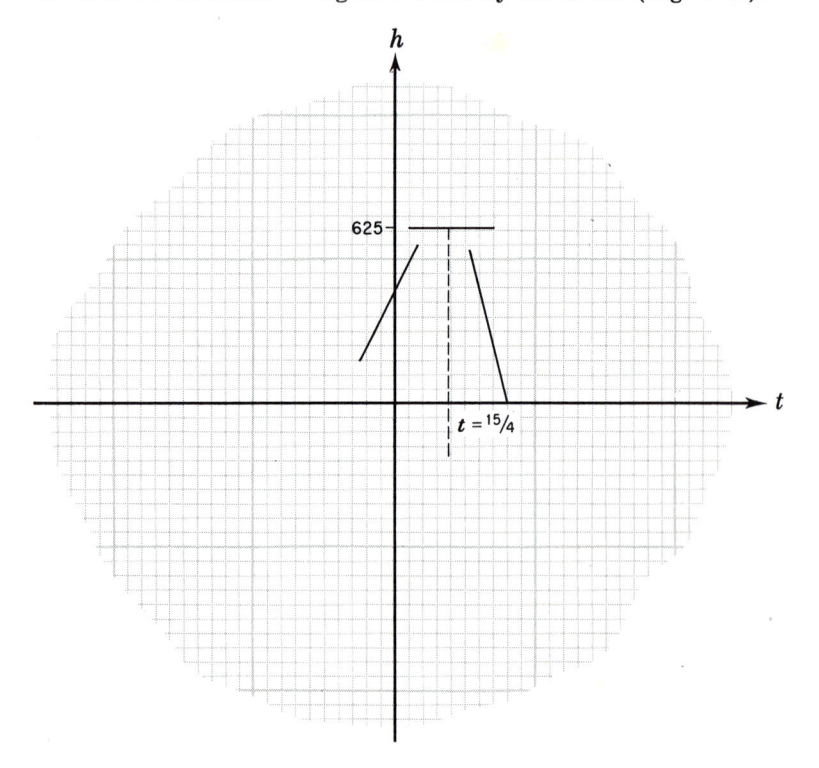

Figure 4-15

If h is graphed as a function of t, a horizontal tangent line will occur when $h' = 0$, that is, when

$$h' = 120 - 32t = 0$$
$$t = \frac{15}{4} \text{ sec.}$$

Sec. 4-7. Applications Involving Maxima and Minima 119

The slope of the tangent line changes from + to 0 to − as t increases through $\frac{15}{4}$. Hence $t = \frac{15}{4}$ corresponds to a maximum value of h. The maximum value is

$$h = 400 + 120(\tfrac{15}{4}) - 16(\tfrac{15}{4})^2 = 625 \text{ ft.}$$

Example 2

A sluice of rectangular cross section (Fig. 4-16) is made from a long strip of tin 40 in. wide by turning up the sides at right angles to the base. What width sides should be turned up to form a sluice of greatest carrying capacity?

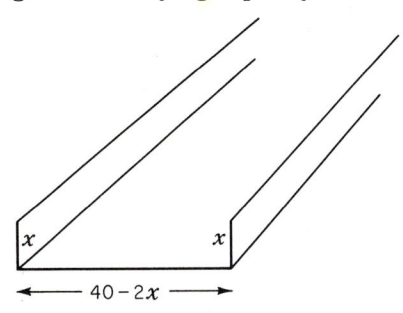

Figure 4-16

The problem is equivalent to finding the trough of greatest cross sectional area.

$$A(x) = x(40 - 2x)$$
$$= 40x - 2x^2.$$

If A is graphed as a function of x, a horizontal tangent line will occur where $A' = 0$.

$$A' = 40 - 4x = 0 \qquad x = 10 \text{ in.}$$

We show that $A(x)$ is a maximum at $x = 10$ by noting that, as x increases through 10, the slope of the tangent line changes from positive to zero to negative. $A(10) = 10(40 - 20) = 200$ is a maximum value for A. The point $(10, 200)$ is a maximum point (Fig. 4-17). If a 10-in. strip is turned up on each edge, the resulting sluice will have the greatest carrying capacity.

Example 3

A farm has a straight fence along one side. The owner wishes to enclose a rectangular plot of ground, using the existing fence as one

120   *Differential Calculus*

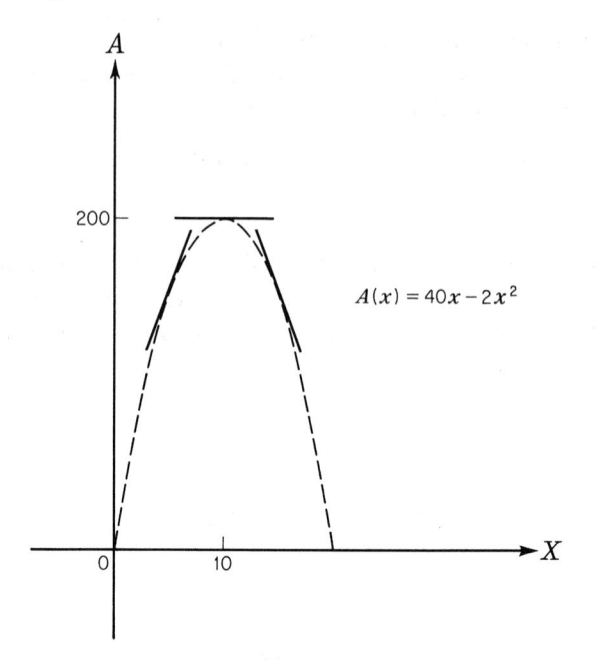

Figure 4-17

side of the plot, and then to divide the plot into two pens with a fence perpendicular to the existing fence. Find the largest total area he may enclose with 300 yd of additional fencing (Fig. 4-18).

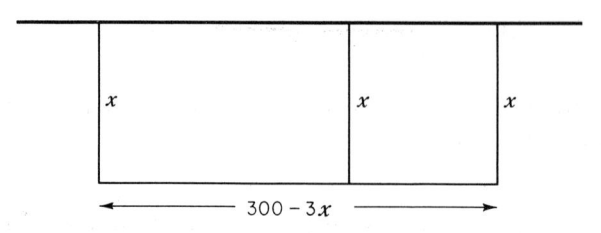

Figure 4-18

$$A = (300 - 3x)x$$
$$= 300x - 3x^2.$$
$$\frac{dA}{dx} = 300 - 6x$$
$$= 6(50 - x)..$$
$$6(50 - x) = 0$$
$$x = 50 \text{ yd}.$$

Sec. 4-7. Applications Involving Maxima and Minima   121

Hence the graph of $A = 300x - 3x^2$ has a horizontal tangent line at $x = 50$.

If $x < 50$, then $50 - x > 0$ and $\dfrac{dA}{dx} = +$.

If $x > 50$, then $50 - x < 0$ and $\dfrac{dA}{dx} = -$.

Hence the slope changes from $+$ to 0 to $-$, indicating a maximum area.

$\left. \begin{array}{r} x = 50 \text{ yd} \\ 300 - 3x = 150 \text{ yd} \end{array} \right\}$ are the dimensions for the maximum area. The maximum area is 7,500 sq yd.

Problem Set 4-7

1. Determine each point at which the curve having as its equation $y = x^3 - 8x^2 + 30x - 7$ has slope 25.

2. Find an equation of a line which is tangent to the curve given by $3x^2 + 4y^2 - xy - 18 = 0$ at the point $(2, -1)$.

3. A closed rectangular box with base twice as long as it is wide is to be built to contain 72 cu ft. Determine its dimensions if the material used in the box is to be a minimum and no allowance is made for the thickness of material along the edges.

4. Find k so that $y = 4x^2 + kx + 9$ will be tangent to the x axis.

5. A tin cylindrical can is to be constructed to contain 125 cu in. Determine the dimensions of a can requiring the least total sheet metal to construct if the circular top and bottom are cut from square pieces of tin with the scraps wasted and the side is cut from large sheets without waste. (Theorem 3, Sec. 4-4, does not apply.)

6. The yearbook committee in a certain school feels that about 600 people will buy yearbooks if the price is $5 each. Their experience suggests that for each 5 cents the price is raised, four of the 600 people will decide not to purchase a yearbook. (a) What price will bring in the greatest total income? (b) If the yearbooks cost $3.50 to produce, what price will bring in the greatest profit? (c) Compare the answers of (a) and (b) and make graphs to explain these facts to the editorial staff, who may understand little mathematics.

7. A telephone answering service for doctors charges each doctor using the service a fee of $3 per month. They have 500 subscribers. The owners believe that for each cent they raise the cost, one doctor will discontinue the service; that is, if the fee is raised to $3.25 per month, 25 of the 500 doctors would stop using the service;

if they raised the fee to $3.30 per month, 30 doctors would drop, etc. On the basis of this assumption, what service fee will bring in the greatest monthly income?

8. A boat company agrees to transport a group of students on an outing for a charge of $6 per student if 75 or fewer students go on the outing. The boat company further agrees that for each student more than the basic 75, they will reduce the charge 5 cents per student *for all students.* What number of students will bring in the largest income to the boat company? How much will each student pay in this case?

9. Jack Trot has a 120-apartment building. His experience leads him to believe that he can rent all 120 apartments at $40 per month for each apartment. He estimates that for each dollar he raises the rent, there will be two empty apartments. That is, if he charges $41, he will only be able to rent 118 of the apartments; if he charges $42 per month, he will only rent 116 apartments, etc. What rental will bring him the greatest income?

10. Given $y = x^3 - 3x^2 - 9x + 5$. Find maximum and minimum points. Make a table and sketch the locus.

11. Calvin Butterball wishes to enclose a rectangular area and divide it into three pens. Calvin decides to use an existing fence for one side of the rectangle and make the cross fences perpendicular to the existing fence as shown in Fig. 4-19. Find the largest total area Cal may enclose with 200 ft of additional fence.

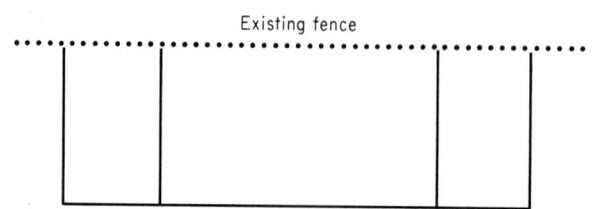

Figure 4-19

12. A box with a square base and open top is to be constructed to hold 4 cu ft. Determine the dimensions of such a box having the least surface area.

13. Silas Hogwinder wishes to enclose a rectangular plot of ground and divide it into four pens by erecting three fences parallel to one side of the rectangle. If Silas has a total of 300 ft of fencing, what is the largest area he may enclose?

4-8. Differentiation of a Product

This theorem is not only a timesaver but provides the basis for an important tool of more advanced mathematics and statistics called *integration by parts*.

Theorem

If $u = u(x)$ and $v = v(x)$ are differentiable functions of x, then

$$\frac{d(u \cdot v)}{dx} = u \frac{dv}{dx} + v \frac{du}{dx}.$$

As usual in proofs involving derivatives, the delta process of Sec. 4-4 is used.
Let $f(x) = u(x) \cdot v(x)$
then $f(x + \Delta x) = u(x + \Delta x) \cdot v(x + \Delta x)$.

Hence, $\Delta f = f(x + \Delta x) - f(x)$
$$= u(x + \Delta x) \cdot v(x + \Delta x) - u(x) \cdot v(x). \quad (1)$$

Now $\Delta u = u(x + \Delta x) - u(x)$; consequently, one easily obtains that $u(x + \Delta x) = u(x) + \Delta u$. Similarly $v(x + \Delta x) = v(x) + \Delta v$.
Substituting in Eq. (1), we obtain

$$\Delta f = [u(x) + \Delta u] \cdot [v(x) + \Delta v] - u(x) \cdot v(x)$$
$$= u(x) \cdot \Delta v + v(x) \cdot \Delta u + \Delta u \cdot \Delta v.$$

$$\frac{\Delta f}{\Delta x} = u(x) \cdot \frac{\Delta v}{\Delta x} + v(x) \cdot \frac{\Delta u}{\Delta x} + \frac{\Delta u}{\Delta x} \cdot \Delta v. \quad (2)$$

$$\lim_{\Delta x \to 0} \frac{\Delta f}{\Delta x} = \lim_{\Delta x \to 0} \left[u(x) \cdot \frac{\Delta v}{\Delta x} + v(x) \cdot \frac{\Delta u}{\Delta x} + \frac{\Delta u}{\Delta x} \cdot \Delta v \right]$$

$$= \lim_{\Delta x \to 0} \left[u(x) \cdot \frac{\Delta v}{\Delta x} + \lim_{\Delta x \to 0} v(x) \cdot \frac{\Delta u}{\Delta x} \right] + \lim_{\Delta x \to 0} \left[\frac{\Delta u}{\Delta x} \cdot \Delta v \right]†$$

$$= \lim_{\Delta x \to 0} u(x) \lim_{\Delta x \to 0} \frac{\Delta v}{\Delta x} + \lim_{\Delta x \to 0} v(x) \lim_{\Delta x \to 0} \frac{\Delta u}{\Delta x}$$

$$+ \lim_{\Delta x \to 0} \frac{\Delta u}{\Delta x} \cdot \lim_{\Delta x \to 0} \Delta v‡$$

$$= u(x) \frac{dv(x)}{dx} + v(x) \frac{du(x)}{dx} + \frac{du(x)}{dx} \cdot 0. \quad (3)$$

† This requires the theorem which asserts that the limit of a sum is equal to the sum of the limits if the latter limits exist.
‡ This requires the theorem which asserts that the limit of a product is equal to the product of the limits if the latter limits exist.

124   *Differential Calculus*

Consequently $\dfrac{d(u \cdot v)}{dx} = u\dfrac{dv}{dx} + v\dfrac{du}{dx}.$

If $\lim\limits_{\Delta x \to 0} \Delta u/\Delta x$ and $\lim\limits_{\Delta x \to 0} \Delta v/\Delta x$ exist, then Δu and Δv must approach zero when Δx approaches zero. (Why is this observation germane?)

Example 1

If $y = (x^3/3 - 9x + 1)(x^2 + 6x + 2)$, find y'.

Use the theorem of this section with $u = \tfrac{1}{3}x^3 - 9x + 1$ and $v = x^2 + 6x + 2$,

$$y' = u\frac{dv}{dx} + v\frac{du}{dx}$$

$$= \left(\frac{x^3}{3} - 9x + 1\right)\frac{d(x^2 + 6x + 2)}{dx} + (x^2 + 6x + 2)\frac{d(\tfrac{1}{3}x^3 - 9x + 1)}{dx}$$

$$= \left(\frac{x^3}{3} - 9x + 1\right)(2x + 6) + (x^2 + 6x + 2)(x^2 - 9)$$

$$= \left(\frac{x^3}{3} - 9x + 1\right) \cdot 2 \cdot (x + 3) + (x^2 + 6x + 2)(x - 3)(x + 3).$$

The factor $x + 3$ appears in each term. Hence,

$$y' = (x + 3)\left[\left(\frac{x^3}{3} - 9x + 1\right)2 + (x^2 + 6x + 2)(x - 3)\right]$$

$$= (x + 3)\left(\frac{5x^3}{3} + 3x^2 - 34x - 4\right).$$

If the factors of y had been "multiplied out," it would not have been so apparent that $x + 3$ is a factor of y'.

Since many applications require the solution of the equation $y' = 0$, having y' in partially factored form is an additional advantage of this theorem.

Example 2

If $f(x) = (5x^2 - 3)(7x^2 + 6)$, find $f'(x)$.

Setting $u = (5x^2 - 3)$ and $v = 7x^2 + 6$, use the theorem of this section.

Sec. 4-8. Differentiation of a Product

$$f'(x) = u\frac{dv}{dx} + v\frac{du}{dx}$$
$$= (5x^2 - 3)(14x) + (7x^2 + 6)(10x)$$
$$= 2x[(5x^2 - 3)7 + (7x^2 + 6)5]$$
$$= 2x(70x^2 + 9).$$

Example 3

Find all points at which $y = (3x^2 - 5x + 4)(2x + 3)$ has slope 9.

Since $\frac{dy}{dx}$ represents the slope, we wish to find values of x at which $\frac{dy}{dx} = 9$.

$$y = (3x^2 - 5x + 4)(2x + 3)$$
$$y' = (3x^2 - 5x + 4)(2) + (2x + 3)(6x - 5)$$
$$= 18x^2 - 2x - 7.$$

Values of x such that $y' = 9$ will be solutions of

$$y' = 9$$
or
$$18x^2 - 2x - 7 = 9$$
$$18x^2 - 2x - 16 = 0$$
$$9x^2 - x - 8 = 0$$
$$(9x + 8)(x - 1) = 0$$
$$x = -\frac{8}{9} \qquad x = 1.$$

Hence, the desired points are $(-\frac{8}{9}, 3{,}212/243)$ and $(1, 10)$.

Problem Set 4-8

1. Find y' if $y = (x^3 - 1)(4x^2 - 2x + 1)$.
2. If $y = (x^3 - 2)(4x^2 + 5x - 7)$, find y'.
3. Determine the slope of $y = (3x^2 - 2x + 5)(4x^2 - 3x)$ at the point on the locus where $x = 2$.
4. If $y = (x^5 - 1)(x^2 - 1)$, find y'.

In Probs. 5 to 8, compute the slope of the line tangent to the given curve at the point of intersection of the given curve C and line L.

5. $C: y = (x^3 + 7x - 4)(x + 2);\ L: x = -2$.
6. $C: y = (x^5 - 4x)(x^2 + 2);\ L: x = 3$.
7. $C: y = (x^3 + 1)(x^7 + 7x^2 + 3x - 5);\ L: x = 1$.
8. $C: y = (x^3 + 2x^2 + 4x + 5)(2x^3 + 4x^2 + 8x - 1);\ L: x = 4$.
9. The line $x = 5$ crosses the curve $y = (3x^2 - 2x)(4x^3 - 5)$ at some point P. Determine the slope of the tangent to the curve at P.

Differential Calculus

10. If $y = (x^2 + 3)(1/x)$, use the theorem of this section to compute y'. HINT: Find $\dfrac{d(1/x)}{dx}$ first, using the delta process.

11. If $y = (x^2 + 5) \cdot 1/(x - 2)$, find y' using the theorem of this section. HINT: Find the derivative of the second factor by the delta process.

In Probs. 12 to 17, find each point at which the given curve has a horizontal tangent (slope zero).

12. $y = (x^2 - 3x + 2)(x^2 + 3x + 1)$.
13. $y = (x^2 - 5x + 4)(x^2 - 5x + 9)$.
14. $y = x^7(x^2 + 4)$. **15.** $y = (ax + b)(ax - b)$.
16. $y = (3x^2 + 4x + 5)(6x^2 + 18x + 13)$.
17. $y = (3x^3 - 4x - 3)(3x^2 + 4x + 1)$.
18. Determine all points at which $y = (3x^2 - 5x + 4)(2x + 3)$ has slope 13.
19. Determine the slope of a line tangent to the curve having equation $y = (x^3 + 4x^2 + 2x - 3) \cdot (x^{15} - 4x^2 - 3x + 1)$ at the point $(0, -3)$.
20. Prove that $\dfrac{dc \cdot v(x)}{dx} = c\dfrac{dv}{dx}$, if c is a constant.

4-9. Differentiation of a Power of a Function

It is possible to multiply out $(t^2 - 5t + 3)^{11}$ and then differentiate the result to determine y' where $y = (t^2 - 5t + 3)^{11}$, but this is a tedious task. The theorem of this section provides a simple method for avoiding this multiplication.

Theorem

$$\frac{d(u^n)}{dx} = nu^{n-1}\frac{du}{dx}$$

if n is a positive integer and $u = u(x)$ a differential function of x.

The theorem can be proved in the manner of Theorem 3, Sec. 4-4, but for variety the theorem of Sec. 4-8 is used instead.

In order to use the theorem

$$\frac{d(u \cdot v)}{dx} = u \cdot \frac{dv}{dx} + v \cdot \frac{du}{dx}$$

Factor $u^n = u \cdot u^{n-1}$. Let $v = u^{n-1}$. Then

$$\frac{d(u^n)}{dx} = u \cdot \frac{du^{n-1}}{dx} + u^{n-1} \cdot \frac{du}{dx}.$$

Sec. 4-9. Differentiation of a Power of a Function

Separate the u^{n-1} of the first term into two factors $u^{n-1} = u \cdot u^{n-2}$ and repeat the process, obtaining

$$\frac{d(u^n)}{dx} = u \cdot \frac{d(u \cdot u^{n-2})}{dx} + u^{n-1} \cdot \frac{du}{dx}$$

$$= u \cdot \left(u \cdot \frac{du^{n-2}}{dx} + u^{n-2} \cdot \frac{du}{dx} \right) + u^{n-1} \cdot \frac{du}{dx}$$

$$= u^2 \cdot \frac{du^{n-2}}{dx} + u^{n-1} \cdot \frac{du}{dx} + u^{n-1} \cdot \frac{du}{dx}$$

$$= u^2 \cdot \frac{d(u \cdot u^{n-3})}{dx} + 2u^{n-1} \cdot \frac{du}{dx}.$$

By repeating the process, we eventually obtain the desired result,

$$\frac{d(u^n)}{dx} = nu^{n-1} \cdot \frac{du}{dx}.$$

This may be sufficient proof to satisfy many readers. However, essentially this proof requires the use of mathematical induction. Students familiar with the process should complete the proof.

Example 1

If $y = u^{11}$, where $u = x^2 - 5x + 3$, find $\frac{dy}{dx}$.

$$\frac{dy}{dx} = \frac{d(u^{11})}{dx} = 11u^{10}\frac{du}{dx}.$$

Since $\frac{du}{dx} = \frac{d(x^2 - 5x + 3)}{dx} = 2x - 5,$

$$\frac{dy}{dx} = 11u^{10}(2x - 5)$$

or $\frac{dy}{dx} = 11(x^2 - 5x + 3)^{10}(2x - 5).$

This method is easier than raising $x^2 - 5x + 3$ to the eleventh power; moreover, the result is partially factored.

Example 2

Find y' where $y = (3x^2 - 7)^5$. Here $y = u^5$ if $u = 3x^2 - 7$. Then

$$y' = 5u^4 \frac{du}{dx}$$

$$= 5(3x^2 - 7)^4 \frac{d(3x^2 - 7)}{dx}$$

$$= 5(3x^2 - 7)^4(6x)$$

$$= 30x(3x^2 - 7)^4.$$

Example 3

If $y = (x^2 - 5x + 3)^5(2x - 5)^3$, find $\dfrac{dy}{dx}$.

Since y is a product, we use

$$\frac{d(u \cdot v)}{dx} = u\frac{dv}{dx} + v\frac{du}{dx}$$

with $u = (x^2 - 5x + 3)^5$ and $v = (2x - 5)^3$. Thus

$$y' = (x^2 - 5x + 3)^5 \frac{d[(2x - 5)^3]}{dx} + (2x - 5)^3 \frac{d[(x^2 - 5x + 3)^5]}{dx}.$$

Then apply the theorem of this section

$$y' = (x^2 - 5x + 3)^5 3(2x - 5)^2 \frac{d(2x - 5)}{dx}$$

$$+ (2x - 5)^3 5(x^2 - 5x + 3)^4 \frac{d(x^2 - 5x + 3)}{dx}$$

$$= (x^2 - 5x + 3)^5 3(2x - 5)^2(2)$$

$$+ (2x - 5)^3 5(x^2 - 5x + 3)^4(2x - 5)$$

$$= (x^2 - 5x + 3)^4(2x - 5)^2[(x^2 - 5x + 3)6 + (2x - 5)^2 5]$$

$$= (x^2 - 5x + 3)^4(2x - 5)^2(26x^2 - 130x + 143).$$

Having a polynomial partially factored may be helpful in applications.

Problem Set 4-9

1. Carry through the proof of this section, using u^5 for u^n and $x^3 - 5x$ for $u(x)$.

2. Find $\dfrac{d[u^7]}{dx}$ where $u = x^3 - 5x + 1$.

3. Find $\dfrac{d[(x^2 - 5x + 2)^4]}{dx}$.

4. Find y' where $y = (x^2 - 5x)^7$.

5. Find y' where $y = (7x^2 - 4x + 3)^{10}$.

6. If $p(x) = (17x^3 - 3)^{15}$, find $\dfrac{dp}{dx}$.

7. If $y = (x - 3)^8$, find $\dfrac{dy}{dx}$.

8. If $y = (t^2 - 2t + 3)^3$, find $\dfrac{dy}{dt}$.

9. Find the maximum and minimum points of the curve and sketch the locus.

$$y = (2x - 5)^3(x^2 - 5x + 3)^5.$$

Sec. 4-10. Derivative of a Composite Function

10. Find the maximum and minimum points, and sketch the locus.
$$h = (3 - 5t + t^2)^{11}.$$

4-10. Derivative of a Composite Function

If $y = 3x^2 + 4x - 9$ and $x = 2t^2 - t$, then it is meaningful either to speak of y as a function of x, or to speak of y as a function of t. (If a value of t is given, then a unique corresponding value of y is determined.) It is quite possible, in this example, to obtain a formula for y in terms of t by substituting $2t^2 - t$ for x in the formula $y = 3x^2 + 4x - 9$, obtaining

$$\begin{aligned} y &= 3(2t^2 - t)^2 + 4(2t^2 - t) - 9 \\ &= 12t^4 - 12t^3 + 11t^2 - 4t - 9. \end{aligned}$$

Then $\dfrac{dy}{dt} = 48t^3 - 36t^2 + 22t - 4$

$$= 2(6t^2 - 3t + 2)(4t - 1).$$

The theorem of this section makes it unnecessary for us to make this substitution, since

$$\begin{aligned} \frac{dy}{dt} &= \frac{dy}{dx} \cdot \frac{dx}{dt} \\ &= \frac{d(3x^2 + 4x - 9)}{dx} \cdot \frac{d(2t^2 - t)}{dt} \\ &= (6x + 4) \cdot (4t - 1) \end{aligned}$$

or, upon substituting $2t^2 - t$ for x,

$$= 2(6t^2 - 3t + 2)(4t - 1)$$

as before.

If $y = f(x)$ and $x = g(t)$, then y may be considered to be a function of t, and $\dfrac{dy}{dt}$ may be a meaningful concept.

Theorem

> If $y = f(x)$ and $x = g(t)$, then if both $\dfrac{dy}{dx}$ and $\dfrac{dx}{dt}$ exist,
> $$\frac{dy}{dt} = \frac{dy}{dx} \cdot \frac{dx}{dt}.$$

Differential Calculus

If $\Delta x \neq 0$ when $\Delta t \neq 0$, then

$$\frac{\Delta y}{\Delta t} = \frac{\Delta y}{\Delta x} \cdot \frac{\Delta x}{\Delta t} \qquad \text{when } \Delta t \neq 0.$$

Since $\dfrac{dx}{dt}$ exists, $\Delta x \to 0$ when $\Delta t \to 0$. (Why?)

Thus
$$\lim_{\Delta t \to 0} \frac{\Delta y}{\Delta t} = \lim_{\Delta t \to 0} \frac{\Delta y}{\Delta x} \cdot \lim_{\Delta t \to 0} \frac{\Delta x}{\Delta t}$$
$$= \lim_{\Delta x \to 0} \frac{\Delta y}{\Delta x} \cdot \lim_{\Delta t \to 0} \frac{\Delta x}{\Delta t}$$
$$\frac{dy}{dt} = \frac{dy}{dx} \cdot \frac{dx}{dt}.$$

If the assumption $\Delta x \neq 0$ when $\Delta t \neq 0$ is not fulfilled, the theorem still holds but a separate proof is required. The proof for this special case will be found in many texts.

Example 1

If $v = \frac{4}{3} \cdot \pi r^3$ and $r = 3t^2 - 5$, find $\dfrac{dv}{dt}$.

$$\frac{dv}{dt} = 4\pi r^2 \frac{dr}{dt}$$

but, since $r = 3t^2 - 5$, $\dfrac{dr}{dt} = 6t$.

Hence, $\dfrac{dv}{dt} = 4\pi r^2 (6t)$
$$= 4\pi (3t^2 - 5)^2 6t.$$

Another type of problem in which the theorems of this chapter come into play is illustrated below.

Example 2

Find $\dfrac{dy}{dx}$ if $x^2 y^3 + x = 5$. It would be possible to solve for y, obtaining $y = \sqrt[3]{\dfrac{5-x}{x^2}}$, and then differentiate this by the delta process. However, this is tedious at best, and it may be *impossible* to solve for y, explicitly. We proceed as follows:

$$x^2 y^3 + x = 5.$$

Sec. 4-10. Derivative of a Composite Function

Differentiate the product x^2y^3, using the theorems of Secs. 4-8 and 4-9.

$$\frac{d(x^2 \cdot y^3)}{dx} = x^2 \cdot \frac{d(y^3)}{dx} + y^3 \cdot \frac{d(x^2)}{dx}$$

$$= x^2 \cdot 3y^2 \frac{dy}{dx} + y^3 \cdot 2x.$$

This method is used to differentiate both members of the equation

$$x^2y^3 + x = 5$$

obtaining $x^2 \cdot 3y^2 \frac{dy}{dx} + y^3 \cdot 2x + 1 = 0.$

Solving for $\frac{dy}{dx}$,

$$\frac{dy}{dx} = \frac{-1 - 2xy^3}{3x^2y^2}, \qquad \text{the desired derivative.}$$

Problem Set 4-10

1. If $y = 3x^2 - 7x + 5$ and $x = 4t^2 - 2$, find $\frac{dy}{dt}$ in terms of t.

2. If $y = x^3 - 4$ and $x = 2t + 11$, find $\frac{dy}{dt}$.

3. Find $\frac{dy}{dx}$ if $x^3y^5 + 3x^2 = 57$.

4. Find $\frac{dy}{dt}$ if $y^2t^3 + 2y - 3t + 7 = 0$.

5. The point $(2, -1)$ lies on the curve $x^2y^5 - 4x^2y + 3x - 18 = 0$. Find the slope of the line tangent to the given curve at $(2, -1)$.

6. If $v = \frac{4}{3}\pi r^3$ and $r = 7t^2 - 5t + 8\pi - 19$, find $\frac{dv}{dt}$.

7. Find $f'(x)$ if

$$f(x) = 45x^9 - 31(x^9 - 3x^3 + 5x - 18)^{27}$$
$$- 41(x^{19} - 3\sqrt{5}\,x^{23} + 5)^{75} + 57.$$

In Probs. 8 to 10, find $\frac{dy}{dx}$.

8. $x^3y^7 - 62x^7y^5 + 4{,}132 = 35\pi.$
9. $x^2y^8 - 37x + 45y^3 + 17 = 0.$
10. $x^7y^6 - 31x^7y^3 - 29x + 52y = 245x^{71}y^5 + 2{,}349{,}876\pi\sqrt{5}.$

4-11. Some Additional Applications

We next turn our attention to further applications. No new concepts or techniques are involved.

Problem Set 4-11

In these problems, you may wish to use the following facts: (1) If $y = 1/x$, $y' = -1/x^2$. (2) If $f(x) = 1/x^2$, $f'(x) = -2/x^3$. These facts may be obtained by use of the delta process; however, you may assume them here. They were derived in Probs. 4 and 5 of Set 4-2.

1. The sum of two numbers is 12. Find the numbers if their product is to be as large as possible. HINT: Let x and $(12 - x)$ be the numbers. The problem is then to determine x and $(12 - x)$ such that $P = x(12 - x)$ is as large as possible *and to prove conclusively that the values given do make the product as large as possible.*

2. Determine a positive number such that the sum of the number and its reciprocal is as small as possible.

3. A closed rectangular box with base twice as long as it is wide is to be built to contain 72 cu ft. Determine its dimensions if the material used in the box is to be at a minimum and no allowance is made for the thickness of the material.

4. (a) Hobart Musson wishes to enclose a rectangular plot of ground and divide it into four pens by erecting three fences parallel to one side of the rectangle. If Hobart has a total of 600 ft of fencing, what is the largest area he can enclose?
(b) Do the four pens need to be of the same size?

5. (a) The Bilgewater Beach is an apartment hotel, containing 125 apartments. The owner's experience leads him to believe that he can rent all 125 apartments at $60 a month for each apartment. He estimates that for each $1 he raises the rent there will be two empty apartments; that is, if he charges $61 a month he will be able to rent only 123 of the apartments. If he charges $62 a month he will rent only 121 apartments, and so on. What rental will bring him the greatest total income?
(b) Rework if there are 110 apartments in all.
(c) Rework if there are only 30 apartments in all.

6. The Norman Bus Company agrees to transport a group of students on an outing for a charge of $6 per student if 75 or fewer

Sec. 4-11. Some Additional Applications

students go on the outing. The bus company further agrees that for each student more than the basic 75 they will reduce the charge 5 cents per student for all the students who are going on the trip. What number of students will bring the largest income to the bus company? How much will each student pay in this case?

7. In the proof of a problem a student uses the statement, "Since a real number plus its square is never negative, it follows that . . . ." Show that the student's proof is not valid.

8. If x is a real number, what is the smallest value of the function $f(x) = x^2 + 1/x^2$?

9. What is the area of the largest isosceles triangle having its equal sides of length 100 cm?

10. (a) A cylindrical container with open top is to have a volume of 125π cu in. Find the dimensions of the container using the smallest total amount of metal.

(b) Generalize the problem of part (a) to a container of volume V. Determine the *ratio* of diameter to height.

11. (a) Find the greatest and least values of the fraction having its numerator 25 greater than the square of its denominator.

(b) Same as (a), but restrict the fraction to be positive.

12. Find the x coordinates of all points at which the given curves $y = x^3 - 2x^2 - 5x + 17$, and $y = \frac{2}{3}x^3 - 3x^2 + 10x - 11$ have parallel tangents corresponding to the same values of x.

13. A block of ice slides down a chute 100 ft long with an acceleration of 16 ft/sec/sec. The distance of the block from the bottom of the chute is given by $s(t) = -8t^2 - 20t + 100$.

(a) At what time does the block reach the bottom of the chute?

(b) How long does it take the block to slide 4 ft; that is, when does $s(t) = 96$?

14. A ball rolls on an inclined plane so that its distance s ft from the bottom of the plane at time t sec is $s(t) = 6t - 2t^2$.

(a) How far up the inclined plane will the ball roll?

(b) At what time does the ball reach the bottom of the inclined plane?

(c) At what times is the ball 1 ft from the bottom of the plane?

15. The point $(0,-4)$ does not lie on the curve $y = x^2$. Nevertheless, there are two lines passing through $(0,-4)$ which are tangent to the curve $y = x^2$. Find the equations of these two lines (Fig.

4-20). HINT: Let the point of tangency be (A,B). Then $B = A^2$. (Why?) Also, the slope of the tangent line is $\dfrac{B+4}{A-0} = 2A$. (Why?)

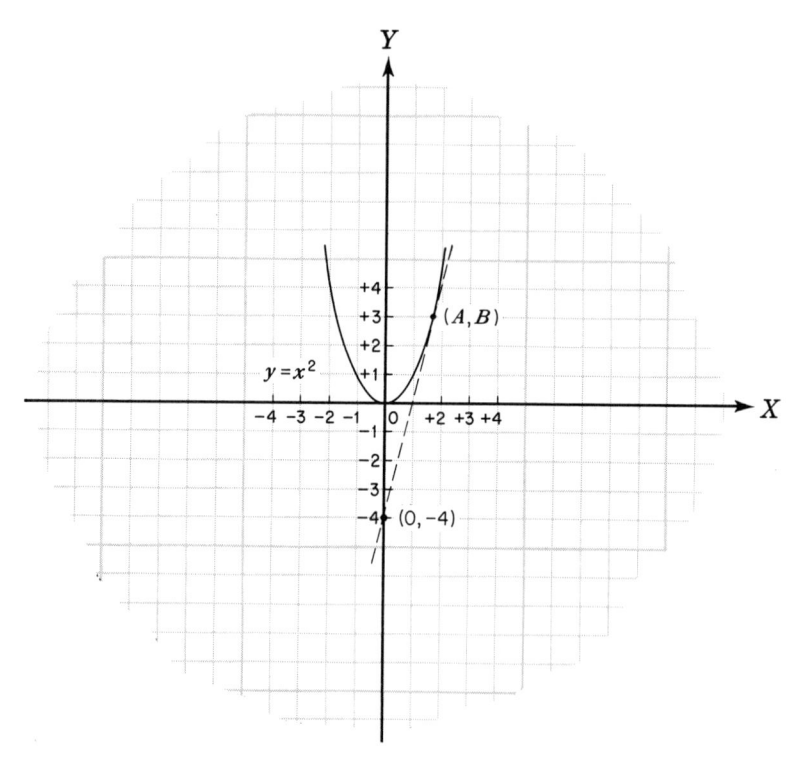

Figure 4-20

16. The height of a projectile shot vertically is given as h where $h = -16t^2 + v_0 t + s_0$ where v_0 is the initial velocity and s_0 is the initial height. If a projectile is shot from ground level and reaches a height of 144 ft in 1 sec, how high will the projectile rise?

17. The cost of fuel for running a steamer, per hour, is proportional to the cube of the speed of the steamer. The fuel costs $20 per hour when the speed is 10 mph. Other expenses for the steamer are $135 per hour.

(a) Find the most economical rate to run the steamer for 27 miles.
(b) For 500 miles.

Sec. 4-11. Some Additional Applications

18. The owner of a large apartment project feels that he can keep all 200 of his apartments filled at a rental of $40 per month per apartment. His experience suggests that if he raises the monthly rental $5 per apartment one apartment will remain vacant; $15 a month and there will be nine empty apartments, and so forth; that is, if he raises the rent $5x$ dollars, then x^2 apartments will remain empty.
(a) What monthly rental will bring in the greatest income?
(b) What percentage of the apartments will be rented at this maximum income?
(c) Over what range of values of x does the function have meaning?
(d) Make graphs and explain your conclusions to a nonmathematical friend.

19. Rework Prob. 18, changing the original assumption of a 200-apartment unit to a 500-apartment unit.

20. Solve Prob. 18, assuming that each empty apartment saves the owner $2 a month on janitorial expenses.

21. A piece of wire 12 ft long is cut into two pieces. One piece is bent into the form of a circle. The other piece is bent into a square.
(a) Where should the cut be made if the *sum* of the areas of the circle and square is to be as small as possible?
(b) Where should the cut be made if the sum of the two areas is to be as large as possible? Justify your answer.

22. A sluice is to be made from a long strip of metal 12 in. wide by folding up 4 in. on each side to form an isosceles trapezoid. Find the width across the top such that the sluice will have a maximum carrying capacity.

23. The stiffness of a beam of rectangular cross section is proportional to the breadth times the cube of the depth of the beam. What is the shape of the stiffest beam which can be cut from a circular log 1 ft in diameter? Use the delta process if necessary.

24. Find the dimensions of the rectangle of maximum area which may be inscribed in a right triangle having legs 4 ft and 6 ft if one corner of the rectangle is to be at the right angle of the triangle.

25. In what x intervals does $y = (x - 2)^2(4x - 5)^3$ increase as x increases?

26. A square sheet of metal, 30 in. on a side, has square pieces cut out of the corners and the edges turned up to form a box with open

136 *Differential Calculus*

top (Fig. 4-21). Determine the dimensions of the box of largest volume which may be so formed.

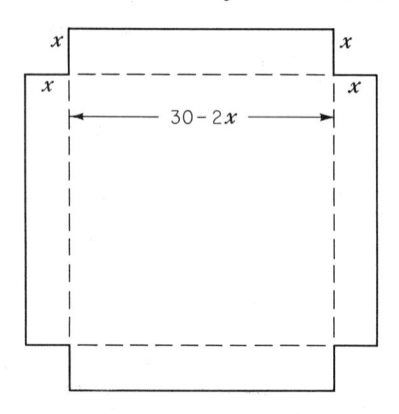

Figure 4-21

In Probs. 27 to 35, sketch the indicated loci, showing all maximum and minimum points. Determine rough approximations of the Y coordinates, not exact values, if desired.

27. $y = 7(2x - 3)^5(x^2 - 3x + 7)^9$.

28. $y = \dfrac{3x^2 - 5x + 7}{x}$.

(a) Work as $u \cdot v$ with $u = 3x^2 - 5x + 7$ and $v = 1/x$.
(b) Work as $y = 3x - 5 + (7/x)$.

29. $y = (479x^2 - 375x + 2{,}193)^{71}$.

30. $y = \pi x^2 - \sqrt{2}\,(x) + 37 - 5\sqrt{17}$.

31. $y = 71(x - 5x^3)^{14}$. **32.** $y = 2 + |3x - 15|$.

33. $y = (4x - 2)^5 x^2$. **34.** $y = 1/x + 1/x^2 - 4x^3$.

35. $y = \begin{cases} 4x - 7 & \text{if } x \geq 3 \\ 8 - x & \text{if } x < 3 \end{cases}$

4-12. Self-test

1. Use the basic definition (delta process) to determine $\dfrac{dy}{dx}$ when $y = \tfrac{3}{2}x + 1$.

2. Determine the equation of a line tangent to the curve $y = x^3 + x^2 - 30x + 5$ such that the tangent line has slope 3.

3. Sketch $y = 4x^3 - 2x^2 - 40x + 3$, showing all maximum and minimum points as well as the approximate intercepts.

Sec. 4-12. Self-test

4. If $y = (x^2 - 4x + 2)^{15} \cdot (2x - 3)^{10}$, find $\dfrac{dy}{dx}$.

5. A stone is projected upward from a tower 144 ft above ground level with an initial velocity of 32 ft/sec/sec upward. Determine the highest point to which the stone ascends and also determine the velocity of the stone as it strikes the ground. (See Prob. 16, page 134.)

6. A square sheet of metal 40 in. on each side has square pieces removed from each corner and the remaining edges turned up to form a box with an open top. Determine the dimension of the box of largest volume which may be so formed.

7. Find the point on the graph $y = x^2$ which is closest to the point $(3, -1)$. Obtain your answer to the nearest tenth only.

5

Extended Theorems of Differentiation

5-1. Negative, Fractional, and Zero Exponents

If n is a positive integer, then b^n is defined as 1 multiplied n times b ($b^n = 1 \cdot \underbrace{b \cdot b \cdot b \cdots b}_{n \text{ factors}}$). Thus $2^5 = 1 \cdot 2 \cdot 2 \cdot 2 \cdot 2 \cdot 2$. Certain "laws of exponents" are derived from this definition in elementary algebra. The more important of these laws are:

I. $a^m \cdot a^n = a^{m+n}$

PROOF, when m and n are positive integers:
$$a^m \cdot a^n = 1 \cdot \underbrace{a \cdot a \cdots a}_{m} \cdot 1 \cdot \underbrace{a \cdot a \cdots a}_{n} = 1 \cdot \underbrace{a \cdot a \cdots a}_{m+n} = a^{m+n}.$$

II. $\dfrac{a^m}{a^n} = \begin{cases} a^{m-n} & \text{if } m > n \\ 1 & \text{if } m = n \\ \dfrac{1}{a^{n-m}} & \text{if } m < n \end{cases}$ where m and n are positive integers.

III. $(a^n)^k = a^{nk}$ where n and k are positive integers.

IV. $(a \cdot b)^n = a^n b^n$ where n is a positive integer.

At this point, symbols (such as 3^{-2}, 5^0, $4^{\frac{1}{2}}$) which involve negative, zero, or fractional exponents *have no meaning. It is desirable to define negative, zero, and fractional exponents so that the above laws hold for such new exponents.* The following definitions satisfy these laws. They cannot be "proved" from the relations above, but it is possible to show that they are consistent with the above definitions. (What does that mean?)

Sec. 5-1. Negative, Fractional, and Zero Exponents

zero exponent
fractional exponent

$b^0 = 1 \quad \text{if } b \neq 0,$

$b^{n/d} = \sqrt[d]{b^n}.$

If d is even, $b^{n/d}$ is defined only for nonnegative b, that is, if d is even, $b \geq 0.$† In any event, n/d must be a positive rational number, that is, both n and d are positive integers.

negative exponent

$b^{-k} = \dfrac{1}{b^k} \quad \text{if } b \neq 0 \quad \text{for } k, \text{ a rational number.}$

A few illustrations may be helpful.

Example 1

$$4^{-\frac{1}{2}} = \frac{1}{4^{\frac{1}{2}}} = \frac{1}{\sqrt{4}} = \frac{1}{2}.\ddagger$$

Example 2

$$(a+b)^{-2} = \frac{1}{(a+b)^2}, \quad \text{providing } (a+b) \neq 0.$$

Example 3

$$\frac{3^5}{3^{-2}} = 3^{5-(-2)} = 3^7 \quad \text{using Law II, extended to negative exponents.}$$

Example 4

$$\frac{\sqrt{x}\,\sqrt[5]{x^2}}{\sqrt[3]{x}}, \quad \text{with } x \neq 0.$$

Fractional exponents are helpful in problems of this type.

$$\frac{x^{\frac{1}{2}} \cdot x^{\frac{2}{5}}}{x^{\frac{1}{3}}} = x^{\frac{1}{2}+\frac{2}{5}-\frac{1}{3}} = x^{\frac{15+12-10}{30}} = x^{\frac{17}{30}} \quad \text{or} \quad \sqrt[30]{x^{17}}$$

if preferred.

Example 5

$(x^{\frac{2}{3}} - 5^{\frac{2}{3}})(x^{\frac{4}{3}} + x^{\frac{2}{3}}5^{\frac{2}{3}} + 5^{\frac{4}{3}}).$ HINT: Recall the well-known identity $(A - B)(A^2 + AB + B^2) = A^3 - B^3.$

---

† If $b^{n/d}$ were defined for $b < 0$, then $(b^{1/d})^n$ might not be equal to $(b^n)^{1/d}$. For example: $[(-1)^{\frac{1}{2}}]^2 = (i)^2 = -1$, while $[(-1)^2]^{\frac{1}{2}} = (1)^{\frac{1}{2}} = 1$. Also 0^0 and $0^{-|k|}$ remain undefined.

‡ The symbol $\sqrt{4}$ means $+2$, not ± 2; although, if $x^2 = 4$, $x = \pm 2$.

Extended Theorems of Differentiation

Hence
$$(x^{\frac{1}{3}} - 5^{\frac{2}{3}})(x^{\frac{2}{3}} + x^{\frac{1}{3}}5^{\frac{2}{3}} + 5^{\frac{4}{3}}) = (x^{\frac{1}{3}})^3 - (5^{\frac{2}{3}})^3 = x - 25.$$

Example 6

$\dfrac{x^{-1} + y^{-2}}{(x+y)^{-2}}.$ This may be written $\dfrac{1/x + 1/y^2\dagger}{1/(x+y)^2}$ and simplified.

However, negative exponents were introduced to *simplify* algebra. Let us not ignore the advantage gained by the introduction of negative exponents. If numerator and denominator of the given expression are multiplied by $xy^2(x+y)^2$, the fraction is unchanged in value. Using the definition of the zero exponent, the simplification is accomplished in two steps.

$$\frac{(x^{-1} + y^{-2})xy^2(x+y)^2}{[(x+y)^{-2}]xy^2(x+y)^2} = \frac{y^2(x+y)^2 + x(x+y)^2}{xy^2}$$
$$= \frac{(y^2 + x)(x+y)^2}{xy^2}$$

Example 7

If $h(x) = 3x^{-1} + 2x^{\frac{1}{3}} - 4$, find $h(8)$.

$h(8) = 3(8^{-1}) + 2(8^{\frac{1}{3}}) - 4 = \frac{3}{8} + 4 - 4 = \frac{3}{8}.$

Recall that in bx^n the n operates on the x only, not the b. Consequently, $2 \cdot 8^{\frac{1}{3}} = 2(8^{\frac{1}{3}}) = 4$.

Example 8

$$\frac{(3x + x^2)x^{-5}}{x^2} - \frac{3x^{-4}}{x} = (3x + x^2)x^{-7} - 3x^{-5}$$
$$= 3x^{-6} + x^{-5} - 3x^{-5}$$
$$= 3x^{-6} - 2x^{-5}.$$

Problem Set 5-1

1. If $h(x) = x^{-\frac{1}{2}} + 3x^2 + 4x^{-2} + 3$, find $h(1)$ and $h(4)$. Does $h(0)$ have meaning? Does $h(3)$ have meaning?

2. If $h(x) = (3x + 3)^{-\frac{1}{2}} + 17x^{-5} + 4^{-\frac{1}{2}} - 3$, find $h(2)$ and $h(26)$ accurate to two decimal places.

3. Simplify $\dfrac{x^{-2} + y^2}{x^{-1} + y^{-1}}.$

4. Simplify $4^{-\frac{1}{2}} + 3^2 - 7 \cdot 2^4 + 3 \cdot 9^{-\frac{1}{2}} - 16^{-2}$ and obtain an approximation accurate to two decimal places.

† The given expression does *not* equal $\dfrac{(x+y)^2}{x+y^2}$. Why not?

Sec. 5-1. Negative, Fractional, and Zero Exponents 141

The expressions in Probs. 5 to 10 arise when the derivatives of certain more complicated expressions are computed. The reader will learn to compute similar derivatives later, but should be able to simplify these fractions now. In each case determine whether the expression on the left may be simplified to that on the right. Also obtain one other simplification of the expression on the left.

5. $\dfrac{(x^2 - x)(2)(2x - 1)(2) - (2x - 1)^2(2x - 1)}{(x^2 - x)^2} = \dfrac{1 - 2x}{x^2(x - 1)^2}$.

6. $\dfrac{x^3 \frac{1}{3}(2x - 1)^{-\frac{2}{3}}(2) - (2x - 1)^{\frac{1}{3}}(3x^2)}{(2x - 1)^{\frac{2}{3}}} = \dfrac{2x^3 - 9x^2(2x - 1)}{3(2x - 1)^{\frac{4}{3}}}$.

7. $\dfrac{x^3(1 - x^2)^{-\frac{1}{2}}}{4x} = \dfrac{1}{4} x^2(1 - x^2)^{-\frac{1}{2}}$.

8. $\dfrac{4\frac{1}{2}(3x^2 + 5)^{-\frac{1}{2}}(6x)}{4(3x^2 + 5)^{\frac{1}{2}}} = \dfrac{3x}{3x^2 + 5}$.

9. $\dfrac{(2x + 7)\frac{1}{2}(4x - 5)^{-\frac{1}{2}}(4) - \sqrt{4x - 5}\,(2)}{(2x + 7)^2} = \dfrac{24 - 4x}{(2x + 7)^2(4x - 5)^{\frac{1}{2}}}$.

10. $\dfrac{(2x + 3)(1) - (x - 3)(2)}{(2x + 3)^2} \cdot (2x + 3)^{-\frac{1}{3}} = \dfrac{9}{(2x + 3)^{\frac{7}{3}}}$.

In Probs. 11 to 26, solve the given equations for real values of the unknown involved. Check for extraneous values.

11. $\dfrac{x^{-1} + 3x}{x} = \dfrac{13}{4}$.
12. $(2x - 5)^{\frac{1}{2}} + 3 = 0$.

13. $4(5x + 7)^{\frac{1}{3}} = 3x$.
14. $(W + 4)^{\frac{1}{2}} + W - 2 = 0$.

15. $3(z + 1)^{-1} + 4(z - 1)^{-1} = 5(1 - z)^{-1}$.

16. $(x - 2)^{\frac{1}{3}} = (x - 6)^{\frac{1}{2}}$.
17. $\dfrac{x^{-1} + 1}{x^{-1} - 1} = 1$.

18. $x^5 = 4x^3$.
19. $x^{-\frac{1}{2}} - (2x)^{-\frac{1}{2}} = 5$.

20. $[x^{\frac{1}{2}} - (2x)^{\frac{1}{2}}]^{-1} = 5$. Compare with Prob. 19.

21. $x(6 - x)^{\frac{1}{2}} = (6 - x)^{\frac{3}{2}}$. *22. $(x^2 - 4x + 3)^{\frac{1}{2}} + 1 = 3x^{-1}$.

23. $(x + 2)^{-\frac{1}{2}}(x + 3)^{-1} + 3 = \frac{91}{30}$.

24. $5x^{-1} + x^{-2} = 6$.
25. $3x^{-2} + 4x^{-1} = -1$.

26. $(x + 2)^{\frac{1}{2}} - 3 = 0$.

In Probs. 27 to 30, show that the expression on the left may be simplified to the expression on the right. These identities arise in hyperbolic function theory, which is used in solving projectile problems involving air resistance, and in electrical circuit theory.

Extended Theorems of Differentiation

27. $\left(\dfrac{e^x + e^{-x}}{2}\right)^2 - \left(\dfrac{e^x - e^{-x}}{2}\right)^2 = 1.$

28. $\left(\dfrac{2}{e^x + e^{-x}}\right)^2 + \left(\dfrac{e^x - e^{-x}}{e^x + e^{-x}}\right)^2 = 1.$

29. $\dfrac{1}{2}(e^{x/2} - e^{-x/2})^2 = \dfrac{e^x + e^{-x}}{2} - 1.$

30. $\left(\dfrac{e^x + e^{-x}}{2}\right)^2 + \left(\dfrac{e^x - e^{-x}}{2}\right)^2 = \dfrac{e^{2x} + e^{-2x}}{2}.$

5-2. Scientific Notation

scientific notation

In applied mathematics it is necessary to use and to compare numbers such as the ionization constants of boric acid, 0.00000000064, and hydrogen cyanide, 0.0000000012. This is easier if *scientific notation* is used. To write a number in *scientific notation*, express it as a number containing *exactly one* nonzero digit to the left of the decimal point times the appropriate power of 10. For example,

$$0.00000000064 = 6.4 \times 10^{-10},$$
$$0.0000000012 = 1.2 \times 10^{-9}.$$

Clearly, 1.2×10^{-9} is the larger. (Why?)

To obtain exactly one digit to the left of the decimal point may require a shift of the decimal point. The number of places the decimal point is shifted indicates the appropriate power of 10.

This notation also indicates the accuracy of a number. For example, to say that the population of a certain city is 30,000 does not indicate whether the number is (a) 30,000 to the nearest 10,000, (b) 30,000 to the nearest 1,000, or (c) 30,000 to the nearest 100. In scientific notation we can indicate this difference with ease (a) 3×10^4, (b) 3.0×10^4, or (c) 3.00×10^4.

In physical problems the sizes of numbers vary over a wide range. The speed of light is 983,570,000 ft/sec while the mass of the hydrogen atom is 0.00000000000000000000000016617 gram. These are more accurately (since tolerance is indicated) and more easily indicated in scientific notation as 9.8357×10^8 ft/sec and 1.6617×10^{-24} gram.

A number has k *significant digits* if, when expressed in scientific notation, there are k digits in the first factor. Thus, 4.70×10^5 has three significant digits.

Sec. 5-2. Scientific Notation

Example 1

Arrange the numbers 37, 6.6×10^{-4}, .0005, 327, 3×10^6, 0, 429,000,000, and $-141,200$ in increasing order of magnitude. Write each number in scientific notation and arrange, to obtain

-1.412×10^5, 0, 5×10^{-4}, 6.6×10^{-4}, 3.7×10, 3.27×10^2, 3×10^6, 4.29×10^8.

Example 2

The larger the ionization constant, the stronger an acid is said to be. A table gives the following ionization constants for weak acids. (a) Which of the acids listed is the weakest? (b) Which is the strongest?

Acid	Ionization constant
Acetic	1.8×10^{-5}
Arsenic	4.5×10^{-3}
Arsenous	2.1×10^{-8}
Benzoic	6.6×10^{-5}
Boric	1.1×10^{-9}

(a) The strongest *of the acids listed* is arsenic acid since it has the largest ionization constant. (b) The weakest acid listed is boric since it has the smallest ionization constant.

One characteristic of an experienced scientist or engineer is the ability to *estimate results*. Scientific notation is useful in obtaining quick estimates. Problem 11 of Set 5-2 requires the solution of

$$\frac{x^2}{0.10 - x} = 2.00 \times 10^{-4}.$$

A chemist would know that x will be very small *compared to* 0.1; hence, for a quick estimate, he would use 0.1 for $(0.1 - x)$. The equation becomes (using the symbol $\cong$ to mean "approximately equal")

$$\frac{x^2}{0.10} \cong 2.0 \times 10^{-4}$$
$$x^2 \cong 2.0 \times 10^{-5}$$
or $\quad x^2 \cong 20 \times 10^{-6}$
$$x \cong \sqrt{20} \times 10^{-3} \cong 4.5 \times 10^{-3}.$$

(Only the positive root is meaningful in the example.)

Extended Theorems of Differentiation

In this particular example the "estimate" is as accurate as the original data will permit. In other problems the estimate may be quite crude, but it still gives a worthwhile check on a more accurately computed result.

Problem Set 5-2

Perform the indicated operations. Estimate your answer *before* doing the arithmetic.

1. $\dfrac{(3.6 \times 10^5)(2.1 \times 10^4)}{8(2.7 \times 10^3)}$.

2. $\dfrac{(2.9 \times 10^4)(6.1 \times 10^{-8})}{\sqrt{4.0 \times 10^{-12}}}$.

3. $\dfrac{(1.1 \times 10^{-9})(2.1 \times 10^{-8})}{4.0 \times 10^{-15}}$.

4. $\dfrac{(6.4 \times 10^{-11})(2.71 \times 10^9)}{0.000000013}$.

5. Solve for y: $1.37 \times 10^5 y + 2.1 \times 10^3 = 7.1 \times 10^3$.

6. Solve for t: $2.1 \times 10^4 t + 1.6 \times 10^{-5} = 30 \times 10^{-6}$.

7. The solubility product constant k_{sp}, at room temperature, of certain compounds is given in the table below. Which compound has the largest k_{sp}? Arrange the compounds in *increasing* order of k_{sp}.

Compound	k_{sp}
Silver sulfide	4.1×10^{-52}
Barium carbonate	8.1×10^{-9}
Mercurous bromide	4.0×10^{-23}
Lead chromate	2.0×10^{-14}
Cadmium hydroxide	1.2×10^{-14}
Cupric iodate	1.4×10^{-7}

8. Will the product of the solubility products of silver sulfide and of cupric iodate be more or less than that of mercurous bromide? This has important application in elementary chemistry.

9. Consult a chemical table and determine the most and the least soluble compounds of those listed. If several different temperatures are considered, use 20°C.

10. Consult a physics text or handbook to determine the speed of light and that of sound. How many times as fast as sound is light? Find W, where W is to the speed of sound as the speed of sound is to the speed of light.

11. Determine the concentration of hydrogen ion in 0.1 molar HCNO (cyanic acid) by solving for X:

$$\frac{X^2}{0.10 - X} = 2.00 \times 10^{-4}.$$

Sec. 5-2. Scientific Notation 145

12. Obtain the concentration of hydrogen ion, X, in a solution containing 0.1 mole HCNO and 0.1 mole NaCNO, by solving the equation:

$$\frac{X(0.10 + X)}{0.10 - X} = 2.00 \times 10^{-4}.$$

Compare the results of Probs. 11 and 12. Chemistry teachers should ask "why."

13. If 100 cc of 0.1 molar ammonium chloride solution is added to 150 cc of 0.1 molar ammonium hydroxide solution, the $(OH)^-$ ion concentration is computed by solving the equation

$$\frac{(0.040 + X)X}{0.060 - X} = 1.8 \times 10^{-5}.$$

Determine the concentration. There is no justification for obtaining your answer to more than two significant digits.

14. Solve the equation $\dfrac{x(0.10 + x)}{(0.025 - x)} = 1.85 \times 10^{-5}$ for x to determine the concentration of H^+ ion when 0.1 mole of solid ammonium hydroxide is added to 1 liter of 0.125 molar acetic acid.

15. Considerations involving the hydrogen atom permit us to compute the mass of an electron as

$$\frac{1.67 \times 10^{-24}}{1.84 \times 10^3 \times 10^3} \text{ kg}.$$

Express this in simpler form.

16. The following expressions arise in computing the electric intensities due to charged particles.

$$A = 9 \times 10^9 \times \frac{12 \times 10^{-9}}{(0.06)^2} \quad \text{newtons/coulomb.}$$

$$B = 9 \times 10^9 \times \frac{12 \times 10^{-9}}{(0.04)^2} \quad \text{newtons/coulomb.}$$

$$C = 9 \times 10^9 \times \frac{12 \times 10^{-9}}{(0.14)^2} \quad \text{newtons/coulomb.}$$

Show that
$A \cong 3.00 \times 10^4$ newtons/coulomb.
$B \cong 6.75 \times 10^4$ newtons/coulomb.
$C \cong 0.55 \times 10^4$ newtons/coulomb.

17. In a simple diode vacuum tube of 250-volt potential, the speed with which electrons are emitted from the cathode and reach the

146 *Extended Theorems of Differentiation*

anode may be computed as

$$v = \frac{2eV}{m} = \frac{2 \times 1.6 \times 10^{-19} \times 250}{9.1 \times 10^{-31}}.$$

Show that $v \cong 8.8 \times 10^{13}$ m/sec.

18. Find when the potential energy V required to produce the deuteron speed attained in a certain cyclotron is

$$V = \frac{1}{2} \times 4.8 \times 10^7 \times (1.8)^2 \times (0.48)^2 \text{ volts}.$$

19. The relative permeability k of a ring of iron of certain dimensions is found to be

$$k = \frac{\left(\dfrac{2.0 \times 10^2}{32}\right)}{1.257 \times 10^{-6}}.$$

Show that $k = 5.0 \times 10^6$.

20. The following equation arises in obtaining the wavelength of light:

$$\frac{1}{x} = 1.097 \times 10^7 \left(\frac{1}{4} - \frac{1}{9}\right).$$

Show that $x = 6.5 \times 10^{-7}$.

21. In each of Probs. 11, 12, 13, and 14, use the assumption that x is small in comparison with the other numbers involved in the problem and that, therefore, expressions such as $(0.06 - x)$ may be replaced by 0.06 without seriously altering the results. Obtain approximate solutions in this way and compare the approximations with the previously computed results.

22. The carbon dioxide cycle is about in balance as far as the inorganic contribution is concerned. (An exchange of about 1×10^8 tons per year is subtracted in the formation of carbonates and is added from geysers, hot springs, and volcanoes.) Photosynthesis, on the other hand, uses up about 6×10^{10} tons per year. Respiration and decay account for a similar addition. What percentage of the total CO_2 used is a result of photosynthesis?

23. (For students who have had elementary chemistry only.) Burning (oxidation) of carboniferous fuels adds 6×10^9 tons of CO_2 per year to the atmosphere. What volume (STP) in cubic miles will be occupied by 6×10^9 tons of CO_2? What will be the

Sec. 5-3. An Extension of the Theorem $d(kx^n)/dx = knx^{n-1}$ 147

volume of O_2 which is used in forming 6×10^9 tons of CO_2 by burning carbon?

5-3. An Extension of the Theorem $\dfrac{d(kx^n)}{dx} = knx^{n-1}$

In Chap. 4, the theorem $\dfrac{d(kx^n)}{dx} = knx^{n-1}$ is proved for n, a *positive integer*. In this section that theorem is extended to include *negative integers* for n, and further extended to the case where n may be any *rational number*. This is a beautiful theorem and, in fact, is even more general than has been indicated here. The remainder of this chapter is devoted to proving and examining the applications of this extended theorem.

Since $x^0 = 1$ if $x \neq 0$, and since the derivative of a constant is zero, the theorem is valid for $n = 0$. (Be sure that you understand this statement before continuing. You should realize that the given sentence is an extension of Theorem 3, Sec. 4-4, and that this extension was not covered in the proof given in Sec. 4-4.)

We next extend the given theorem to include negative integral exponents

$$\frac{d(kx^{-m})}{dx} = -kmx^{-m-1},$$

where $-m$ is a negative integer and k is a constant.

PROOF: If $y = kx^{-m}$, then $x^m y = k$. By using the formula of Sec. 4-8, $\dfrac{d(u \cdot v)}{dx} = u \cdot \dfrac{dv}{dx} + v \cdot \dfrac{du}{dx}$, with $u = x^m$ and $v = y$, we obtain

$$\frac{d(x^m \cdot y)}{dx} = \frac{d(k)}{dx}$$

$$x^m \cdot \frac{dy}{dx} + y \cdot mx^{m-1} = 0.$$

Solving for $\dfrac{dy}{dx}$,

$$\frac{dy}{dx} = \frac{-y \cdot mx^{m-1}}{x^m}$$

$$= \frac{-ym}{x}.$$

Extended Theorems of Differentiation

Since we were given $y = kx^{-m}$, this is equivalent to

$$\frac{dy}{dx} = \frac{-kx^{-m} \cdot m}{x}$$

$$\frac{d(kx^{-m})}{dx} = -kmx^{-m-1}.$$

This follows the general rule of Theorem 3, Sec. 4-4, with $n = -m$. Combining the results obtained thus far, we have

$$\boxed{\frac{d(kx^n)}{dx} = knx^{n-1} \quad \text{for all integers } n \text{ (positive, zero, negative) and every constant } k.}$$

Example 1

Find y' where $y = 7/x^2$. It is no longer necessary to compute the derivative of $y = 7/x^2$ by the delta process, since if

$$y = 7/x^2 = 7 \cdot x^{-2}$$
$$y' = 7(-2)x^{-2-1}$$
$$= -14x^{-3}.$$

Example 2

If $B(x) = \dfrac{(4x^2 - 7x + 5)^5}{(x^2 - 3x + 1)^3}$, find $\dfrac{dB}{dx}$.

Since $B(x) = (4x^2 - 7x + 5)^5(x^2 - 3x + 1)^{-3}$

we use $\dfrac{d(u \cdot v)}{dx} = u \cdot \dfrac{dv}{dx} + v \cdot \dfrac{du}{dx}$, with $u = (4x^2 - 7x + 5)^5$ and $v = (x^2 - 3x + 1)^{-3}$. Thus

$$\frac{dB}{dx} = (4x^2 - 7x + 5)^5 \cdot (-3)(x^2 - 3x + 1)^{-4}(2x - 3)$$
$$+ (x^2 - 3x + 1)^{-3} \cdot 5(4x^2 - 7x + 5)^4(8x - 7).$$

Problem Set 5-3

In Probs. 1 to 10, compute the derivative of the given function.

1. $y = (4x^3 - 5x + 6)^{-3}$.
2. $y = (14x^7 - 5x^3 + 11)^{-5}$.
3. $F(x) = 7(x^2 - 4x + 3)^{15} \cdot (2x^3 - 5x + 1)^{-8}$.
4. $G(x) = \dfrac{7(x^2 - 4x + 3)^{15}}{(2x^3 - 5x + 1)^8}$.

Sec. 5-4. *A Further Extension of the Theorem* $d(kx^n)/dx = knx^{n-1}$ 149

5. $p(x) = 4x^2 - 5x^{-3} + 2x - 11x^{-5}$.

6. $y(x) = \dfrac{3x^2 - 5x + 7}{x}$. **7.** $f(x) = (x^2 - 5x + 3)^{-4}$.

8. $y = (2x^5 - 7x + 5)^{-4} \cdot (x^2 - 2x + 1)$.
9. $y = 1/x^4$.
10. $y = 7/x^4 + \sqrt{11}\,(x^2) - 31\sqrt{\pi^{-2}}$.
11. (a) Does the curve $y = 7/x^5$ ever have a tangent line whose slope is positive?
(b) Does the curve $y = 7/x^6$ ever have a tangent line whose slope is positive?
12. Find a point at which the curve $y = 4/x^2$ has slope -9.

13. Compute $\dfrac{d\left[\dfrac{x^2 - 5x + 1}{2x + 3}\right]}{dx} = \dfrac{d[(x^2 - 5x + 1) \cdot (2x + 3)^{-1}]}{dx} = ?$

14. Prove the following formula for the derivative of a quotient of two differentiable functions of x, $u = u(x)$, $v = v(x)$.

$$\frac{d(u/v)}{dx} = \frac{v\dfrac{du}{dx} - u\dfrac{dv}{dx}}{v^2}.$$

HINT: $u/v = u \cdot v^{-1}$.

15. The point $(2,3)$ lies on the graph $xy = 6$. Find the points at which the line tangent to $xy = 6$ at the point $(2,3)$ crosses the x and y axes.

*16. Find the length of the *shortest* line segment which is tangent to $xy = 6$ and has its end points on the coordinate axes.

5-4. *A Further Extension of the Theorem* $\dfrac{d(kx^n)}{dx} = knx^{n-1}$

The extension of Theorem 3, Sec. 4-4, to the case where the exponent is a rational number† will be undertaken next.

Since (by Prob. 20, Set 4-8) $\dfrac{d[c \cdot v(x)]}{dx} = c \cdot \dfrac{d[v(x)]}{dx}$, we need only show that

---

† The reader is reminded that a rational number is a number of the form A/B where A and B are integers and $B \neq 0$. Since $B = 1$ is a valid choice, the integers (positive, negative, zero) are included among the rational numbers along with "the fractions."

Extended Theorems of Differentiation

$$\frac{d(x^{p/q})}{dx} = \frac{p}{q} \cdot x^{(p/q)-1} \qquad \text{where } p \text{ and } q \text{ are integers, } q \neq 0.$$

Let $\quad y = x^{p/q}$
then $\quad y^q = x^p \quad$ (Why?)

$$\frac{d(y^q)}{dx} = \frac{d(x^p)}{dx}$$

$$qy^{q-1}\frac{dy}{dx} = px^{p-1}. \qquad \text{(What theorems are used?)}$$

Solving for $\dfrac{dy}{dx}$,

$$\frac{dy}{dx} = \frac{px^{p-1}}{qy^{q-1}}.$$

Since $y = x^{p/q}$,

$$y^{q-1} = (x^{p/q})^{q-1} = x^{p(q-1)/q} = x^{(pq-p)/q} = x^{p-(p/q)}.$$

Upon making this substitution for y^{q-1} in $\dfrac{dy}{dx}$, we have

$$\frac{dy}{dx} = \frac{px^{p-1}}{qx^{p-(p/q)}} = \frac{p}{q}x^{p-1-[p-(p/q)]}$$

$$\frac{d(x^{p/q})}{dx} = \frac{p}{q}x^{(p/q)-1} \qquad \text{as desired.}$$

We have now established the almost unbelievably general theorem

$$\frac{d(kx^n)}{dx} = knx^{n-1}, \text{ where } n \text{ is } \textit{any rational number} \text{ (positive,}$$
negative, zero) and k is any constant.

The theorem given above is, as we have mentioned, even more general than this. Actually, the theorem is valid for any *real* exponent n. However, we cannot prove this at present; in fact, we are not even ready to define what x^n means in some cases. For example, what do the symbols $5^{\sqrt{3}}$ or 7^{π} mean? In order to define such symbols, it is necessary either to consider convergent sequences or to use the logarithmic function. We are not ready to consider either of these alternatives.

The student should *not* feel that he is now, at last, free to forget the delta process. It is true that many problems on which we formerly were forced to use the delta process may now be solved

Sec. 5-4. A Further Extension of the Theorem $d(kx^n)/dx = knx^{n-1}$

by the use of the general theorem given above. However, in later work we shall wish to take the derivative of the sin x and the derivative of 2^x and of other functions which are not rational functions. In order to do this, and for other reasons, the student should bear in mind at all times that *these short cuts are merely timesavers*, and that *the fundamental notion of differentiation is to find the limit of a certain fraction whose numerator and denominator each approach zero*.

Example 1

Find y' if $y = \sqrt{3x + 7}$
$$y = (3x + 7)^{\frac{1}{2}}$$
$$y' = \tfrac{1}{2}(3x + 7)^{-\frac{1}{2}}(3).$$

How did the factor 3 come into the picture?

Example 2

Find the slope of $y = \dfrac{\sqrt{4x + 1}}{(x^2 + 2x - 6)^3}$ at the point where the curve crosses the line $x = 2$.

$$y = (4x + 1)^{\frac{1}{2}} \cdot (x^2 + 2x - 6)^{-3}.$$

Using $\dfrac{d(u \cdot v)}{dx} = u \cdot \dfrac{dv}{dx} + v \cdot \dfrac{du}{dx}$ where one has $u = (4x + 1)^{\frac{1}{2}}$ and $v = (x^2 + 2x - 6)^{-3}$, we obtain

$$y' = (4x + 1)^{\frac{1}{2}} \cdot (-3)(x^2 + 2x - 6)^{-4}(2x + 2)$$
$$+ (x^2 + 2x - 6)^{-3} \cdot \tfrac{1}{2}(4x + 1)^{-\frac{1}{2}}(4).$$

Upon substituting $x = 2$,

$$y'(2) = (9)^{\frac{1}{2}} \cdot (-3)(2)^{-4}(6) + (2)^{-3} \cdot \tfrac{1}{2} 9^{-\frac{1}{2}}(4) = \cdots = \frac{-79}{24},$$

the desired slope.

Problem Set 5-4

In Probs. 1 to 10, find the derivative of the given function.
1. $y = \sqrt{3x - 7}$.
2. $y = x\sqrt{3x - 7}$.
   (a) Do as $y = u \cdot v$ with $u = x$, $v = (3x - 7)^{\frac{1}{2}}$.
   (b) Do as $y = \sqrt{3x^3 - 7x^2}$.

Extended Theorems of Differentiation

3. $F(x) = 4(3x^2 - 7x + 5)^{\frac{3}{2}}$. 　　4. $G(t) = \dfrac{\sqrt{1-2t}}{4-7t^2}$.

5. $h(t) = 12t^{-1} - 4\sqrt{17}$.
6. $e(t) = (5t^2 - 3t + 2)^{-\frac{1}{3}}(4t-3)^2$.
7. $e(t) = \dfrac{(4t-3)^2}{\sqrt[3]{5t^2 - 3t + 2}}$.　　HINT: Note Prob. 6.

8. $f(x) = \dfrac{x}{\sqrt{3-5x}}$.
9. $v(x) = 11(4x-7)^{13} + \sqrt{\pi^2 - 5}$.
10. $u(x) = 21\sqrt{17}\,(4x-3)^{-\frac{3}{4}} + \dfrac{7x}{(x-1)^2} + 91\sqrt{17\pi}$.

11. A tin cylindrical can is to be constructed to contain 125 cu in. Determine the dimensions of a can requiring the least total sheet metal to construct if the circular top and bottom are cut from square pieces of tin *with the scraps wasted* and the side is cut from a large sheet without waste.

12. A voltaic cell has constant emf and constant internal resistance r. The work done by this cell in sending a steady current through an external circuit of resistance x is given by

$$W = \frac{kx}{(x+r)^2} = kx(x+r)^{-2}$$

where k and r are constants. For what value of x is W a maximum?

13. John Garnett has the monopoly of the sale of student pins, which cost him $3 each. If the number sold varies inversely as the cube of the selling price, what selling price will yield the greatest total profit to John?

14. Determine intervals in which $y = (x+6)(x-1)^3$ is increasing as x increases.

15. The weight W of hot gas passing up a chimney in a given time is given by

$$W = \frac{k\sqrt{0.96T - T_0}}{T}$$

where T denotes the (variable) temperature of the gas, T_0 is the constant temperature of the outside air, and k is a constant. What value of T will force a maximum weight of gas through the chimney in the given time interval?

Sec. 5-4. *A Further Extension of the Theorem* $d(kx^n)/dx = knx^{n-1}$ 153

16. An electron moves on the path $y = x^2 - \frac{9}{2}$. At what position is the electron nearest the origin? What is this minimum distance?

17. An electron travels along the path represented by $x^2 = y$. How close does the electron come to an accelerator placed at (3,0)?

18. A particle travels along the curve $xy = 5$. How close does it come to the point (3,0)?

19. Find the minimum cost of the material for a box with a square base which contains 16 cu ft if the material for the top and bottom costs $2 per square foot and that for the sides costs $1 per square foot.

20. (a) What is the smallest number which can be found by adding a positive number to its reciprocal?
(b) Rework (a) without the restriction to "positive" numbers.

21. Find the dimensions of the cylinder of largest volume which can be inscribed in a right circular cone of height 24 in. and radius of the base 8 in.

22. Find the volume of the largest right circular cylinder which can be inscribed in a sphere of radius 24 in.

23. A sheet of paper is to contain 18 sq in. of sensitized emulsion. The margins at the top and bottom of the emulsion are to be 2 in. each and a 1-in. margin is to be maintained at each side of the emulsion. Find the dimensions of the smallest piece of paper which can be used.

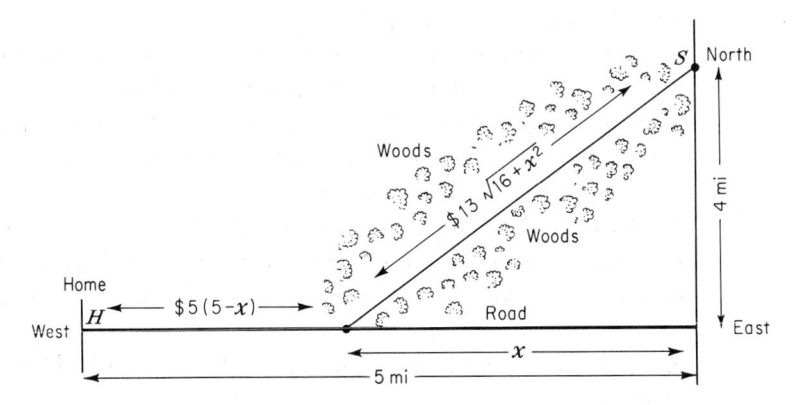

Figure 5-1

24. John wishes to run a telephone line from his home to a point 5 miles east and 4 miles north of his home (Fig. 5-1). A road running *EW* passes his home. Along the road the line costs $5 per

mile. Through the woods north of the road the cost is $13 per mile.
(a) Find the most economical route; that is, where should John start angling through the woods?
(b) Make appropriate comments on the domain of definition of the function which rule out the possibility that x may be negative.

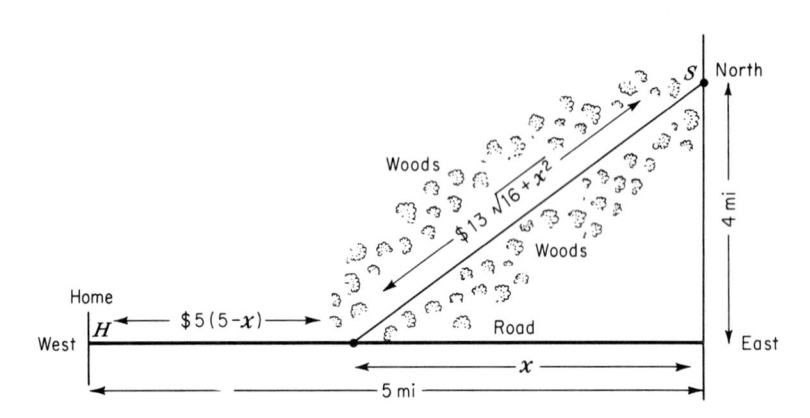

Figure 5-1

25. Rework Prob. 24 if the costs are $40 and $50 per mile, respectively.

26. (a) Rework Prob. 25 if the costs are $5 per mile along the road and $50 per mile through the woods.
(b) $50 per mile along the road and $30 per mile through the woods.

27. The intensity of light at any point varies inversely as the square of the distance between the point and the center of the light source. Two lights A and B are 6 ft apart. Light A has an intensity six times as great as that of light B.
(a) How far from A on line segment AB is the intensity least?
(b) Light A is 6 in. in radius and light B 4 in. in radius, and the point may not be taken inside of these radii. Where on line AB is the light intensity greatest?

28. Find the equation of the line tangent to $y^2x^3 - 4x^2y^2 + 7xy^3 = 6$ at $(2,1)$.

In Probs. 29 to 33, use the rule $\dfrac{d(u/v)}{dx} = \dfrac{v\dfrac{du}{dx} - u\dfrac{dv}{dx}}{v^2}$ derived in Prob. 14, Set 5-3, to determine the derivative of the given function.

29. $f(x) = \dfrac{(2x-1)^2}{x^2-x}$. (See Prob. 5, Set 5-1.)

30. $g(x) = \dfrac{x^3}{\sqrt[3]{2x-1}}$. (See Prob. 6, Set 5-1.)

31. $h(x) = \dfrac{\sqrt{4x-5}}{(2x+7)}$. (See Prob. 9, Set 5-1.)

32. $\dfrac{x-3}{2x+3}$.

33. $\dfrac{\sqrt{5x+7}}{\sqrt[3]{2x+9}}$.

5-5. Self-test

1. Describe, in words, the exact nature of the "extensions of previous theorems" made in this chapter.

2. Determine $\dfrac{dy}{dx}$ where

(a) $y = \sqrt{4x-3}$.
(b) $y = (\sqrt{x^2-5x+1})^3 \sqrt[4]{(2x-3)^5}$.
(c) $2xy = 5$.

3. Let $u = u(x)$ and $v = v(x)$ be differentiable functions of x. Using the formula for the derivative of a product, derive the formula for

$$\frac{d(u/v)}{dx} = \frac{d(u \cdot v^{-1})}{dx}.$$

4. Sketch the curve $y = \dfrac{\sqrt{4+x^2}}{x^2-9}$, showing all maximum and minimum points as well as approximate intercepts.

5. Determine the slope of $y = 3\sqrt{2x+1} \cdot (5x-4)^{\frac{3}{2}} + 17$ at that point where the curve crosses the line $x = 4$.

6

Rates of Change

6-1. Rate of Change

derivative

It is now within our power to compute the derivative of every *rational function*.† The applications have emphasized that the derivative of $f(x)$ at a given point $[a, f(a)]$ is the slope of the line tangent to $y = f(x)$ at $[a, f(a)]$. This is a valid interpretation, but by no means the only interpretation of the derivative. Fundamentally, *a derivative is a rate of change*. Before discussing the derivative further, let us review the meaning of rate of change.

6-2. Average Rate of Change

Consider a particle P moving in a straight line where the distance from a fixed point 0 (origin) is $s = s(t)$, some function of time t. The distance of P from 0 at time t_1 is $s(t_1)$ and at time $t_1 + \Delta t$ the distance is $s(t_1 + \Delta t)$. The *average velocity* or *average rate of change of s with respect to time t* over the interval from $t = t_1$ to $t = t_1 + \Delta t$ is given by

average velocity

$$\text{Average velocity} = \frac{\Delta s}{\Delta t} = \frac{s(t_1 + \Delta t) - s(t_1)}{\Delta t}.$$

Example 1

A man travels south from Chicago. At 2 P.M. he is 135 miles from Chicago, while at 3:30 P.M. he is 210 miles from Chicago. Deter-

† A rational function is a quotient of two polynomials. Polynomial is defined in Sec. 4-5.

Sec. 6-2. Average Rate of Change

mine his average rate of change of distance with respect to time (average velocity) during this period.

Here $t_1 = 2$, $t_1 + \Delta t = 3.5$. Hence

$$\Delta t = 1.5 \qquad s(t_1) = s(2) = 135 \qquad s(t_1 + \Delta t) = s(3.5) = 210.$$

$$\text{Average velocity} = \frac{s(t_1 + \Delta t) - s(t_1)}{\Delta t} = \frac{210 - 135}{1.5} = \frac{75}{1.5}$$

$$= 50 \text{ mph}.$$

Example 2

A ball is thrown downward from the top of a tower. One second after it is thrown, the ball is 224 ft from the ground. Three seconds after it is thrown the ball is 76 ft from the ground. Determine the average velocity of the ball during this period.

$$t_1 = 1 \qquad t_1 + \Delta t = 3 \qquad \Delta t = 2$$
$$s(t_1) = s(1) = 224 \quad \text{and} \quad s(t_1 + \Delta t) = s(3) = 76.$$

Hence

$$\text{Average velocity} = \frac{s(t_1 + \Delta t) - s(t_1)}{\Delta t} = \frac{s(3) - s(1)}{2}$$

$$= \frac{76 - 224}{2} = \frac{-148}{2} = -74 \text{ ft/sec}.$$

(Why is the sign negative?)

The reader is cautioned that this does *not* mean that the ball was falling at 74 ft/sec during the entire 2 sec. It means that *if* the ball had fallen at 74 ft/sec during the period from $t = 1$ to $t = 3$, it would have covered the same distance.

Problem Set 6-2

1. Find the average velocity of a ball which falls 64 ft in 2 sec.
2. If the ball of Prob. 1 falls 144 ft in 3 sec, find its average velocity.
3. Find the average velocity during the third second of the ball of Probs. 1 and 2. Does the average velocity during the first three seconds equal the average of the velocity during the third second and the first two seconds?
4. If $s(t) = 8 + 10t - t^2$, find the average velocity from $t = 2$ to $t = 5$.
5. If $s(t) = 10t^2 + 4t$, find the distance Δs between $t = 2$ and $t = 7$. Also find the average velocity in this interval.

158   *Rates of Change*

6. If $s(t) = 6t^2 + 2$, find the average velocity in the intervals $0 \leq t \leq 1$ and $10 \leq t \leq 11$.

7. If a motorist makes a trip at an average speed of 30 mph and returns over the same route at an average speed of 60 mph, what is his average speed for the entire trip? The answer is *not* 45 mph.

8. Math teachers should generalize Prob. 7, using the concept of *harmonic mean*.

6-3. Velocity

In Example 1, Sec. 6-2, we obtained 50 mph as the average rate of change of distance or average velocity. This does not mean that the man was going 50 mph as he passed through the town of McLean. He may have been going much faster, or he may have been going 15 mph, or he may have stopped for dinner and hence for part of the time had a speed of zero mph. He may have traveled at each of these speeds, while passing through McLean. Instantaneous speed is a concept which is quite familiar but hard to formulate without using the idea of limit. Try it and see. Average speed is easy, but what is instantaneous speed?

There is an important difference between *average* rate of change of distance with respect to time (average velocity) and *instantaneous* rate of change of distance with respect to time (velocity or speedometer reading). This instantaneous rate of change,† or velocity $v(t_1)$, is defined as the limit of the average velocity as the intervals Δt approach zero; that is,

velocity
$$\text{Velocity at time } t_1 = v(t_1) = \lim_{\Delta t \to 0} \left(\frac{\Delta s}{\Delta t} \right)$$
$$= \lim_{\Delta t \to 0} \frac{s(t_1 + \Delta t) - s(t_1)}{\Delta t}.$$

The velocity at time $t = t_1$ may be thought of as obtained by taking the limit of the average velocities on smaller and smaller intervals of time Δt surrounding $t = t_1$. The distance $\Delta s = s(t + \Delta t) - s(t)$ also becomes smaller and smaller of course, but the fraction $\frac{\Delta s}{\Delta t}$ may

---

† One of the famous paradoxes of Zeno concerns an arrow in flight which supposedly can never move since at each instant it must be somewhere and if it is somewhere, it is not in motion. A discussion of this and other paradoxes will be found in *Mathematics and the Imagination* by Kasner and Newman, pp. 37ff.

Sec. 6-3. Velocity

easily have a finite limit. The astute reader will note that the right-hand member of this equation is exactly the definition of $\frac{d[s(t)]}{dt}$ evaluated at $t = t_1$. Hence velocity at time $t_1 = v(t_1) = \frac{d[s(t)]}{dt}$ evaluated at $5 = t_1$.

If a point moves on an axis so that its directed distance from the origin at time t is given by $s(t) = t^4 - 2t^2 + 1$, then $\frac{ds}{dt} = s'(t) = 4t^3 - 4t$ is the instantaneous rate of change of the distance with respect to time. We use the word *velocity* for this concept. In general,

Velocity at time t is $v(t) = s'(t)$.

speed The sign of $v(t)$ determines whether $s(t)$ is increasing [$v(t) > 0$] or decreasing [$v(t) < 0$]. The term *speed* is used when this distinction is not made; that is, speed = |velocity|.

The velocity $v(t)$ is itself a function of t, and we may compute the instantaneous rate of change of $v(t)$, obtaining $v'(t)$. The rate of *acceleration* change of velocity with respect to time is the *acceleration*.

$$\text{Acceleration} = a(t) = \frac{dv(t)}{dt} = v'(t).$$

In our example

$$s(t) = t^4 - 2t^2 + 1$$
$$v(t) = s'(t) = 4t^3 - 4t$$
and $a(t) = 12t^2 - 4$ at time t.

The symbol $s''(t)$ may be used for $a(t)$, since $a(t)$ is the derivative of the derivative of $s(t)$.

Example 1

A particle moves along the s axis. The directed distance of the particle from the origin is given by $s(t) = t^4 - 4t^3 + 8t - 3$. Determine the position, velocity, and acceleration of the particle when $t = 2$.

Since $s(t) = t^4 - 4t^3 + 8t - 3$,
$$v(t) = s'(t) = 4t^3 - 12t^2 + 8.$$

[Do *not* divide by four; $v(t)$ does *not* equal $t^3 - 3t^2 + 2$. (Why not?)] and $a(t) = v'(t) = 12t^2 - 24t$. (Why not $t^2 - 2t$?)
At $t = 2$,

$$s(2) = 16 - 32 + 16 - 3 = -3 \text{ units of } s \text{ from the origin}$$
$$(\text{viz.}, -3 \text{ ft}).$$
$$v(2) = 32 - 48 + 8 = -8 \text{ units of } s \text{ per } t \text{ unit (viz., } -8 \text{ ft/sec}).$$
$$a(2) = 48 - 48 = 0 \text{ (units of } s \text{ per } t \text{ unit) per } t \text{ unit}$$
$$(\text{viz.}, 0 \text{ ft/sec/sec}).$$

Special attention must be given to the units in applied problems.

Example 2

Determine all positions of the particle of Example 1 when the velocity is 8 units of s per t unit.

Since $v(t) = 4t^3 - 12t + 8$,
we set $v(t) = 4t^3 - 12t + 8 = 8$
or $4t^3 - 12t^2 = 0$
$4t^2(t - 3) = 0$
$t = 0, \quad t = 0, \quad t = 3$.

At these times $v(t) = 8$. The desired positions are given by $s(0) = -3$ and $s(3) = -6$.

Example 3

A stone is thrown upward into the air from the top of a tower 400 ft high (see Example 1, Sec. 4-7). The height h, in feet, of the stone from the ground t sec after it is thrown is

$$h(t) = 400 + 120t - 16t^2.$$

(a) Determine the velocity of the stone 2 sec after it is thrown.

(b) Determine the velocity of the stone when it strikes the ground. Work out the answers yourself before you look below.

(a) Since $v(t) = h'(t) = 120 - 32t$

$$v(2) = 120 - 32(2) = 56 \text{ ft/sec}.$$

(b) The stone strikes the ground when $h(t) = 0$. In Sec. 4-7 it was shown that $t = 10$ is the desired solution of the equation $h(t) = 400 + 120t - 16t^2 = 0$. The problem requests the velocity at $t = 10$, namely, $v(10)$

$$v(10) = 120 - 32(10) = -200 \text{ ft/sec}.$$

Sec. 6-3. Velocity 161

The negative velocity indicates the distance from the ground $h(t)$ is decreasing; that is, the stone is going downward.

Problem Set 6-3

In Probs. 1 to 5, determine the position, velocity, and acceleration (at the indicated time) of a particle which moves along the s axis and whose directed distance from the origin is given by $s(t)$. Describe the motion of the particle for $-100 \leq t \leq 100$.

1. $s(t) = t^2 - 3t + 5$ at $t = 1$, $t = 7$, and $t = 11$.
2. $s(t) = t^4 + 4t - 9$ at $t = 1$, $t = 0$, and $t = -3$.
3. $s(t) = 64t - 16t^2$ at $t = 0$, $t = 5$, and $t = 10$.
4. $s(t) = 17t$ at $t = 0$, $t = 5$, and $t = 7$.
5. $s(t) = 11$ at $t = 0$, $t = 5$, and $t = 15$.

In Probs. 6 to 10, determine the times at which the particle has the given velocity or acceleration and determine the distances of the particle from the origin at these times.

6. $s(t) = 10t^3 + 7t^2 + 3$ where $a(t) = 2$.
7. $s(t) = 12t^3 - 3t + 5$ where $v(t) = 6$.
8. $s(t) = t^4$ where $v(t) = 32$.
9. $s(t) = 3t^3 - 7t + 2$ where $a(t) = 36$.
10. $s(t) = 17 + t - t^2$ where $v(t) = 9$.

11. A hockey puck is struck at time $t = 0$. The distance in feet of the puck from the spot at which it was struck is $s(t) = 72t - 3t^2$, where t is in seconds. How far did the puck travel before coming to rest? In what t interval does the function have meaning in this problem?

12. A stone is thrown upward from the top of a tower 100 ft above the ground with an initial velocity of 64 ft/sec. The height h of the stone above the ground t sec after it is thrown is given by $h(t) = -16t^2 + 64t + 100$. Determine the velocity of the stone as it passes a window 20 ft above ground level.

13. From a point 1,200 ft above the earth's surface, a stone is thrown upward with a speed of 160 ft/sec. Find the impact velocity if the height of the stone at any time t after the stone is thrown is

$$h(t) = -16t^2 + 160t + 1{,}200.$$

14. A block of ice slides down a chute 100 ft long with an acceleration of 16 ft/sec/sec. The distance of the block from the bottom

of the chute is given by
$$s(t) = -8t^2 - 20t + 100.$$
Find the velocity of the block at the bottom of the chute; that is, when $s(t) = 0$.

15. How fast is the block of Prob. 14 sliding when it has slid 4 ft?

16. A ball rolls on an inclined plane so that its distance s ft from the bottom of the plane at time t sec is $s(t) = 6t - 2t^2$. How far up the inclined plane will the ball roll?

17. At what time and with what speed does the ball of Prob. 16 reach the bottom of the inclined plane?

18. What is the speed of the ball of Prob. 16 when it is 1 ft from the bottom of the plane?

19. John Garnett is coasting down a long hill on his bike. His distance (in feet) from the top of the hill at any time t in seconds is given by $s(t) = 5t^2 + 7t$. After John has coasted 160 ft from the top of the hill, he strikes a large stone and falls from his bike. Determine his velocity at the moment he strikes the stone. Express this velocity in miles per hour. Do you think it likely that John was hurt in the fall?

20. If John had fallen when he had coasted only half as far as in Prob. 19, with what velocity would he have fallen?

6-4. General Rate of Change

The remarks of Sec. 6-3 on rate of change are not limited to time rate of change of distance. Other rates of change are handled in a similar manner. The change in volume with respect to time, for example, is $dV/dt = \lim_{\Delta t \to 0} \Delta V/\Delta t$.

Example 1

Hydrogen is being forced into a spherical balloon at the constant rate of 1,000 cu in./min. How fast is the radius of the balloon increasing when the balloon is 2 ft in radius?

The volume V of a sphere is $V = \frac{4}{3}\pi r^3$. We are given $\frac{dV}{dt} = 1{,}000$ cu. in./min and are asked to find $\frac{dr}{dt}$ when $r = 2$ ft $= 24$ in. Then, since $r = r(t)$, using Sec. 4-10, we have
$$\frac{dV}{dt} = \frac{dV}{dr} \cdot \frac{dr}{dt} = \frac{d\frac{4}{3}\pi r^3}{dr} \cdot \frac{dr}{dt} = 4\pi r^2 \frac{dr}{dt}.$$

Sec. 6-4. General Rate of Change

Substituting the known values into this equation, we have

$$1{,}000 = 4\pi(24)^2 \frac{dr}{dt}.$$

$$\frac{dr}{dt} = \frac{1{,}000}{4\pi(24)^2} = 0.14 \text{ in./min.}$$

Example 2

The horizontal V-shaped trough illustrated in Fig. 6-1 is 4 ft long with vertical ends and has a 90° angle between its sides. Water is poured into the trough at the rate of 500 cu in./min. At what rate is the depth of the water increasing when the water is 5 in. deep?

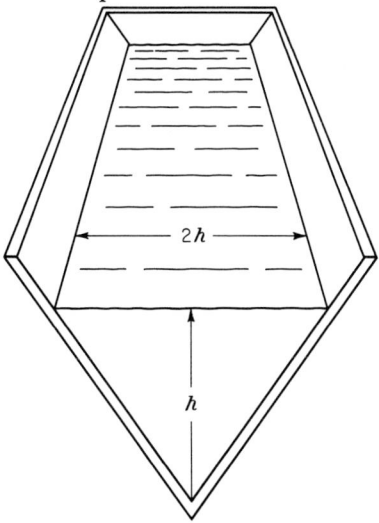

Figure 6-1

Let V = the volume of water in the trough and $h = h(t)$ the depth of the water in inches. The *width* of the water surface is $2h$ when the depth is h. (Why?) The cross-section area is h^2 sq in. and

$$V = 4(12)h^2 \text{ cu in.}$$
$$V = 48h^2 \text{ cu in.}$$

The problem states that $\dfrac{dV}{dt} = 500$ cu in. per minute and requests $\dfrac{dh}{dt}$ when $h = 5$ in.

$$\frac{dV}{dt} = \frac{dV}{dh} \cdot \frac{dh}{dt} = \frac{d(48h^2)}{dh} \cdot \frac{dh}{dt} = 96h \frac{dh}{dt}.$$

Substituting the known values, we obtain

$$500 = 96(5) \frac{dh}{dt}$$

and $\dfrac{dh}{dt} = \dfrac{500}{96(5)} = 1.04$ in./min.

Problem Set 6-4

1. Gas is being forced into a spherical balloon at the rate of 400 cu in./min. How fast is the radius of the balloon increasing when the radius is 5 in.?
2. Same as Prob. 1 for a 10-in. radius.
3. Gas is forced into a spherical balloon at the rate of 600 cu in./min. How fast is the radius of the balloon increasing when it encloses 288π cu in. of gas?
4. At what rate is the surface area of the balloon of Prob. 1 changing when $r = 5$ in.?
5. At what rate is the surface area of the balloon of Prob. 3 changing at the instant in question?
6. Determine the rate at which the depth of the water is changing in the trough of Example 2, Sec. 6-4, when the water is 1 in. deep.
7. Same as Prob. 6 when the water is 10 in. deep.

Discuss and sketch the functions given in Probs. 8 to 15. List all maximum, minimum, and inflection points and the approximate intercepts.

8. $y = 2x^3 - 7x + 5$.
9. $y = 3x^2 - 4x + 2/x$.
10. $y = 2x^2 + 11$.
11. $y = x^3 - 12x$.
12. $y = 4x^3 - 108$.
13. $y = 12 - x^3$.
14. $y = 5 + 3x - 10x^3$.
15. $y = -7 - 2x + x^2$.

16. A steamer is rented for an excursion for 100 passengers at $10 per passenger providing 100 passengers purchase tickets. If fewer than 100 tickets are sold, the rental remains at $1,000. The owners agree to reduce all fares 5 cents per fare for each person beyond the basic 100. (a) How much is the maximum rental the owners may expect on the excursion? (b) How many passengers will need to go to obtain this maximal rental? (c) Assuming that the fire laws will not permit the steamer to carry more than 180 passengers, what will be the smallest gross income the owners may receive for a legal load? (d) Sketch a graph of income I vs. number of passengers x. Note that, for $0 < x \leq 100$, I is constant.

Sec. 6-4. General Rate of Change

17. An exporter can ship a cargo of 100 tons today at a profit of $5 a ton. By waiting, he can add 20 tons per week to the shipment, but the profit on all that he exports will be reduced 25 cents per ton per week. How long will it be profitable to wait?

18. A light is placed at ground level 50 ft from a high wall. A man 5 ft tall starts at the light and walks directly toward the wall at the rate of 8 ft/sec. How fast is the top of his shadow moving down the wall when he is halfway there?

19. The demand for an article varies inversely as the $\frac{5}{2}$ power of the selling price. If the articles cost $1 each to manufacture, determine the selling price which will produce the greatest profit.

20. The amount of candy and pop an individual huckster can sell at a football game decreases as the number of hucksters increases. If 100 hucksters can average $80 each in sales at a big game, and if for each additional huckster added the average per huckster drops 10 cents, find the number of hucksters for a maximum total income for the concession owner. How much does each huckster sell at this maximum? How much larger is the maximum total income than if 100 hucksters were employed?

21. John Garnett is in the hospital recovering from a bike accident. A doctor is inflating a spherical balloon to please John. If hydrogen is supplied at 100 cu in./min, how long will it take to inflate the balloon to a radius of 10 in.? How fast is the radius increasing when the balloon is 5 in. in radius? 10 in. in radius? Just before it bursts at 12.3 in. in radius?

22. Show that $y = x^3 - 6x^2 - 15x + 8$ is neither rising nor falling at the points where $x = -1$ and $x = 5$.

23. Consider the curve of Prob. 22. Tell whether the curve is rising or falling as x increases on each of the intervals (a) $x < -1$, (b) $-1 < x < 5$, (c) $5 < x$.

24. Determine the critical points and the intervals in which $y = (x + 6)(x - 1)^3$ is increasing as x increases.

25. A tin container is to be constructed in the form of a right circular cylinder containing 27 cu in. volume. If the top and bottom of the can are cut from square sheets and the corner pieces are wasted, find the dimensions of the container requiring the least tin.

26. Bill McFurson wishes to fence in a rectangle of ground in his pasture and then divide the rectangle into three (not necessarily equal) pens, by erecting two fences parallel to the ends of the rectangle. If the three pens enclose a total area of 1,800 sq ft, determine the smallest amount of fencing Bill can use to do the job.

27. Sand is being poured on the ground, forming a conical pile with its altitude equal to two-thirds of the radius of its base. If the sand is falling at the rate of 12 cu ft/sec, how fast is the altitude increasing at the time when the volume is 100 cu ft?

28. Find the slope of the line tangent to the curve whose implicit equation is $x^3y^2 - 3x^5y^4 + 70x + 7{,}560 = 0$ at the point $(2, -3)$.

29. If $V = \frac{4}{3}\pi r^3$ where $r = r(t)$, find $\frac{dV}{dt}$ in terms of r and $\frac{dr}{dt}$.

30. If $v = s^5 + 17$ where $s = 3t^2 - 5$, find $\frac{dv}{dt}$ in terms of s and $\frac{ds}{dt}$.

31. If $W = 4H^3S$ where $H = 3S^2 - 5$, find $\frac{dW}{dS}$ in terms of H, S, and $\frac{dH}{dS}$.

32. Evaluate the derivatives in Prob. 30 at $t = 2$.

33. Evaluate the derivatives in Prob. 31 at $s = 3$.

34. Work Prob. 30 by first substituting $s = 3t^2 - 5$ into the equation $v = s^5 + 17$, obtaining $v = (3t^2 - 5)^5 + 17$ and then differentiating.

6-5. Self-test

1. A box slides in a straight line until it comes to rest. Its distance $s(t)$ from a certain spot is given by $s(t) = 20t - 5t^2$ during the time the box is in motion.

(a) When and where does the box come to rest?

(b) What is the domain of definition of the function $s(t)$?

2. The height $h(t)$ of a stone thrown upward from the top of a tower is given by $h(t) = -16t^2 + 64t + 95$, where t is the time in seconds after the stone is thrown.

(a) How high was the tower?

(b) How high did the stone go?

(c) How fast was the stone going when it went past a spot 15 ft above ground level?

(d) Approximately when did the stone strike the ground?

3. Rework Example 2, Sec. 6-4, using a triangle having a 60° angle between the sides.

4. At what rate is the volume of a spherical balloon increasing if its radius is increasing at the rate of 2 in./sec, and its radius is 5 in.?

7
Integral Calculus

sum

7-1. Σ Notation

In mathematics it is often necessary to add elements together. It is usual to use the capital Greek letter sigma Σ to denote a *sum*. Σ is the Greek equivalent of S, the first letter of the word "sum."

$$\sum_{k=1}^{50} k = 1 + 2 + 3 + 4 + \cdots + 49 + 50.$$

$$\sum_{i=1}^{n} b_i = b_1 + b_2 + b_3 + \cdots + b_n.$$

$$\sum_{n=1}^{5} 1/n = \tfrac{1}{1} + \tfrac{1}{2} + \tfrac{1}{3} + \tfrac{1}{4} + \tfrac{1}{5}.$$

$$\sum_{x=2}^{4} x^2 = 2^2 + 3^2 + 4^2 = 29.$$

The notation $\sum_{i=0}^{10} a_i$ is read "the summation of a_i from $i=0$ to $i=10$" and is shorthand for the expression

$$a_0 + a_1 + a_2 + a_3 + a_4 + a_5 + a_6 + a_7 + a_8 + a_9 + a_{10}.$$

A polynomial (Sec. 4-5) may be expressed neatly as

$$\sum_{i=0}^{n} a_i x^i = a_0 + a_1 x + a_2 x^2 + \cdots + a_n x^n.$$

The reader should show that, if

$$f(x) = \sum_{i=0}^{n} a_i x^i \quad \text{then} \quad f'(x) = \sum_{i=1}^{n} i a_i x^{i-1}.$$

167

168 *Integral Calculus*

Example 1

Find $\sum_{x=1}^{5} (x^2 + 3)$.

$$\sum_{x=1}^{5} (x^2 + 3) = (1^2 + 3) + (2^2 + 3) + (3^2 + 3) + (4^2 + 3) + (5^2 + 3)$$
$$= 4 + 7 + 12 + 19 + 28 = 70.$$

Example 2

Find $\sum_{k=0}^{3} \frac{\sqrt{4+k}}{2}$.

$$\sum_{k=0}^{3} \frac{\sqrt{4+k}}{2} = \frac{\sqrt{4+0}}{2} + \frac{\sqrt{4+1}}{2} + \frac{\sqrt{4+2}}{2} + \frac{\sqrt{4+3}}{2}$$
$$= \tfrac{1}{2}(2 + \sqrt{5} + \sqrt{6} + \sqrt{7}).$$

Example 3

Find the smallest value of n such that $\sum_{i=1}^{n} i(i+1) > 15$.

$$\sum_{i=1}^{n} i(i+1) = 1 \cdot 2 + 2 \cdot 3 + 3 \cdot 4 + 4 \cdot 5 + \cdots + n(n+1).$$

For $n = 2$, $\sum_{i=1}^{2} i(i+1) = 1 \cdot 2 + 2 \cdot 3 = 8$.

For $n = 3$, $\sum_{i=1}^{3} i(i+1) = 1 \cdot 2 + 2 \cdot 3 + 3 \cdot 4 = 20$.

For $n < 3$ the sum is less than 15, while for $n \geq 3$ the sum is greater than 15. The desired value, therefore, is $n = 3$.

Example 4

Find $\sum_{x=2}^{4} \frac{5 \cdot 4 \cdot 3 \cdots (5 - x + 1)}{1 \cdot 2 \cdot 3 \cdots x}$.

$$\sum_{x=2}^{4} \frac{5 \cdot 4 \cdot 3 \cdots (5-x+1)}{1 \cdot 2 \cdot 3 \cdots x} = \frac{5 \cdot 4}{1 \cdot 2} + \frac{5 \cdot 4 \cdot 3}{1 \cdot 2 \cdot 3} + \frac{5 \cdot 4 \cdot 3 \cdot 2}{1 \cdot 2 \cdot 3 \cdot 4}$$
$$= 10 + 10 + 5 = 25.$$

Sec. 7-1. Σ Notation

Example 5

Find $\sum_{i=1}^{4} (x_i - 2)^2$ where $x_1 = 1$, $x_2 = 5$, $x_3 = 0$, $x_4 = 8$.

$$\sum_{i=1}^{4} (x_i - 2)^2 = (1 - 2)^2 + (5 - 2)^2 + (0 - 2)^2 + (8 - 2)^2$$
$$= 1 + 9 + 4 + 36 = 50.$$

A study of the proofs of the following theorems and examples will help the reader gain familiarity with the Σ notation.

Theorem 1

If c is a constant, then $\sum_{i=1}^{n} c = nc$.

We use the definition of the notation directly.

$$\sum_{i=1}^{n} c = c + c + \cdots + c = nc.$$

Theorem 2

$$\sum_{i=1}^{n} cx_i = c \sum_{i=1}^{n} x_i.$$

$$\sum_{i=1}^{n} cx_i = cx_1 + \cdots + cx_n = c(x_1 + x_2 + \cdots + x_n)$$
$$= c \sum_{i=1}^{n} x_i.$$

Theorem 3

$$\sum_{i=1}^{n} (x_i + y_i - z_i) = \sum_{i=1}^{n} x_i + \sum_{i=1}^{n} y_i - \sum_{i=1}^{n} z_i.$$

$$\sum_{i=1}^{n} (x_i + y_i - z_i) = (x_1 + y_1 - z_1) + (x_2 + y_2 - z_2)$$
$$+ \cdots + (x_n + y_n - z_n)$$
$$= (x_1 + x_2 + \cdots + x_n) + (y_1 + y_2$$
$$+ \cdots + y_n) - (z_1 + z_2 + \cdots + z_n)$$
$$= \sum_{i=1}^{n} x_i + \sum_{i=1}^{n} y_i - \sum_{i=1}^{n} z_i.$$

Next consider a theorem which is not as obvious as those just proved.

Integral Calculus

Theorem 4

The sum of the positive integers from 1 to n is $\displaystyle\sum_{x=1}^{n} x = \frac{n + n^2}{2}$.

The expression $\displaystyle\sum_{x=1}^{n} [x^2 - (x-1)^2]$ will be expanded in two different ways, one of which contains the expression $\displaystyle\sum_{x=1}^{n} x$. The expansions are equated, and the resulting equation is solved for $\displaystyle\sum_{x=1}^{n} x$. The first method of expansion is by the definition, which yields:

$$\sum_{x=1}^{n} [x^2 - (x-1)^2] = (1^2 - 0^2) + (2^2 - 1^2) + (3^2 - 2^2)$$
$$+ (4^2 - 3^2) + \cdots + [(n-1)^2 - (n-2)^2]$$
$$+ [n^2 - (n-1)^2] = n^2. \qquad \text{(Why ``= n^2''?)} \quad (1)$$

However, from another viewpoint we express the same expansion as

$$\sum_{x=1}^{n} [x^2 - (x-1)^2] = \sum_{x=1}^{n} (x^2 - x^2 + 2x - 1)$$
$$= \sum_{x=1}^{n} (2x - 1)$$
$$= \sum_{x=1}^{n} (2x) - \sum_{x=1}^{n} 1 \qquad \text{by Theorem 3.} \quad (2)$$

$$\sum_{x=1}^{n} [x^2 - (x-1)^2] = 2\left[\sum_{x=1}^{n} (x)\right] - n$$
$$\text{by Theorems 2 and 1.} \quad (3)$$

From Eqs. (1) and (3) we obtain

$$2\left[\sum_{x=1}^{n} (x)\right] - n = n^2$$

which has solution $\displaystyle\sum_{x=1}^{n} x = \frac{n + n^2}{2}$, as desired. This, of course, is

Sec. 7-1. Σ Notation

the familiar formula expressing the sum of an arithmetic sequence ("progression") as the number of terms, n, times the average, $(1 + n)/2$, of the first and last terms

$$\sum_{x=1}^{n} x = n \cdot \frac{1+n}{2} = n \cdot \left(\frac{a+l}{2}\right)$$

where a is the first term and l is the last term.

Problem Set 7-1

1. Find $\displaystyle\sum_{x=1}^{7} x^2$.

2. Find $\displaystyle\sum_{x=1}^{5} (2x + 1)$.

3. Find $\displaystyle\sum_{x=1}^{5} (2x) + 1$. Compare with Prob. 2.

4. Find $4 \displaystyle\sum_{n=1}^{3} n^2$.

5. Find $\displaystyle\sum_{k=0}^{4} (k + 1)^2$.

6. Find $\displaystyle\sum_{x=1}^{10} (x - 5)^2$.

7. Find $\displaystyle\sum_{x=1}^{5} (x - 3.5)^2$.

8. Find $\displaystyle\sum_{i=1}^{5} (x_i - 4)$, $x_1 = 1$, $x_2 = 7$, $x_3 = 4$, $x_4 = 2$, $x_5 = 6$.

9. Find $\displaystyle\sum_{i=1}^{5} x_i^2$ for x_i as given in Prob. 8.

10. Find $\displaystyle\sum_{i=1}^{5} (x_i - 4)$ for x_i as given in Prob. 8.

11. Find $\displaystyle\sum_{i=1}^{5} (x_i - 4)^2$ for x_i as given in Prob. 8.

12. Find the smallest value of n such that $\displaystyle\sum_{x=1}^{n} i^2 > 20$.

13. Find the smallest value of n such that $\displaystyle\sum_{x=1}^{n} (x - 4)^2 > 30$.

172   *Integral Calculus*

14. For what range of values of n will $\sum_{x=1}^{n} x^2$ lie between 10 and 50?

15. Find $\sum_{x=1}^{4} \dfrac{4 \cdot 3 \cdots (4 - x + 1)}{1 \cdot 2 \cdots x}$. (See Example 4.)

16. Write the first four terms of $\sum_{x=1}^{n} a_x$, where $a_x = \dfrac{1}{x(x+1)}$. Find $\sum_{x=1}^{n} a_x$. HINT: $a_x = \dfrac{1}{x} - \dfrac{1}{x+1}$.

17. Find $\sum_{x=1}^{n} x^2$, the sum of the square of the positive integers from 1 to n. HINT: Express $\sum_{x=1}^{n} [x^3 - (x-1)^3]$ in two ways and apply Theorem 4. *Ans.* $\tfrac{1}{6}n(n+1)(2n+1)$.

18. Find $\sum_{x=1}^{n} x^3$. See hint of Prob. 17. *Ans.* $\tfrac{1}{4}n^2(n+1)^2$.

19. Prove or disprove: $\sum_{x=1}^{n} [x^2(x-1)] = \left[\sum_{x=1}^{n} x^2\right] \cdot \left[\sum_{x=1}^{n} (x-1)\right]$.

20. Find $\sum_{i=2}^{7} \dfrac{5 \cdot 4 \cdot 3 \cdots (5-i)}{1 \cdot 2 \cdot 3 \cdots (i-1)}$.

7-2. $\lim_{x \to \infty} f(x)$

Section 3-2 discussed the meaning of $\lim_{x \to b} f(x) = L$ and noted that this was equivalent to stating that, "for each $\epsilon > 0$, there exists a δ such that if $0 < |x - b| < \delta$, then $|f(x) - L| < \epsilon$." In other words, if x is near b but not equal to b, then $f(x)$ is near L. We have also discussed, informally, what happens to certain function values, say $2/x$, as x becomes very large. This concept is formalized using the *limit notation*. The expression $\lim_{x \to \infty} f(x) = L$ is read "the

Sec. 7-3. Area

$\lim_{x \to \infty} f(x)$

limit of $f(x)$ as x *increases without bound* is L." Do not use the word "infinity" for the symbol ∞. To do so would substitute a word for a concept. To say that $\lim_{x \to \infty} f(x) = L$ means that for each $\epsilon > 0$ there exists a number N such that, for all $x > N$, $|f(x) - L| < \epsilon$. That is, $f(x)$ becomes and remains as close ($\epsilon > 0$) to L as desired, for x sufficiently large.

It is in this sense that we state that $\lim_{n \to \infty} \frac{k}{n} = 0$.

7-3. Area

Before studying the derivative, we discussed the definition of a line tangent to a curve at a given point P. The concept was a familiar one, but it proved difficult to give an exact definition of "a line tangent to a curve at a point P" until the notation of a "limiting position" was introduced. This leads, eventually, to the derivative

$$\frac{df}{dx} = \lim_{\Delta x \to 0} \frac{\Delta f}{\Delta x} = \lim_{\Delta x \to 0} \frac{f(x + \Delta x) - f(x)}{\Delta x}$$

The reader may discover that although the notion of area is familiar, a precise definition of area is somewhat elusive. The area of a rectangle is easily defined, in the manner of Euclid, as the length times the breadth. Since each rectangle may be broken into two congruent right triangles, we then derive the formula for the area of a right triangle as one-half the base times the altitude. The area of nonright triangles is quickly determined by dropping a perpendicular from the vertex to the opposite side, thus breaking the nonright triangle into two right triangles. The reader should show that this definition leads to the formula

$A = \frac{1}{2} \cdot$ (the length of the base) $\cdot$ (the altitude)

no matter whether the foot of the perpendicular strikes the opposite side inside or outside of the triangle (Figs. 7-1 and 7-2). Since every polygonal figure can be broken up into triangles, a formula for the area of a polygon follows.

Figure 7-1   **Figure 7-2**

This is about as far as we can go with the concepts discussed so far. Every school child knows that the area of a circle is πr^2. However, this formula does not follow easily from the above definition of area. Before continuing with this section, ponder carefully how you would prove that the area of a circle is πr^2.

The ancient mathematician Archimedes was faced with a similar dilemma. One possible solution is to define the area of a circle, which is undefined at this stage, as the limiting value of the area of an inscribed, *regular* polygon of n sides as the number of sides increases without bound, if such a limit exists. This definition yields a result which agrees with our intuitive notion of area. However, another possible definition of area of a circle, which would also agree with our intuitive notions, would be to define the area of a circle as the limiting value of an area of an inscribed polygon of n sides as the number of sides increases without bound and as the length of the longest side approaches 0, if the limit exists.

In the second definition the polygon need not be a regular polygon (that is, have all sides of the same length). It is quite conceivable that this definition might give a different value for the area of a circle. Two other possible definitions arise by considering circumscribed polygons in place of inscribed polygons. We have, then, four reasonable definitions for the area of a circle. They *may* all yield the same result or they may yield different results. Note that each of these definitions makes use of the notion of a limit.

By using the notion of a limit, Archimedes defined not only the area of a circle but also the area of a segment of a parabola. We shall not use Archimedes' methods because the more powerful methods of analytic geometry are much simpler. The area of the parabola, rather than that of the circle, has been chosen here since the resulting algebra is easier. The area of the circle involves the area of inscribed or circumscribed polygons, and uses the formulas

$$A = \frac{n}{2} r^2 \sin\left(\frac{2\pi}{n}\right) \qquad \text{area of inscribed polygon.}$$

$$A = nr^2 \tan\left(\frac{\pi}{n}\right) \qquad \text{area of circumscribed polygon.}$$

In order to take the limit of either of these expressions as n increases without bound, it is necessary to examine certain trigonometric limits. Although it is within our ability to do so at this point, let

Sec. 7-3. Area 175

us instead concentrate on the main issue, namely, the definition of area of a closed figure whose boundary does not consist entirely of straight-line segments.

Consider the graph of $y = x^2$ shown in Fig. 7-3. This parabola, the x axis, and the line $x = 1$ enclose a region of the xy plane with which we wish to associate a number called *area*.

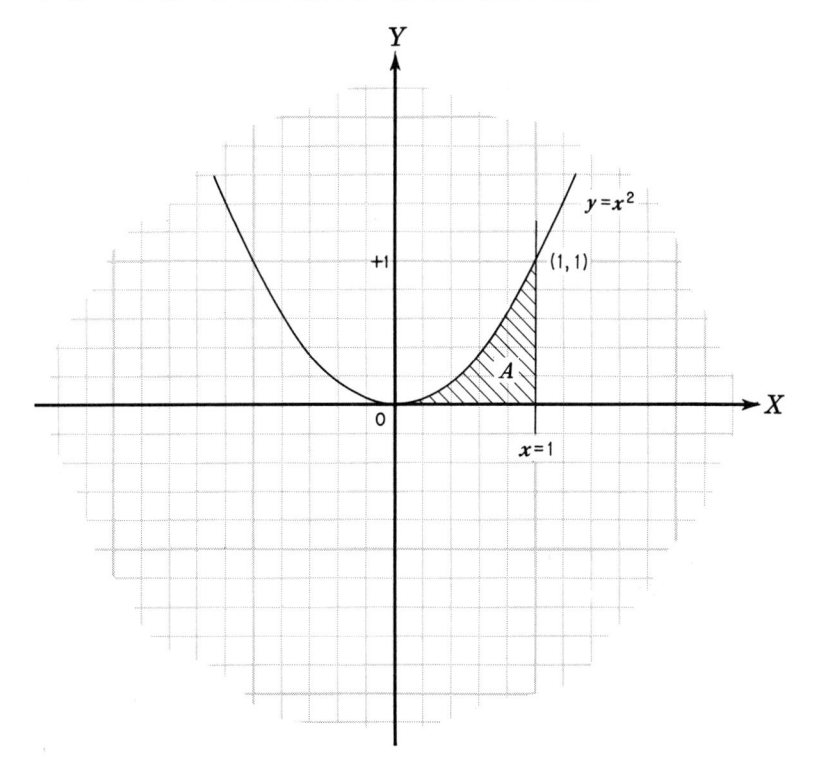

Figure 7-3

Let us divide the x axis between 0 and 1 into n equal intervals by placing points at 0, $1/n$, $2/n$, $3/n$, ..., $n - 1/n$, 1 (Fig. 7-4). Construct a series of rectangles with these intervals as bases and an upper corner of the rectangle on the curve $y = x^2$. Figure 7-4 will occur if the rectangles are constructed to lie below the curve $y = x^2$; Fig. 7-5 will occur if the rectangles are so taken that the curve lies under the rectangles.

If we are to give meaning to the idea of area under the parabola, the area should be greater than or equal to the area obtained in Fig.

176 Integral Calculus

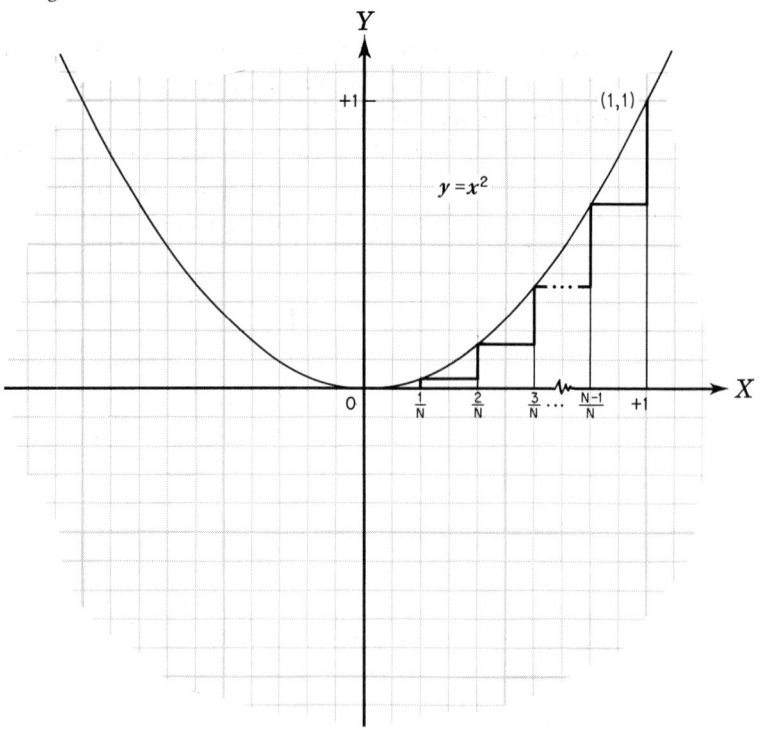

Figure 7-4

7-4 and less than or equal to the area obtained in Fig. 7-5. If the area desired is denoted by the letter A, we have the following inequality:

{Area of inscribed polygon (Fig. 7-4)} $< A$ (area under parabola)
$<$ {area of circumscribed polygon (Fig. 7-5)}

$$\left[\left(\frac{0}{n}\right)^2 + \left(\frac{1}{n}\right)^2 + \left(\frac{2}{n}\right)^2 + \cdots + \left(\frac{n-1}{n}\right)^2\right]\left(\frac{1}{n}\right) < A$$

$$< \left[\left(\frac{1}{n}\right)^2 + \left(\frac{2}{n}\right)^2 + \left(\frac{3}{n}\right)^2 + \cdots + \left(\frac{n-1}{n}\right)^2 + \left(\frac{n}{n}\right)^2\right]\left(\frac{1}{n}\right).$$

The squared terms come from the ordinates of $y = x^2$ at the place where the rectangle crosses the curve. The width of each rectangle is $1/n$. Factoring $1/n^2$ from each term,

$$\left(\frac{1}{n}\right)^3 [0^2 + 1^2 + \cdots + (n-1)^2] < A$$

$$< \left(\frac{1}{n}\right)^3 \cdot [1^2 + 2^2 + 3^2 + \cdots + (n-1)^2 + n^2].$$

Sec. 7-3. Area

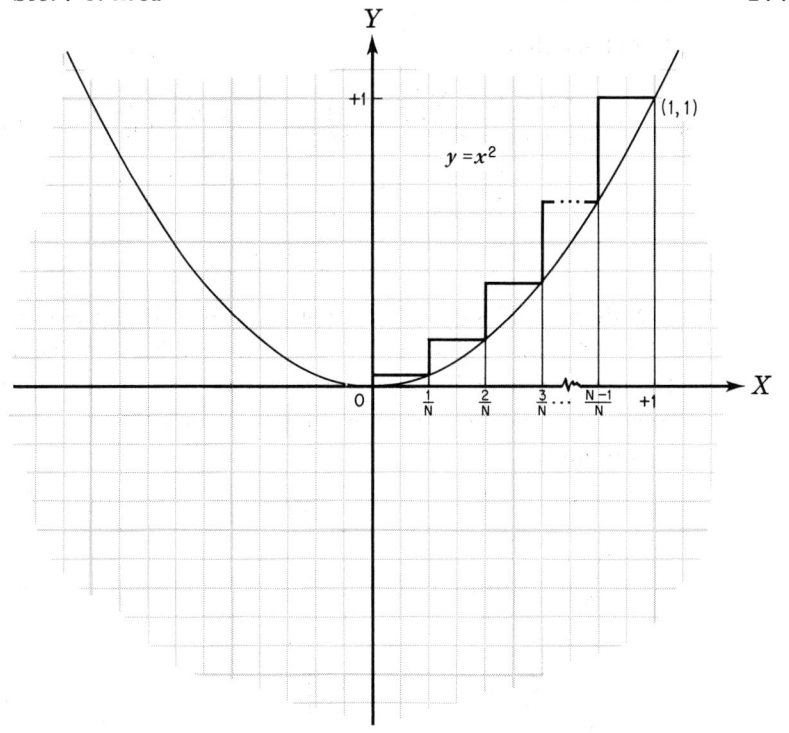

[Figure: graph of $y = x^2$ with rectangles from 0 to 1 divided at $\frac{1}{N}, \frac{2}{N}, \frac{3}{N}, \ldots, \frac{N-1}{N}, +1$, point $(1,1)$ marked]

Figure 7-5

Problem 17 of Set 7-1 shows that

$$1^2 + 2^2 + \cdots + k^2 = \sum_{i=1}^{k} i^2 = \frac{k(k+1)(2k+1)}{6}.$$

Therefore, the given inequality reduces to

$$\left(\frac{1}{n^3}\right)\frac{(n-1)(n)(2n-1)}{6} < A < \left(\frac{1}{n^3}\right)\frac{n(n+1)(2n+1)}{6}.$$

which is equivalent to

$$\frac{(n-1)(2n-1)}{6n^2} < A < \frac{(n+1)(2n+1)}{6n^2}$$

or $\quad \dfrac{1}{6}\left(1 - \dfrac{1}{n}\right)\cdot\left(2 - \dfrac{1}{n}\right) < A < \dfrac{1}{6}\left(1 + \dfrac{1}{n}\right)\cdot\left(2 + \dfrac{1}{n}\right).$

If we permit n to increase without bound, the limit of $(1/n)$ is zero.

$$\lim_{n \to \infty} 1/n = 0.$$

178   *Integral Calculus*

It follows that

$$\lim_{n \to \infty} \left[\frac{1}{6}\left(1 - \frac{1}{n}\right)\left(2 - \frac{1}{n}\right)\right] \leq A \leq \lim_{n \to \infty} \left[\frac{1}{6}\left(1 + \frac{1}{n}\right)\left(2 + \frac{1}{n}\right)\right]$$

$$\frac{1}{6}(1 - 0)(2 - 0) \leq A \leq \frac{1}{6}(1 + 0)(2 + 0)$$

$$\frac{1}{3} \leq A \leq \frac{1}{3}.$$

Note that $\leq$ signs have crept in, after taking the limit, in place of the $<$ signs. This is a result of taking a limit. The limit of the fraction $1/n$, as n increases, is zero, but none of the fractions themselves is zero. We conclude that the area under the parabola is $\frac{1}{3}$ square unit.

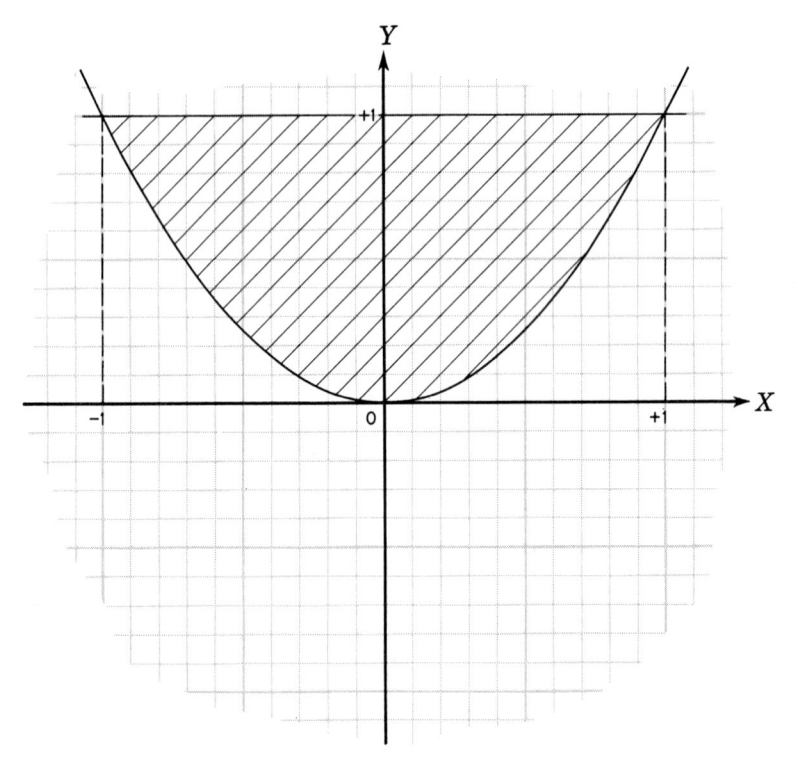

Figure 7-6

The area of the segment of the parabola which Archimedes found (Fig. 7-6) is obtained by taking the area of the rectangle (namely,

Sec. 7-3. Area 179

two units) and subtracting from it twice the area just found. The area of the segment of the parabola $y = x^2$, bounded by the line $y = 1$, is $2 - 2(\frac{1}{3}) = \frac{4}{3}$ square units. We have therefore solved, using modern analytical methods, a problem which sorely taxed the best brains of ancient Greece.

Problem Set 7-3

Use the method described in this section to compute as many of the following areas as you can by determining upper and lower bounds for the area in terms of rectangles and then taking the limit of these bounds as the width of the rectangle approaches zero.

1. $y = 5x$ between $x = 0$, $x = 1$, and the x axis. Show that the area so obtained has a limit which agrees with the area of a triangle obtained by the methods of ordinary geometry. If this were not true the method described would be unsatisfactory.
2. $y = 5x + 4$ between $x = 0$ and $x = 1$. HINT: The width of the rectangles is $6/n$ rather than $1/n$. (Why?)
3. $y = 5x + 4$ between $x = 0$ and $x = 6$.
4. $y = x^2$ between $x = 10$ and $x = 11$.
5. $y = 3x^2$ between $x = 0$ and $x = 1$.
6. $y = 3x^2$ between $x = 0$ and $x = 2$. HINT: The width of the rectangles is now $2/n$ rather than $1/n$.
7. $y = x - x^2$ and the x axis. Note that this curve intercepts the x axis at $x = 0$ and $x = 1$. The area is $\frac{1}{6}$ square units.
8. $y = x^2$, the line $y = 2$, and the line $x = 0$.
9. Discuss, in a short paragraph, the meaning of area and explain why the area bounded by a parabola and a line needs definition other than that of a polygonal area.
10. Examine a high school or junior high school text and comment on the definition, if any, given for the area of a circle. If possible, find a proof that this area is πr^2 and discuss its weaknesses, if any.

7-4. The Spring Problem

In physics you may have learned that the amount of work done when a constant force moves a body a certain distance is defined as the product of the magnitude of the force by the distance moved.

work

Integral Calculus

If the force is not constant, what then is the amount of *work*? To be specific, let us consider a spring which satisfies Hooke's law. In this case the force is proportional to the amount of stretch from its normal length, $f(s) = ks$. If we make a proper choice of units or select a proper spring, we may be lucky enough to have k, the constant of the spring, equal to 1. Let us assume, in order to simplify the arithmetic involved in this problem, that we have made this lucky choice and $F(s) = s$. If the original length of the spring is 12 ft and we pull the spring out to 17 ft in length, we have stretched it a distance s of 5 ft. As we start to pull the spring, the amount of stretching is small and the force $F(s)$ necessary to stretch it is also small. As the spring is pulled out, s increases and so does the necessary force F.

The force is *not* constant but is a function of the amount of stretch. Therefore, the old definition of work as force times distance does not apply. Hence, to find W, we need to extend our definition of work, just as previously we needed to extend the definition for area when the bounding curve was not composed of straight-line segments. Since this is a physical problem, you may well have an intuitive feeling concerning the meaning of "work" to stretch the spring.

Whatever your intuitive idea of work may be, you will almost certainly agree that the work done is more than $0 \cdot 5 = 0$ ft-lb, which is the amount of work that would have been done had the minimum force been applied during the entire 5-ft stretch. Furthermore, you are apt to agree that the work done will be less than would have been required had the maximum force been required during the entire 5-ft stretch, that is, $5 \cdot 5 = 25$ ft-lb. We thus have bounds for the amount of work W.

$$0 \leq W \leq 25.$$

Let us now obtain a better approximation. Suppose we consider the stretching as being done in two steps. That is, we first pull the string out to $14\frac{1}{2}$-ft length ($2\frac{1}{2}$-ft stretch) doing an amount of work W_1 and then to 17-ft length doing an amount of work W_2, so that stretching is done in two $\frac{5}{2}$ steps. In the first step, the amount of work W_1 is greater than $0 \cdot \frac{5}{2} = 0$ ft-lb and less than $\frac{5}{2} \cdot \frac{5}{2} = \frac{25}{4}$ ft-lb. (Why?)

Sec. 7-4. The Spring Problem

$$0 \cdot \frac{5}{2} \leq W_1 \leq \frac{5}{2} \cdot \frac{5}{2}$$

$$0 \leq W_1 \leq \frac{25}{4}.$$

In the second step the work done is W_2, which is greater than $\frac{5}{2} \cdot \frac{5}{2} = \frac{25}{4}$ and less than $5 \cdot \frac{5}{2} = \frac{25}{2}$ ft-lb, since the maximum force is 5 lb and the distance is $\frac{5}{2}$ ft.

$$\frac{5}{2} \cdot \frac{5}{2} \leq W_2 \leq 5 \cdot \frac{5}{2}$$

$$\frac{25}{4} \leq W_2 \leq \frac{25}{2}.$$

Since $\quad W = W_1 + W_2$

$$0 + \frac{25}{4} \leq W_1 + W_2 \leq \frac{25}{4} + \frac{25}{2}$$

$$\frac{25}{4} \leq W \leq \frac{75}{4}$$

$$6.25 \leq W \leq 18.75.$$

This is a better bound on the total work W than was obtained before. If the stretching process had been done in five equal steps, the bounds would be

$$10 \leq W \leq 15.$$

You are asked to carry out this computation in Prob. 1 of Set 7-4. If the stretching had been done in 50 steps, the bounds would be

$$12.25 \leq W \leq 17.75.$$

If 100 steps had been used, the bounds would be

$$12.375 \leq W \leq 12.625.$$

If the 5-ft stretch is considered in 200 consecutive steps, each of $\frac{5}{200}$ units length, the resulting bounds are

$$12.4375 \leq W \leq 12.5625.$$

Perhaps it is evident that our project is to obtain a general formula for the amount of work done. The method employed is to divide the interval into n parts, obtaining upper and lower bounds for W_k

182 **Integral Calculus**

on each part. These bounds are then summed to obtain upper and lower bounds for the total work W.

$$\sum_{k=1}^{n} \text{(lower bounds on each interval)} \leq W$$
$$\leq \sum_{k=1}^{n} \text{(upper bounds on each interval)}.$$

We now take the limit in the inequality as the number in intervals n increases without bound and the length of each interval approaches zero. If the interval is divided into n parts (each of length $5/n$), the amount of force required at the beginning of the kth interval is $(k - 1)$ times $5/n$. The amount of force required at the end of the kth interval is k times $5/n$. In other words,

$$\sum_{k=1}^{n} \frac{(k-1)5}{n} \cdot \frac{5}{n} \leq W \leq \sum_{k=1}^{n} \frac{k \cdot 5}{n} \cdot \frac{5}{n}.$$

Upon passing to the limit

$$\lim_{n \to \infty} \sum_{k=1}^{n} \frac{(k-1)5}{n} \cdot \frac{5}{n} \leq W \leq \lim_{n \to \infty} \sum_{k=1}^{n} \frac{k \cdot 5}{n} \cdot \frac{5}{n}.$$

$$\lim_{n \to \infty} \frac{25}{n^2} \sum_{k=1}^{n} (k-1) \leq W \leq \lim_{n \to \infty} \frac{25}{n^2} \sum_{k=1}^{n} k$$

$$\lim_{n \to \infty} \frac{25}{n^2} \cdot \frac{n^2 - n}{2} \leq W \leq \lim_{n \to \infty} \frac{25}{n^2} \cdot \frac{n^2 + n}{2}$$

using Theorem 4, Sec. 7-1.

$$\lim_{n \to \infty} \left(\frac{25}{2} - \frac{25}{2n} \right) \leq W \leq \lim_{n \to \infty} \left(\frac{25}{2} + \frac{25}{2n} \right)$$

$$\frac{25}{2} \leq W \leq \frac{25}{2}.$$

Hence $W = \dfrac{25}{2} = 12.5$. (Why?)

If the upper and lower bounds for W had not had a common limit, our conclusion would not have been clear. However, when this common limit exists, then the common limit is *defined* to be the work done by the variable force.

Sec. 7-4. The Spring Problem 183

This extension of the "static" definition of work involves the limit concept, just as the concept of limit was involved in defining an area whose bounds were not straight lines, or an instantaneous velocity, or the slope of a tangent.

Use the following notation to define the work done, W, in moving a body from $s = a$ to $s = b$, against a variable force $f(s)$.

$(x_i \text{ min}) = $ a value of x giving the smallest value of $f(x)$ in the ith interval.

$(x_i \text{ max}) = $ a value of x giving the greatest value of $f(x)$ in the ith interval.

$x_i = $ the length of the ith interval.

Then the work W done by moving from $s = a$ to $s = b$ against a force $f(s)$ satisfies the following inequalities:

$$\sum_{i=1}^{n} f(x_i \text{ min}) \cdot \Delta x_i \leq W \leq \sum_{i=1}^{n} f(x_i \text{ max}) \cdot \Delta x_i.$$

If the limits of the two extremities in the above inequality are identical, we shall then define this common value as the amount of work done.

The student who notes a similarity between problems of work and problems of area is quite correct. The concepts are closely related.

Problem Set 7-4

1. Carry out the computation of the stretching process on the spring mentioned on page 181 for five steps, and show that $10 \leq W \leq 15$.
2. Carry out the process of approximating an area described in Sec. 7-3, for the area under the curve $y = x$ for x between 1 and 5. Use $n = 5$ in your approximation. Note the similarity between this problem and Prob. 1.
3. Evaluate the limit of the two sums given as an approximation of W for the function $f(s) = s$ where the number of intervals n increases without bound. In other words, rework the text example.
4. Evaluate the limit of the sum obtained for the area under $y = x$ between $x = 0$ and $x = 5$. Show that this result is the same as that obtained in Prob. 3.

184   *Integral Calculus*

7-5. Some Remarks on $\lim\limits_{n\to\infty} \sum\limits_{i=1}^{n} f(\xi_i)\,\Delta x_i$

We have discussed several functions in which

$$\lim_{n\to\infty} \sum_{i=1}^{n} f(\min x_i)\,\Delta x_i = \lim_{n\to\infty} \sum_{i=1}^{n} f(\max x_i)\,\Delta x_i = L$$

where L is the common value of the limits given. If ξ_i is any point in the ith interval, then

$$\sum_{i=1}^{n} f(\min x_i)\,\Delta x_i \leq \sum_{i=1}^{n} f(\xi_i)\,\Delta x_i \leq \sum_{i=1}^{n} f(\max x_i)\,\Delta x_i$$

and, since the outer two functions have the same limit L as n increases without bound, it follows that

$$\lim_{n\to\infty} \sum_{i=1}^{n} f(\min x_i)\,\Delta x_i = \lim_{n\to\infty} \sum_{i=1}^{n} f(\xi_i)\,\Delta x_i$$

$$= \lim_{n\to\infty} \sum_{i=1}^{n} f(\max x_i)\,\Delta x_i = L.$$

Actually there are many functions which have the property that the limit $\lim\limits_{n\to\infty} \sum\limits_{i=1}^{n} f(\xi_i)\,\Delta x_i$, where ξ_i is some point in the interval Δx_i, exists and is independent of the choice of the point ξ_i.

7-6. The Definite Integral

Theorem

Let $f(x)$ be a single-valued, continuous function on the (closed) interval $a \leq x \leq b$. Divide the interval from a to b into n subintervals (not necessarily of equal length). In each subinterval x_i select a point ξ_i. Then

$$\lim_{\substack{n\to\infty \\ \max \Delta x_i \to 0}} \sum_{i=1}^{n} f(\xi_i)\,\Delta x_i$$

exists and is independent of the choice of ξ_i and of Δx_i as long as the length of the maximum Δx_i approaches zero as n becomes large without bound. [This follows from the conditions placed on the function $f(x)$, but the proof is *not* simple.]

Sec. 7-6. The Definite Integral

definite integral

This limit, when it exists, is known as the *definite integral* of $f(x)$ from a to b and is written

$$\int_a^b f(x)\,dx = \lim_{\substack{n\to\infty \\ \max \Delta x_i \to 0}} \sum_{i=1}^n f(\xi_i)\,\Delta x_i.$$

The notation of the elongated S, standing for sum, in the integral sign is due to the mathematician, Baron Gottfried Wilhelm von Leibnitz (German, 1646–1716), a contemporary of Jolliet and Scarlatti.

Example 1

Perform the indicated integration:

$$\int_0^3 (3x + 4)\,dx.$$

This is equivalent to finding the area of the trapezoid bounded by $y = 0$, $x = 0$, $y = 3x + 4$, and $x = 3$ (Fig. 7-7).

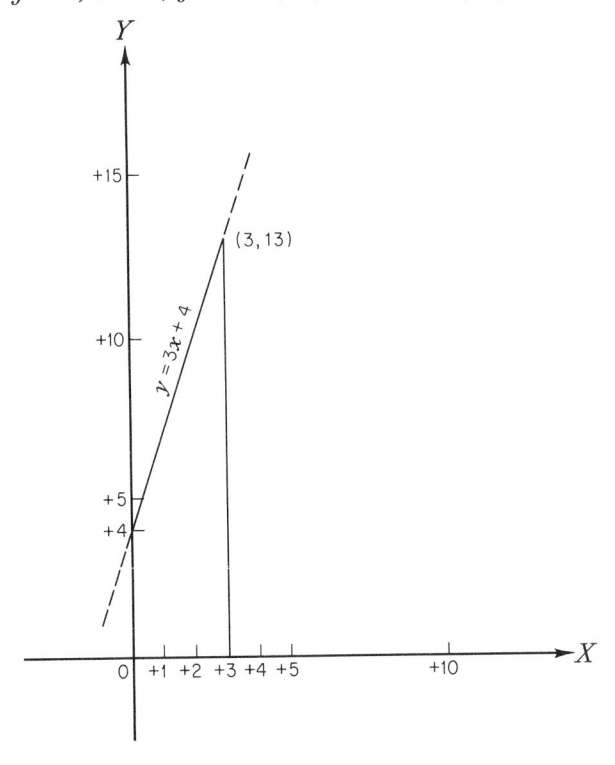

Figure 7-7

186   *Integral Calculus*

If we use equal subintervals $\Delta x_i = 3/n$ and select the right end point $i(3/n)$ in each subinterval as the point ξ_i, then we obtain

$$\int_0^3 (3x + 4)\, dx = \lim_{\substack{n \to \infty \\ \Delta x_i = \frac{3}{n} \to 0}} \sum_{i=1}^n \left[3\left(\frac{3i}{n}\right) + 4\right] \cdot \frac{3}{n}$$

$$= \lim_{n \to \infty} \left(\frac{27}{n^2} \sum_{i=1}^n i + \frac{12}{n} \sum_{i=1}^n 1\right)$$

$$= \lim_{n \to \infty} \left(\frac{27}{n^2} \cdot \frac{n^2 + n}{2} + \frac{12}{n} \cdot n\right)$$

$$= \lim_{n \to \infty} \left(\frac{27}{2} + \frac{27}{2n} + 12\right)$$

$$= \lim_{n \to \infty} \left(\frac{51}{2} + \frac{27}{2n}\right) = \frac{51}{2}.$$

Hence: $\int_0^3 (3x + 4)\, dx = \frac{51}{2}.$

By the method of integration, the area of the trapezoid is $\frac{51}{2}$ square units. This value is the same as that given by the ordinary Euclidean definition. Prove this.

Example 2

$$\int_0^1 (1 - x^2)\, dx.$$

Again taking equal subintervals $\Delta x_i = 1/n$ and letting $\xi_i = i/n$, the point at the extreme right of each subinterval,

$$\int_0^1 (1 - x^2)\, dx = \lim_{n \to \infty} \sum_{i=1}^n \left(1 - \frac{i^2}{n^2}\right) \frac{1}{n}$$

$$= \lim_{n \to \infty} \left(\frac{1}{n} \sum_{i=1}^n 1 - \frac{1}{n^3} \sum_{i=1}^n i^2\right)$$

$$= \lim_{n \to \infty} \left\{\frac{1}{n} \cdot n - \frac{1}{n^3} \left[\frac{1}{6} n(n+1)(2n+1)\right]\right\}$$

by Prob. 17, Set 7-1

$$= \lim_{n \to \infty} \left(1 - \frac{2n^3 + 3n^2 + n}{6n^3}\right)$$

$$= \lim_{n \to \infty} \left(1 - \frac{1}{3} - \frac{1}{2n} + \frac{1}{6n^2}\right)$$

$$= \frac{2}{3}.$$

Sec. 7-6. *The Definite Integral*

Sketch the curve, $y = 1 - x^2$, and show that $\frac{2}{3}$ is a reasonable value for the area under the curve between $x = 0$ and $x = 1$.

Example 3

Find the finite area bounded by $y = x^3$, $y = 0$, $x = 0$, $x = 2$ (Fig. 7-8).

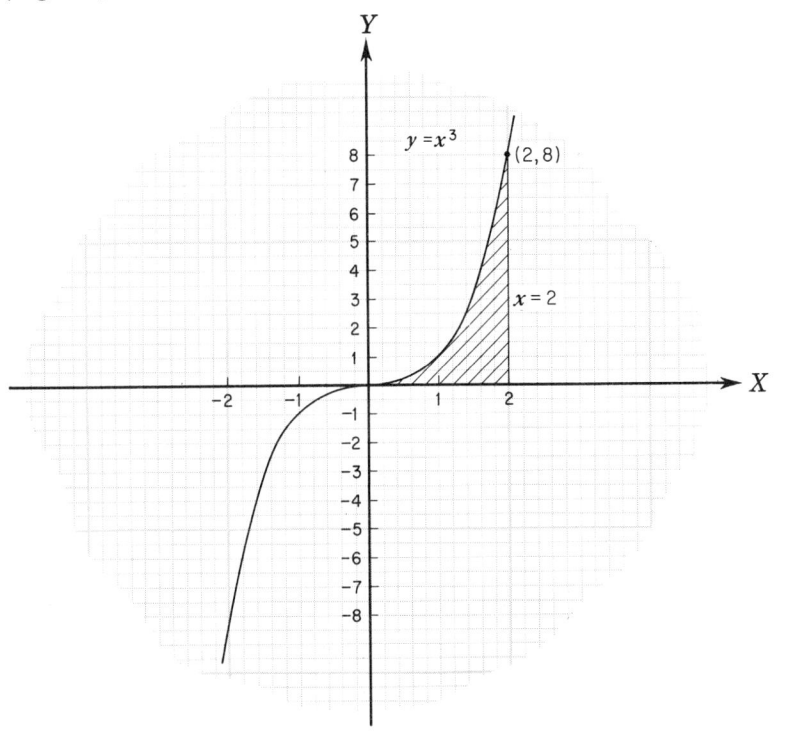

Figure 7-8

$$\int_0^2 x^3 \, dx = \lim_{n \to \infty} \sum_{i=1}^{n} \left(\frac{2i}{n}\right)^3 \frac{2}{n} \quad \text{(Why?)}$$

$$= \lim_{n \to \infty} \frac{16}{n^4} \sum_{i=1}^{n} i^3$$

$$= \lim_{n \to \infty} \left(\frac{16}{n^4} \cdot \frac{n^4 + 2n^3 + n^2}{4}\right)$$

$$= \lim_{n \to \infty} \left(4 + \frac{8}{n} + \frac{4}{n^2}\right)$$

$$= 4.$$

Problem Set 7-6

Evaluate the following integrals, using the definition of this section. Sketch the functions and shade the area which is equal to the integral.

1. $\int_0^1 (3x + 5)\, dx.$ 2. $\int_0^3 x^2\, dx.$ 3. $\int_0^4 5x\, dx.$

4. $\int_0^2 (2x^2 - x)\, dx.$ 5. $\int_0^2 (x^3 - 4)\, dx.$

6. $\int_0^1 x^4\, dx.$ HINT: $\sum_{i=1}^{n} i^4 = \dfrac{6n^5 + 15n^4 + 10n^3 - n}{30}.$

7. $\int_4^5 x\, dx.$ Check by Euclidean geometry.   8. $\int_4^5 x^2\, dx.$

9. $\int_2^5 x^3\, dx.$ 10. $\int_0^2 (x - 1)\, dx.$ Try to interpret your answer geometrically.

11. $\int_0^k 1\, dx.$ 12. $\int_0^k x\, dx = \dfrac{k^2}{2}.$ 13. $\int_0^k x^2\, dx = \dfrac{k^3}{3}.$

Prove the statements given in Probs. 14 through 20.

14. $\int_0^k x^3\, dx = \dfrac{k^4}{4}.$ 15. $\int_0^k x^4\, dx = \dfrac{k^5}{5}.$ HINT. See Prob. 6.

16. $\int_a^b f(x)\, dx + \int_b^c f(x)\, dx = \int_a^c f(x)\, dx.$

17. $\int_a^b [f(x) + g(x)]\, dx = \int_a^b f(x)\, dx + \int_a^b g(x)\, dx.$

18. $\int_a^b [f(x) - g(x)]\, dx = \int_a^b f(x)\, dx - \int_a^b g(x)\, dx.$

19. $\int_a^b kf(x)\, dx = k \int_a^b f(x)\, dx.$

20. Find $\int_0^3 |x - 1|\, dx.$

7-7. Some Preliminary Theorems on Integrals

The following theorems are basic in dealing with integrals. Each may be proved either directly from the definition, or by more subtle

Sec. 7-7. Some Preliminary Theorems on Integrals

means, depending upon your own ingenuity. They were proved in Probs. 16, 17, 18, and 19 of Set 7-6.

Theorem 1

$$\int_a^b f(x)\, dx + \int_b^c f(x)\, dx = \int_a^c f(x)\, dx.$$

Theorem 2

$$\int_a^b [f(x) \pm g(x)]\, dx = \int_a^b f(x)\, dx \pm \int_a^b g(x)\, dx.$$

Theorem 3

$$\int_a^b kf(x)\, dx = k \int_a^b f(x)\, dx.$$

At this point in our study of the derivative, a nice formula for the derivative of kx^n made life much simpler. It would be delightful if we could prove a similar formula here.

By use of	We have shown that
$\sum_{i=1}^{n} 1 = n$	$\int_0^k 1\, dx = k$
$\sum_{i=1}^{n} i = \dfrac{n^2 + n}{2}$	$\int_0^k x\, dx = \dfrac{k^2}{2}$
$\sum_{i=1}^{n} i^2 = \dfrac{n(n+1)(2n+1)}{6}$	$\int_0^k x^2\, dx = \dfrac{k^3}{3}$
$\sum_{i=1}^{n} i^3 = \dfrac{n^4 + 2n^3 + n^2}{4}$	$\int_0^k x^3\, dx = \dfrac{k^4}{4}$
$\sum_{i=1}^{n} i^4 = \dfrac{6n^5 + 15n^4 + 10n^3 - n}{30}$	$\int_0^k x^4\, dx = \dfrac{x^5}{5}$

We might (quite correctly) suspect that $\int_0^k x^m dx$ could be evaluated by use of a similar formula for $\sum_{i=1}^{n} i^m$.

Even though it is possible, by methods similar to those of Probs. 17 and 18 of Set 7-1, to build up a formula for $\sum_{i=1}^{n} i^k$ for $k = 1, 2, 3,$

190 *Integral Calculus*

4, ... (any given value), the *general* formula for $\sum_{i=1}^{n} i^m$ is not known. Fortunately, an even more powerful theorem than the theorems for $\int_0^k x^m \, dx$ will come to our rescue. Before continuing, let us gain a bit more proficiency in the manipulation of integrals. It should be pointed out that the variable of integration is *not* an essential part of the notation; that is

$$\int_a^b f(x) \, dx = \int_a^b f(t) \, dt = \int_a^b f(z) \, dz = \int_a^b f(n) \, dn, \text{ etc.}$$

Problem Set 7-7

To obtain the integrals of the given functions in Probs. 1 to 6, *without* using the lim Σ process, use the theorems of this section and the nine "facts" given.

GIVEN: $\int_0^1 1 \, dx = 1 \qquad \int_0^1 x \, dx = \frac{1}{2} \qquad \int_0^1 x^2 \, dx = \frac{1}{3}.$

$\int_1^2 1 \, dx = 1 \qquad \int_1^2 x \, dx = \frac{3}{2} \qquad \int_1^2 x^2 \, dx = \frac{7}{3}.$

$\int_2^3 1 \, dx = 1 \qquad \int_2^3 x \, dx = \frac{5}{2} \qquad \int_2^3 x^2 \, dx = \frac{19}{3}.$

1. $\int_0^3 (3x^2 - 5x + 2) \, dx.$ HINT:

$\int_0^3 (3x^2 - 5x + 2) \, dx = 3 \int_0^3 x^2 \, dx - 5 \int_0^3 x \, dx + 2 \int_0^3 1 \, dx,$

by Theorems 2 and 3. However, by Theorem 1 we have

$\int_0^3 f(x) \, dx = \int_0^1 f(x) \, dx + \int_1^2 f(x) \, dx + \int_2^3 f(x) \, dx,$

2. $\int_1^3 (3x^2 - 5) \, dx.$ \qquad **3.** $\int_0^2 (9x^2 - 4x) \, dx.$

4. $\int_0^3 (4t^2 - 3t + 6) \, dt.$ \qquad **5.** $\int_1^3 3(91z - 14z^2) \, dz.$

6. If you are also given that $\int_2^4 x^2 \, dx = \frac{56}{3}$, can you determine $\int_3^4 x^2 \, dx$?

Sec. 7-8. Fundamental Theorem of Integral Calculus

7. Prove Theorem 1.   **8.** Prove Theorem 2.
9. Prove Theorem 3.

10. Speculate on the possible meaning of $\int_2^1 x \, dx$.

7-8. Fundamental Theorem of Integral Calculus

The area under the continuous curve $y = f(t)$ between $t = a$ and $t = b$ is defined as $A = \int_a^b |f(t)| \, dt$ (Fig. 7-9a). If $f(t)$ is never negative between $t = a$ and $t = b$, then $A = \int_a^b f(t) \, dt$. If $f(t)$ is sometimes negative (Fig. 7-9b), then the word "under" is interpreted to mean the area between the curve and the t axis.

Think of the left-hand boundary $t = a$ as fixed while the right-hand boundary $t = b$ is permitted to move. Then as the right-hand boundary line changes, corresponding areas of A will be determined. In fact, if the right-hand boundary is taken at some value t_1, a corresponding area $A(t_1)$ is determined.

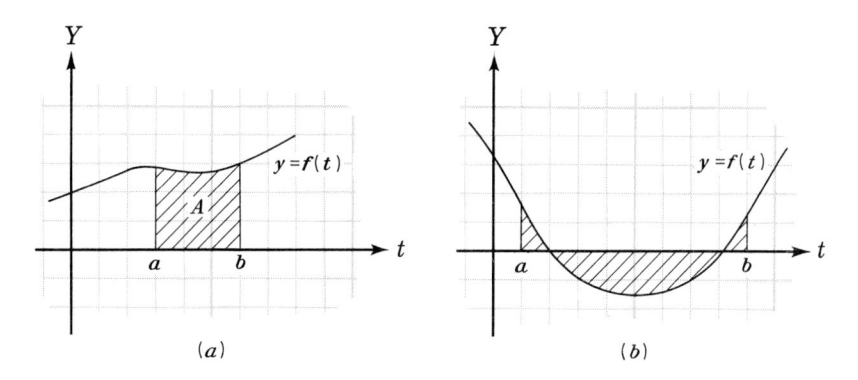

Figure 7-9

Since for each value x a corresponding value $A(x)$ is determined, Area = $A(x)$ is a function of x. (See Fig. 7-10.) (Describe the domain of definition and range of this function.)

Since $A(x)$ is a function of x, it *may* have a derivative with respect to x. It will turn out that computing this derivative, in general, will be one of the most profitable things we have done in this course.

192 *Integral Calculus*

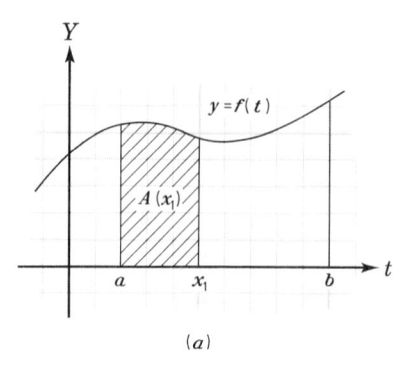

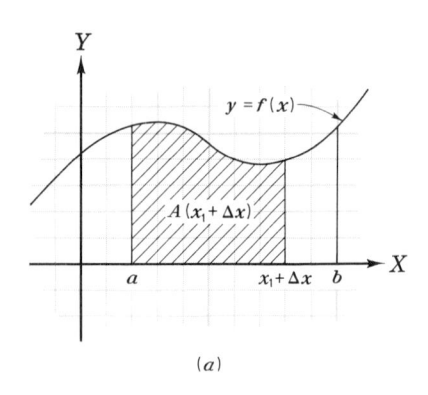

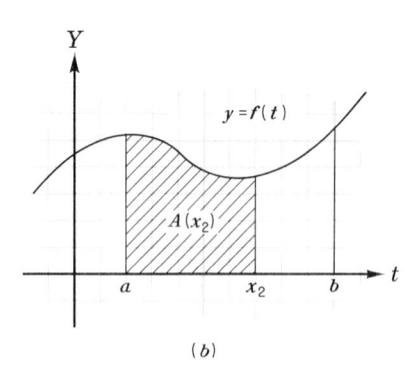

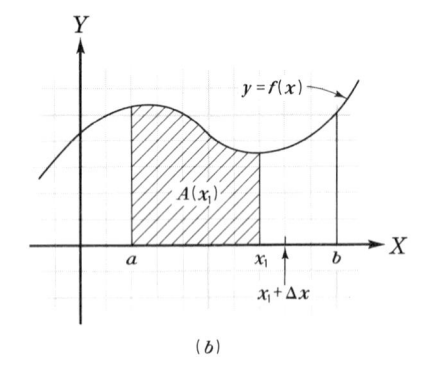

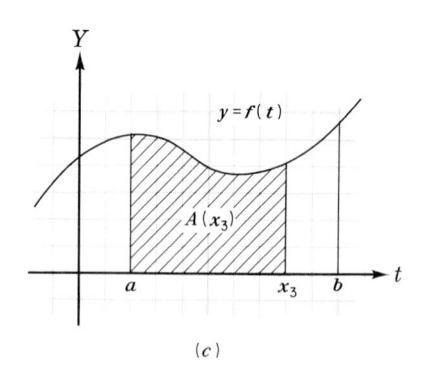

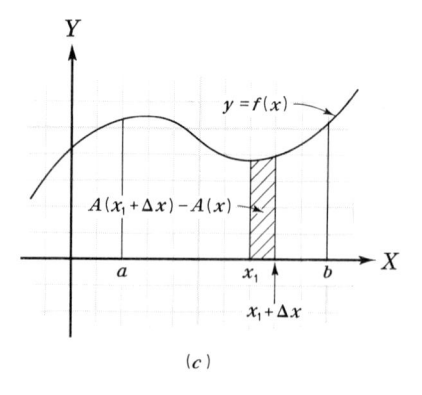

Figure 7-10

Figure 7-11

Sec. 7-8. Fundamental Theorem of Integral Calculus

We return to the delta process to find $\dfrac{dA}{dx}$. We must compute

$$\left.\frac{dA}{dx}\right|_{x=x_1} = \lim_{\Delta x \to 0} \frac{A(x_1 + \Delta x) - A(x_1)}{\Delta x}.$$

The numerator $A(x_1 + \Delta x) - A(x_1)$ represents the area of a small strip between $t = x_1$ and $t = x_1 + \Delta x$, as Fig. 7-11c suggests.

Now, $A(x_1 + \Delta x) - A(x_1)$ is greater than or equal to the area of a rectangle having width Δx and having height equal to the smallest value of $f(t)$ in the interval $x_1 \leq t \leq x_1 + \Delta x$. Call this value $f(\min \xi)$. Furthermore, $A(x_1 + \Delta x) - A(x_1)$ is less than or equal to the area of a rectangle having width Δx and height equal to the largest value of $f(t)$ in the interval $x_1 \leq t \leq x_1 + \Delta x$. Call this value $f(\max \xi)$. Examine Fig. 7-12 on page 194 with care before continuing.

Thus $f(\min \xi)\Delta x \leq A(x_1 + \Delta x) - A(x_1) \leq f(\max \xi)\Delta x$

If $\Delta x > 0$, we may divide by Δx without disturbing the inequalities,† to obtain

$$f(\min \xi) \leq \frac{A(x_1 + \Delta x) - A(x_1)}{\Delta x} \leq f(\max \xi).$$

Taking limits as $\Delta x \to 0$,

$$\lim_{\Delta x \to 0} f(\min \xi) = \lim_{\Delta x \to 0} \frac{A(x_1 + \Delta x) - A(x_1)}{\Delta x} = \lim_{\Delta x \to 0} f(\max \xi).$$

Since $f(\min \xi)$ and $f(\max \xi)$ are each values of $f(t)$ with t on the interval $x_1 \leq t \leq x_1 + \Delta x$, it follows that as $\Delta x \to 0$, they each must approach $f(x_1)$ as a limiting value. (Why? Consider the meaning of the statement "$f(t)$ is continuous at $t = x_1$.") Thus

$$f(x_1) = \lim_{\Delta x \to 0} \frac{A(x_1 + \Delta x) - A(x_1)}{\Delta x} = f(x_1).$$

We conclude:

$$\left.\frac{dA}{dx}\right|_{\text{at } x = x_1} = \lim_{\Delta x \to 0} \frac{A(x_1 + \Delta x) - A(x_1)}{\Delta x} = f(x_1).$$

Recall that, even though the derivative is computed at a point $[x_1, f(x_1)]$, it is usual to drop the subscripts and write:

† If $\Delta x < 0$, the inequalities are reversed, but the same limiting conclusion is reached.

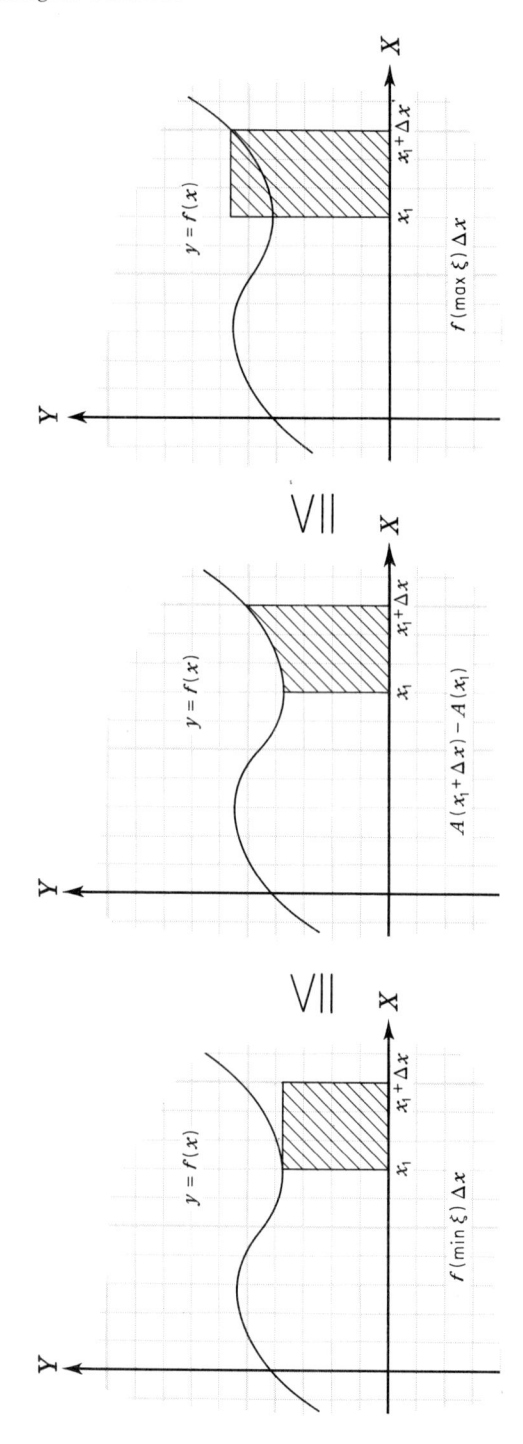

Figure 7-12

Sec. 7-8. Fundamental Theorem of Integral Calculus   195

fundamental theorem of integral calculus

Theorem

If $A(x) = \int_a^x f(t)\, dt$ then $\dfrac{dA(x)}{dx} = f(x)$.

Note that $\int_a^x f(t)\, dt$ is a function of x, not a function of t.

The reader may wish to speculate at this point as to why this theorem is of such great importance. Before entering this discussion, let us consider a few examples of the theorem's meaning.

Example 1

If $A(x) = \int_0^x t^2\, dt$, find $\dfrac{dA}{dx}$.

The theorem states $\dfrac{dA}{dx} = x^2$. $A(x)$ represents the area shaded in Fig. 7-13. We may interpret this as stating that as x changes, the

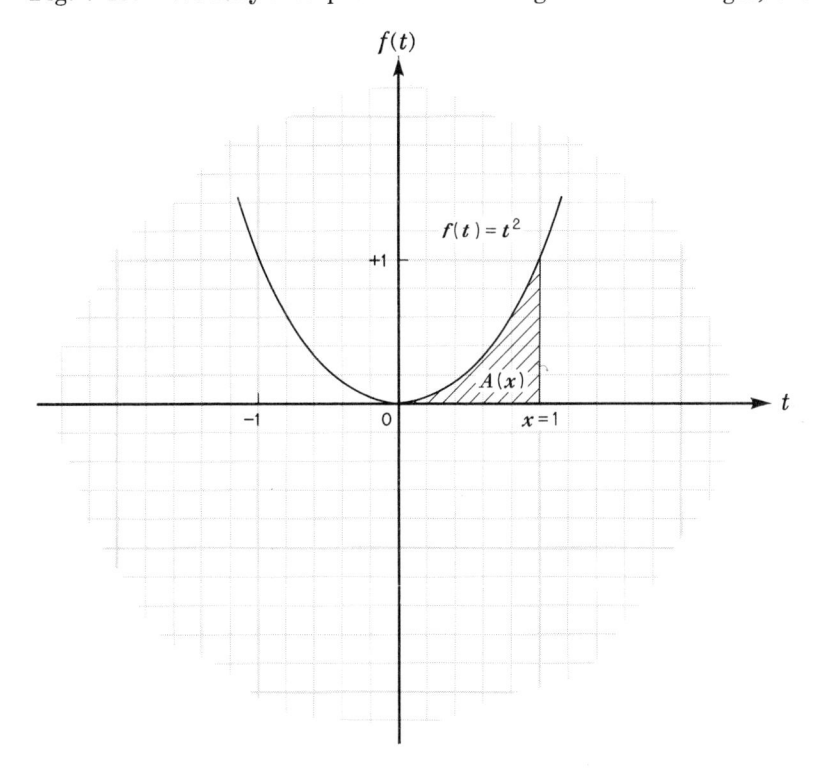

Figure 7-13

Integral Calculus

rate of change of $A(x)$ with respect to the change in x is (x^2); that is, at $x = 2$, $A(x)$ is changing four times as fast (in square units) as x is changing (in linear units).

Example 2

If $A(x) = \int_0^x (t^2 + 3) \, dt$, find $\dfrac{dA(x)}{dx}$.

The theorem states $\dfrac{dA(x)}{dx} = x^2 + 3$.

The student should make sketches and remarks similar to those given after Example 1.

The true importance of the fundamental theorem of integral calculus does not lie in finding derivatives of $A(x)$.

The theorem provides, in many cases, an easy way of *evaluating the integral*, without having to use the "lim Σ" definition. Since, if $A(x) = \int_a^x f(t) \, dt$, $\dfrac{dA}{dx} = f(x)$, it follows that *if* we can find a function whose derivative is $f(x)$, then this function *might* be $A(x)$. It also *might not be*. Problem 35, Set 4-5, states that if $\dfrac{dH(x)}{dx} = 0$, then $H(x) = c$, a constant. This result may be used to prove that if $\dfrac{df(x)}{dx} = \dfrac{dg(x)}{dx}$, then $f(x) = g(x) + c$ for some constant c. Prove it now: Let $H(x) = f(x) - g(x)$ and form $\dfrac{dH}{dx}$.

From this we conclude that,

Theorem

> If a function $I(x)$ can be found such that $dI/dx = f(x)$, then $\int_a^x f(x) \, dx = I(x) + c$. (Why?)

Before showing how to determine the value of the constant c, we give several examples.

Example 3

$\int_1^x 3t^2 \, dt = x^3 + c$, since $\dfrac{dx^3}{dx} = 3x^2 = f(x)$.

Sec. 7-8. Fundamental Theorem of Integral Calculus

Example 4

$$\int_0^x t^4 \, dt = \frac{1}{5} x^5 + c. \qquad \text{(Why?)}$$

Example 5

$$\int_3^x \sqrt{t} \, dt = \int_3^x t^{\frac{1}{2}} \, dt = \frac{2}{3} x^{\frac{3}{2}} + c. \qquad \text{(Why?)}$$

Example 6

$$\int_a^x \frac{dt}{t^3} = \int_a^x t^{-3} \, dt = -\frac{1}{2} x^{-2} + c.$$

In general, since $k \neq -1$, $\dfrac{d\left[\dfrac{x^{k+1}}{k+1}\right]}{dx} = x^k$, it follows from the theorem of this section that

$$\int_a^x t^k \, dt = \frac{x^{k+1}}{k+1} + C \qquad \text{for rational } k \neq -1.$$

The constant C will be determined in the next section.
Using the theorems of Sec. 7-7, that is,

$$\int_a^x [f(t) \pm g(t)] \, dt = \int_a^x f(t) \, dt \pm \int_a^x g(t) \, dt$$

and $\int_a^x k \cdot f(t) \, dt = k \int_a^x f(t) \, dt$

it is possible to integrate more complicated functions.

Example 7

$$I = \int_1^x (t^3 - 7t^{\frac{1}{2}} + 5 - t^{-8}) \, dt = \frac{x^4}{4} + C_1 - \frac{14}{3} x^{\frac{3}{2}} + C_2 + 5x$$
$$+ C_3 + \frac{1}{7} x^{-7} + C_4.$$

Since $C_1 + C_2 + C_3 + C_4$ is still a constant, we write C to represent their sum, giving

$$I = \int_1^x (t^3 - 7t^{\frac{1}{2}} + 5 - t^{-8}) \, dt = \frac{x^4}{4} - \frac{14}{3} x^{\frac{3}{2}} + 5x + \frac{1}{7x^{-7}} + C.$$

Example 8

$$\int_2^x (z^2 - 4z + 3\sqrt{z})\, dz = \frac{x^3}{3} - 2x^2 + 2x^{\frac{3}{2}} + C.$$

Before continuing, we define $\int_b^a f(t)\, dt = -\int_a^b f(t)\, dt$, for $a < b$, and $\int_a^a f(t)\, dt = 0$. Note that these are matters of definition, and that the definitions are consistent with the theorems of Sec. 7-7.

Problem Set 7-8

1. Make sketches and remarks similar to those given in Example 1 for each of Examples 2 to 6.

2. $\int_1^x (t^4 - 2t)\, dt.$

3. $\int_0^x (t - \sqrt{t})\, dt.$

4. $\int_4^x (t - \sqrt{t})\, dt.$ Use shaded sketches to show the difference between the integrals of Probs. 3 and 4. There is a difference, an important one.

5. $\int_1^x (t^{-3} + \sqrt{t} - 5)\, dt.$

6. $\int_2^x (z^2 - 3z + 4)\, dz.$

7. $\int_1^x (z^2 + z)\, dz.$

8. $\int_3^x (4z - 2)\, dz.$

9. $\int_2^x (w^2 - 5w)\, dw.$

10. $\int_2^x (x^2 - 5x)\, dx.$

11. Draw sketches showing the areas under consideration in each of Probs. 1 to 5.

12. Draw sketches showing the areas under consideration in each of Probs. 6 to 10.

13. Explain why the definitions given just before this problem set are reasonable definitions.

14. Prove the theorem: If $\dfrac{df(x)}{dx} = \dfrac{dg(x)}{dx}$, then $f(x) = g(x) + C$ for some constant C. Discuss any necessary restrictions on f and g. Also state and prove the converse of this theorem.

7-9. Integration Continued

It is not, in general, difficult to determine the value of the constant C in the formula

Sec. 7-9. Integration Continued

$$\int_a^x f(t)\, dt = I(x) + C \quad \text{where} \quad \frac{dI(x)}{dx} = f(x).$$

Since the variable of integration is immaterial, it is common practice (although admittedly open to some confusion) to write x in place of t under the integral sign,

$$A(x) = \int_a^x f(x)\, dx = I(x) + C \quad \text{where}$$

$$\frac{dI(x)}{dx} = f(x) = A'(x).$$

In this notation, the x in the upper limit and in $I(x)$ are not the same as the x in $f(x)\,dx$. Which type of x is the x of $A(x)$?

We next determine the constant C in

$$A(x) = \int_a^x A'(x)\, dx = \int_a^x f(x)\, dx = I(x) + C,$$

where $\dfrac{dI(x)}{dx} = f(x)$.

The value of C is determined by observing that $A(x)$ is the area between the ordinates $x = a$ and $x = x$.

When $x = a$, $A(a) = 0$. (Why?)

Consequently,

$$A(a) = 0 = I(a) + C \quad \text{and} \quad C = -I(a).$$

Thus, in general

$$A(x) = I(x) - I(a),$$

where $I(x)$ is a function such that $\dfrac{dI(x)}{dx} = f(x)$.

The area $A = A(b)$, under $y = f(x)$ between $x = a$ and $x = b$, is given by $A = A(b) = I(b) - I(a)$, where $I'(x) = f(x)$. We define the symbol $I(t)\Big|_a^b$ as

$$\int_a^b f(x)\, dx = \int_a^b f(t)\, dt = I(t)\Big|_a^b = I(b) - I(a).$$

The area under the *nonnegative* function $y = f(t) \geq 0$ between

$t = a$ and $t = b$, $a < b$, is given by $A = \int_a^b f(t)\,dt = I(b) - I(a)$.
Why is $f(t)$ restricted to be a nonnegative function?

Since the t does not appear in the result, it is once again appropriate to point out that the variable of integration is merely a "dummy variable," since

$$\int_a^b f(t)\,dt = \int_a^b f(x)\,dx = \int_a^b f(z)\,dz = \int_a^b f(\phi)\,d\phi = \int_a^b f(y)\,dy.$$

If $f(t)$ is always negative in the interval $a \leq t \leq b$, then in the sum, $\lim_{n \to \infty} \sum_{i=1}^{n} f(\xi_i)\,\Delta t_i = \int_a^b f(t)\,dt$, is negative, since each $f(\xi_i)$ is negative. The area is defined to be the integral of the absolute value of $f(t)$; that is, $A(x) = \int_a^b |f(t)|\,dt,\ a \geq b$.

If $f(t)$ assumes both positive and negative values in the interval $a \leq t \leq b$, then some of the rectangular areas will be added while others will be subtracted. Hence, in finding areas, care must be taken to separate the region above and below the axis. This is because we really seek not $\int_a^b f(t)\,dt$, but rather $\int_a^b |f(t)|\,dt$ which is *not*, in general, *the same as* $\left|\int_a^b f(t)\,dt\right|$. (Why not? Use sketches.) In "work" problems and many other applications in which we actually seek $\lim_{n \to \infty} \sum_{i=1}^{n} f(\xi_i)\,\Delta x$ rather than $\lim_{n \to \infty} \sum_{i=1}^{n} |f(\xi_i)|\,\Delta x$, no such precaution need be taken. The integral automatically compensates for "sinks and sources" or "work done by and work being done on" corresponding to negative and positive values of $f(t)$. In each case, *the fundamental theorem is still valid.*

It is common practice to omit the limits of integration a and x and write

$$\int x\,dx = \frac{x^2}{2} + C$$

or $\int (5x^2 + 7)\,dx = \left(\frac{5}{3}\right)x^3 + 7x + C.$

indefinite integral This integral is called an *indefinite integral*, since no limits of integration are specified. The constant C is used to show this.

Sec. 7-9. Integration Continued

Example 1

Determine the area under $y = x^3 - 4x$ between $x = 3$ and $x = 5$.

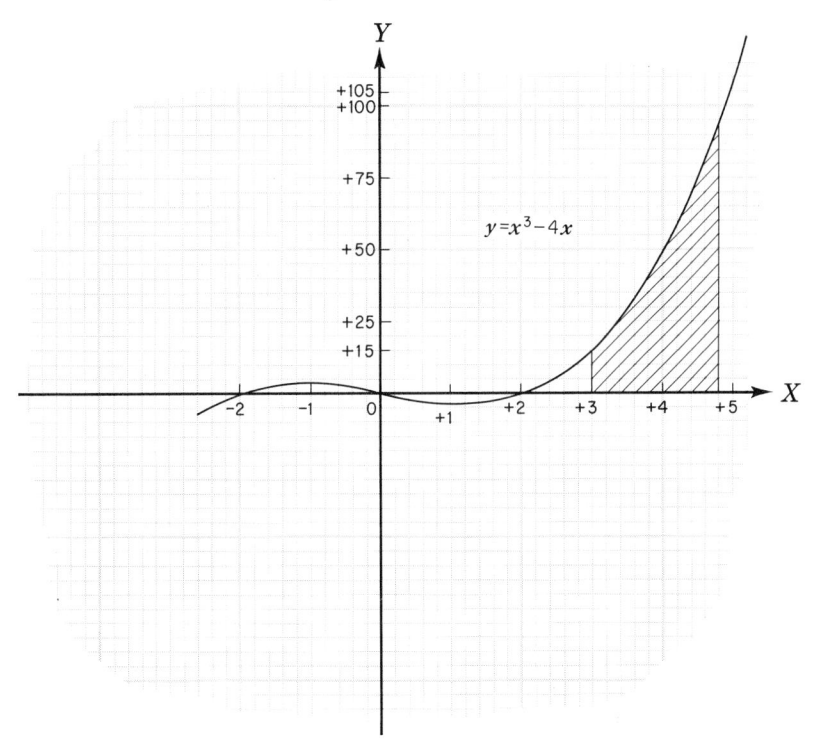

Figure 7-14

A sketch of the locus (Fig. 7-14) shows that $f(x) = x^3 - 4x$ does not change sign in the interval $3 \leq x \leq 5$. Then, since

$$\frac{d\left[\dfrac{x^4}{4} - 2x^2\right]}{dx} = x^3 - 4x,$$

$$A = \int_3^5 (x^3 - 4x)\, dx = \left.\frac{x^4}{4} - 2x^2\right|_3^5 = \left(\frac{625}{4} - 50\right)$$
$$- \left(\frac{81}{4} - 18\right) = \frac{544}{4} - 32 = 104 \text{ square units.}$$

Example 2

Find the area bounded by $y = -x^2 - 2x + 3$ and the x axis. The locus has x intercepts -3 and $+1$ (Fig. 7-15).

202 Integral Calculus

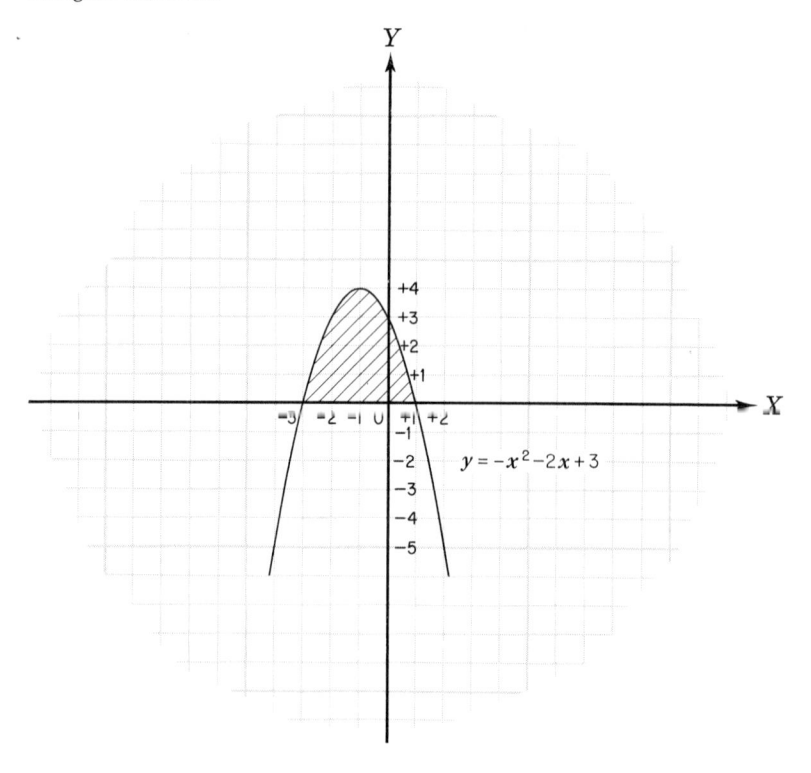

Figure 7-15

$f(x) = -x^2 - 2x + 3$ is continuous and nonnegative in the interval $-3 \leq x \leq 1$.

$$A = \int_{-3}^{1} (-x^2 - 2x + 3)\, dx = \left. \frac{x^3}{3} - x^2 + 3x \right|_{-3}^{1}$$

$$= \left(-\frac{1^3}{3} - 1^2 + 3 \right) - \left[-\frac{(-3)^3}{3} - (-3)^2 + 3(-3) \right]$$

$$= \left(\frac{5}{3} \right) - (-9) = \frac{35}{3} \text{ square units.}$$

Example 3

Determine the area bounded by the x axis and $y = x^3 - 4x$ between the ordinates $x = -1$ and $x = 3$ (Fig. 7-16). This is the function considered in Example 1, but the interval is different.

Since $f(x) = x^3 - 4x$ changes sign at $x = 0$ and at $x = 2$, within the interval $-1 \leq x \leq 3$, the area will be found by adding the

Sec. 7-9. Integration Continued 203

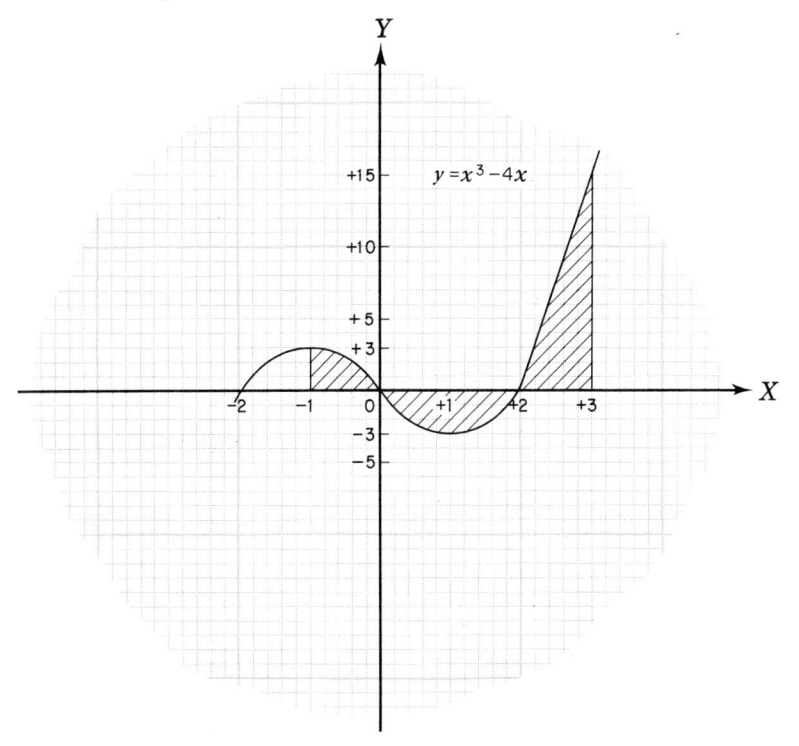

Figure 7-16

absolute values of the integrals in the three intervals:

$$-1 \leq x \leq 0 \qquad 0 \leq x \leq 2 \qquad 2 \leq x \leq 3,$$

in each of which the sign of $f(x)$ does not change.

$$A = \int_{-1}^{0} (x^3 - 4x)\, dx + \left| \int_{0}^{2} (x^3 - 4x)\, dx \right| + \int_{2}^{3} (x^3 - 4x)\, dx$$

$$= \left(\frac{x^4}{4} - 2x^2 \right)\Big|_{-1}^{0} + \left| \left(\frac{x^4}{4} - 2x^2 \right)\Big|_{0}^{2} \right| + \left(\frac{x^4}{4} - 2x^2 \right)\Big|_{2}^{3}$$

$$= 0 - \left[\frac{(-1)^4}{4} - 2(-1)^2 \right] + \left| \left[\frac{2^4}{4} - 2(2)^2 \right] - 0 \right|$$

$$\qquad + \left[\frac{3^4}{4} - 2(3)^2 \right] - \left[\frac{2^4}{4} - 2(2)^2 \right]$$

$$= -\left(\frac{1}{4} - 2 \right) + |(4 - 8)| + \left(\frac{81}{4} - 18 \right) - (4 - 8)$$

$$= -\left(-\frac{7}{4} \right) + |(-4)| + \left(\frac{9}{4} \right) - (-4) = \frac{7}{4} + 4 + \frac{9}{4} + 4$$

$$= 12 \text{ square units.}$$

Note that this area is *not* equal to $\int_{-1}^{3} (x^3 - 4x)\, dx$, which has the value of 4.

Problem Set 7-9

In Probs. 1 to 10, determine, by integration, the area bounded by the following curves and the x axis between the ordinates indicated. In each case, sketch the curve and shade the area found.

1. $y = 5x$ between $x = 0$ and $x = 4$. Show that the integral gives the same value for the area of this triangle as may be found by elementary geometry.
2. $y = 5x$ between $x = 3$ and $x = 5$. Compare the result obtained upon integration with that obtained by geometry.
3. $y = 5x$ between $x = 2$ and $x = 7$.
4. $y = 5x + 4$ between $x = 0$ and $x = 4$. Also compute the area by geometry. Compare with Prob. 1.
5. $y = 5x + 7$ between $x = -2$ and $x = 3$. Also find this area by geometry.
6. $y = 3x^2$ between $x = 0$ and $x = 2$. This area cannot be found by the methods usually considered in high school plane geometry.
7. $y = 3x^2$ between $x = 1$ and $x = 3$.
8. $y = x^3 + 8$ between $x = -1$ and $x = 4$. HINT: To aid in the sketch of $y = x^3 + 8$, consider its x and y intercepts, and also values of x for which the tangent is horizontal.
9. $y = x^2 - 3x$ between $x = -1$ and $x = 4$. HINT: See Example 8.
10. $y = 5x - x^2$ between $x = -2$ and $x = 3$.
11. Determine the area bounded by $y = -x^2 + 4x$ and the x axis.
12. Determine the area bounded by $y = x^2 - 4x$ and the x axis. Compare this problem, both as to sketch and as to result, with Prob. 11.
13. Find the area bounded by the x axis and the function $y = f(x)$ between $x = 0$ and $x = 3$ if

$$f(x) = \begin{cases} x^2 & \text{for } x \leq 2 \\ -x + 6 & \text{for } x > 2. \end{cases}$$

HINT: Sketch the locus and consider separate integrals over the intervals $0 \leq x \leq 2$ and $2 \leq x \leq 3$.

14. Find the area bounded by the x axis and $y = f(x)$ between $x = 1$ and $x = 5$, if $f(x)$ is the function defined in Prob. 13.

Sec. 7-9. Integration Continued

15. Determine the area bounded by the x axis and $y = g(x)$ between $x = 2$ and $x = 5$ if

$$g(x) = \begin{cases} 2x + 3 & \text{if } x \leq 3. \\ -x + 12 & \text{if } x > 3. \end{cases}$$

Also determine this area by geometry.

16. The same as Prob. 15 between the ordinates $x = -2$ and $x = 7$.

17. In computing the moment of inertia of a circular disk (radius a) with respect to its center (a wheel with respect to its axle) in rectangular cartesian coordinates, we must compute the integral of an integral

$$I_0 = \int_{-a}^{a} \left[\int_{-\sqrt{a^2-x^2}}^{\sqrt{a^2-x^2}} (x^2 + y^2) \, dy \right] dx$$

$$= \int_{-a}^{a} \left[2x^2 \sqrt{a^2 - x^2} + \frac{2}{3} (a^2 - x^2)^{\frac{3}{2}} \right] dx.$$

This is, at best, a difficult integration to perform and is beyond our present knowledge. To compute this moment of inertia, *using polar coordinates*, we evaluate the integral $I_0 = \int_0^a 2\pi r^3 \, dr$. This we may do, even with our present slight knowledge of integration. This is one illustration of the importance of polar coordinates. Evaluate the latter integral to obtain I_0.

Integrate:

18. $\int (x^2 - 4x + 3) \, dx.$ **19.** $\int (3x^5 - \sqrt{x}) \, dx.$
20. $\int (\sqrt{x^3} - x^{-5} + 4) \, dx.$

Explain why the following statements are meaningless:

21. $\int_0^2 (t^3 - t^{-2}) \, dt.$ **22.** $\int_{-1}^{7} (x + \sqrt{x}) \, dx.$ **23.** $\int_1^4 \frac{x \, dx}{x - 2}.$

24. We have, at present, no formula which will evaluate $\int_1^x \frac{dt}{t}$, but it is quite possible for you to make sketches which show that the integral is meaningful. Do so.

25. Which is larger $\int_1^3 \frac{dx}{x}$ or $\int_{11}^{13} \frac{dx}{x}$? Explain your conclusion, using the graph of $y = 1/x$.

26. If $I(x)$ is a function having a formula which is known but is too long to write down here, and if its domain of definition is the real numbers in the interval $-1 \leq x < 100$ while its range is the set of real numbers, and if $\int f(x)\, dx = I(x) + C$, find a symbolic representation for each of the following which is meaningful. State why some are without meaning.

(a) $\int_3^7 f(x)\, dx.$ (b) $\int_{-1}^1 f(x)\, dx.$ (c) $\int_{10}^{99} f(z)\, dz.$

(d) $\int_{50}^{100} f(x)\, dx.$ (e) $\int_4^{17} f(z)\, dz.$ (f) $\int_{-4}^7 f(t)\, dt.$

(g) $\int_{30}^{300} f(x)\, dx.$

27. (a) Give an example of an integral which has meaning but which you cannot, at present, evaluate.
(b) Find a pair of numbers A, B, such that the value of the integral of part (a) lies between A and B.

28. Give an example of three functions $f_1(x)$, $f_2(x)$, and $f_3(x)$ such that

(a) $\int_1^3 f_1(x)\, dx$ is negative.   (b) $\int_1^3 f_2(x)\, dx$ is positive.

(c) $\int_1^3 f(x)\, dx$ is zero but $f(1) \neq f(2)$.

29. Give an example of a function $g(x)$ such that $\int_1^3 g(x)\, dx$ has meaning but $\int_1^5 g(x)\, dx$ does not, at present, have meaning.

30. (a) Give an example of a function $q(x)$ such that $\int_1^2 q(x)\, dx$ is meaningless.
(b) Explain why $q(x) = 3x^{-1}$ does *not* satisfy the conditions of part (a), even though we cannot evaluate $\int_1^2 3x^{-1}\, dx$ by formula at present.

7-10. Extensions of the Definite Integral

The definite integral $\int_a^b f(x)\, dx$ has been defined for $f(x)$, a single-valued function which is continuous on the (closed) interval *Riemann* $a \leq x \leq b$. This integral is known as the *Riemann definite integral*, *definite* *integral* after the mathematician Riemann (German, 1826–1866). It is possible to make many extensions of the Riemann integral which

Sec. 7-10. Extensions of the Definite Integral

are not only useful but essential in today's science. It will not be possible for us to study these extensions in this short course, but several will be mentioned as part of your "general mathematical education."

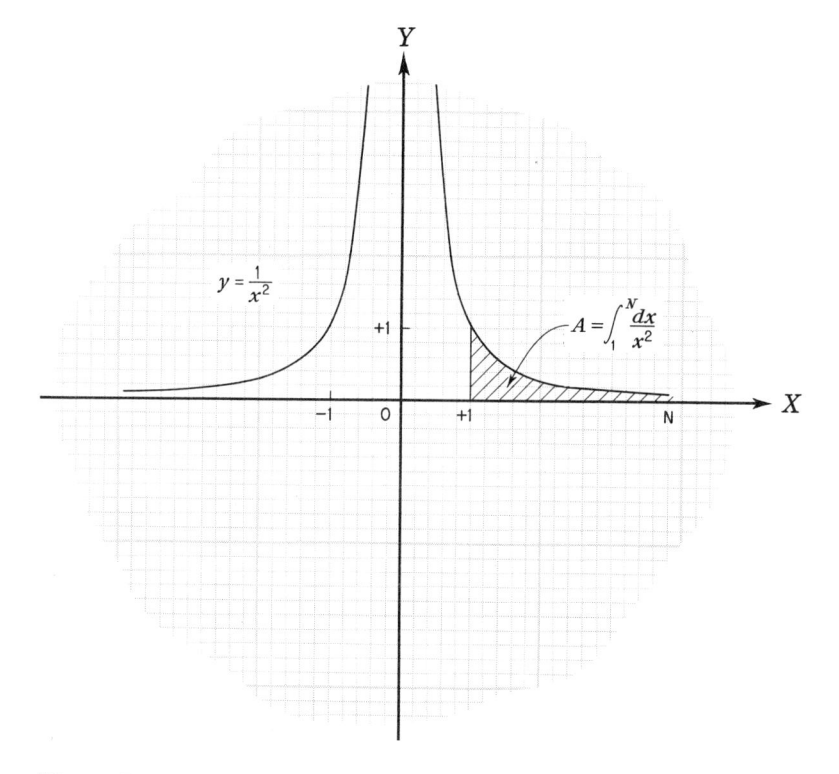

Figure 7-17

The curve $y = 1/x^2$ is shown in Fig. 7-17 with the area represented by $\int_1^N \frac{dx}{x^2}$ shaded. For each value of $N > 1$, we obtain

$$A(N) = \int_1^N \frac{dx}{x^2} = \int_1^N x^{-2}\, dx = -x^{-1}\Big|_1^N = -\frac{1}{N} + 1$$

$$= 1 - \frac{1}{N}.$$

The question now arises, "What happens as the end point N is taken farther and farther to the right, that is, as N increases without bound?"

Integral Calculus

From the sketch of $y = 1/x^2$, it is apparent that, for each value of N, a specific area $A(N) = (1 - 1/N)$ is determined. It is not at all geometrically obvious, however, that as N increases, the area $A(N)$ becomes arbitrarily close to 1 square unit. However,

$$\lim_{N \to \infty} A(N) = \lim_{N \to \infty} \left(1 - \frac{1}{N}\right) = 1,$$

and we conclude that for any predetermined degree of tolerance ($\epsilon > 0$), it is possible to find a value K such that if $N > K$, then $A(N)$ is nearer 1 square unit than was required (that is, if $N > K$, $|A(N) - 1| < \epsilon$).

It is usual to extend our definition of area again to include this case. We write

$$\int_a^\infty f(x)\, dx = \lim_{N \to \infty} \int_a^N f(x)\, dx$$

if the latter limit exists. The area under the curve $y = 1/x^2$ for $x > 1$ is a *finite* area (namely, 1) even though the graph of $y = 1/x^2$, $x > 1$ "goes on forever." This is quite similar to the reasoning used in geometric sequences (progressions) such as

$$1 + \frac{1}{2} + \frac{1}{4} + \frac{1}{8} + \frac{1}{16} + \frac{1}{32} + \cdots + \frac{1}{2^n} + \cdots$$

$$= \lim_{N \to 0} \left[\frac{1 - (\frac{1}{2})^N}{1 - \frac{1}{2}}\right] = 2.$$

The left member contains infinitely many individual terms but is said to actually equal 2 (not *approximately* equal), since for any predetermined degree of tolerance ($\epsilon > 0$) there is a number K such that the sum of the first N terms, for any number $N > K$, is nearer 2 than was required.

(That is, if $N > K$, $|1 + \frac{1}{2} + \frac{1}{4} + \cdots + \frac{1}{2^{N-1}} - 2| < \epsilon$.)

The symbol $\int_0^3 \frac{dx}{\sqrt{x}}$ is without meaning since $y = 1/\sqrt{x}$ is not continuous at $x = 0$. (See Fig. 7-18.)

We define

$$\int_0^3 \frac{dx}{\sqrt{x}} = \lim_{\beta \to 0^+} \int_{0+\beta}^3 \frac{dx}{\sqrt{x}},$$

Sec. 7-10. Extensions of the Definite Integral

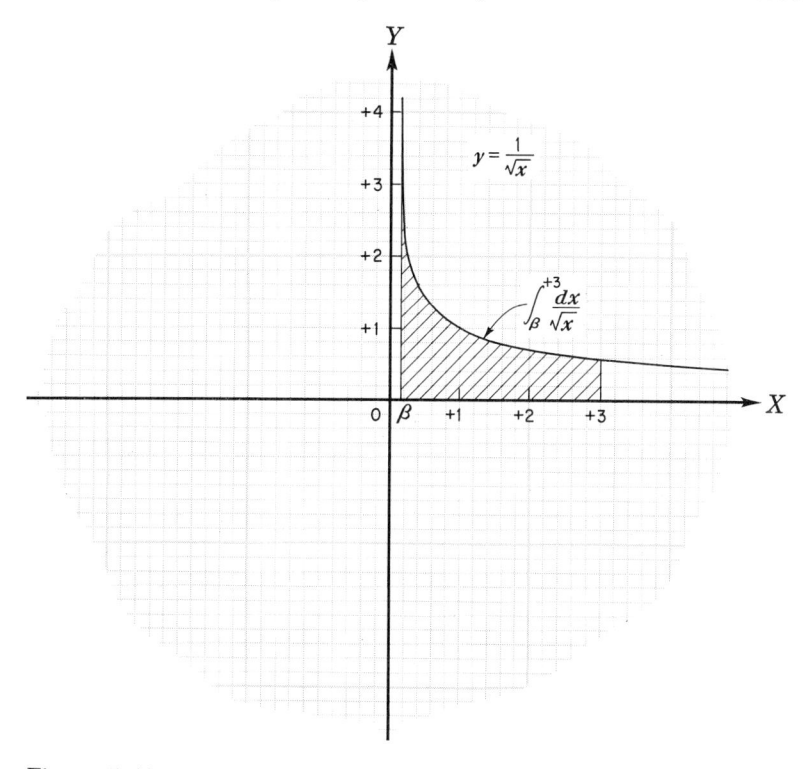

Figure 7-18

if the latter limit exists. The symbol $\beta \to 0^+$ implies $\beta > 0$ as $\beta \to 0$. In this fashion it is possible again to extend our definition of integral and of area.

$$\int_0^3 \frac{dx}{\sqrt{x}} = \lim_{\beta \to 0^+} \int_{0+\beta}^3 x^{-\frac{1}{2}} dx = \lim_{\beta \to 0^+} 2x^{\frac{1}{2}} \Big|_\beta^3$$
$$= \lim_{\beta \to 0^+} (2\sqrt{3} - 2\sqrt{\beta}) = 2\sqrt{3}.$$

The reader may feel that the curves $y = 1/x^2$ (Fig. 7-17) and $y = 1/\sqrt{x}$ (Fig. 7-18) look much alike for $x > 0$. However,

$$\int_1^\infty \frac{dx}{x^2} = 1 \qquad \text{while} \qquad \int_1^\infty \frac{dx}{\sqrt{x}} = \lim_{N \to \infty} (2\sqrt{N} - 2)$$

which *does not exist*. Also

$$\int_0^3 \frac{dx}{x^2} = \lim_{\beta \to 0^+} \int_{0+\beta}^3 \frac{dx}{x^2} = \lim_{\beta \to 0^+} \left(\frac{1}{3} - \frac{1}{\beta}\right)$$

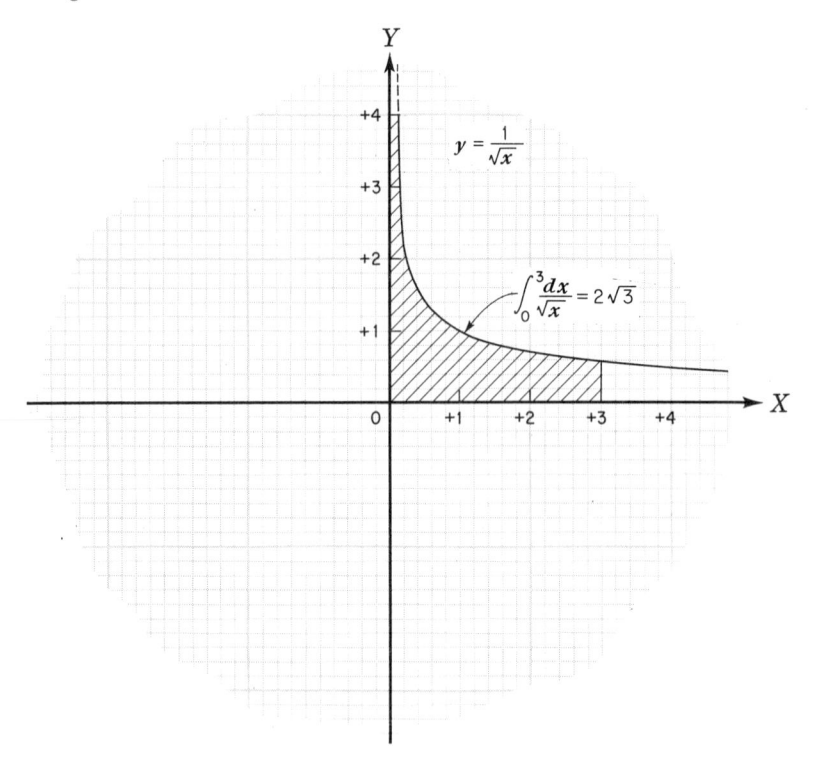

Figure 7-19

which *does not exist while* $\int_0^3 \dfrac{dx}{\sqrt{x}} = 2\sqrt{3}$. Thus the two curves have some striking dissimilarities.

It is also possible to extend the definition of $\int_a^b f(x)\, dx$ to cover certain cases in which points of discontinuity occur *inside* the range of the integration limits, say at c, where $a < c < b$. Then we define

$$\int_a^b f(x)\, dx = \lim_{\beta \to 0^+} \int_a^{c-\beta} f(x)\, dx + \lim_{\delta \to 0^+} \int_{c+\delta}^b f(x)\, dx$$

providing *each* of the limits exists. This makes $\int_{-1}^2 \dfrac{dx}{\sqrt{|x|}}$ meaningful but fails to give meaning to $\int_{-1}^2 \dfrac{dx}{\sqrt{x}}$ or to $\int_{-1}^2 \dfrac{dx}{x^2}$. (Why?) Each of the latter integrals remains without meaning.

In many problems in probability and statistics it is necessary to

Sec. 7-11. Techniques of Integration   211

define an integral for step functions such as $\int_1^{10} f(x)\,dx$, where $f(x)$ is the function given in Fig. 7-20.

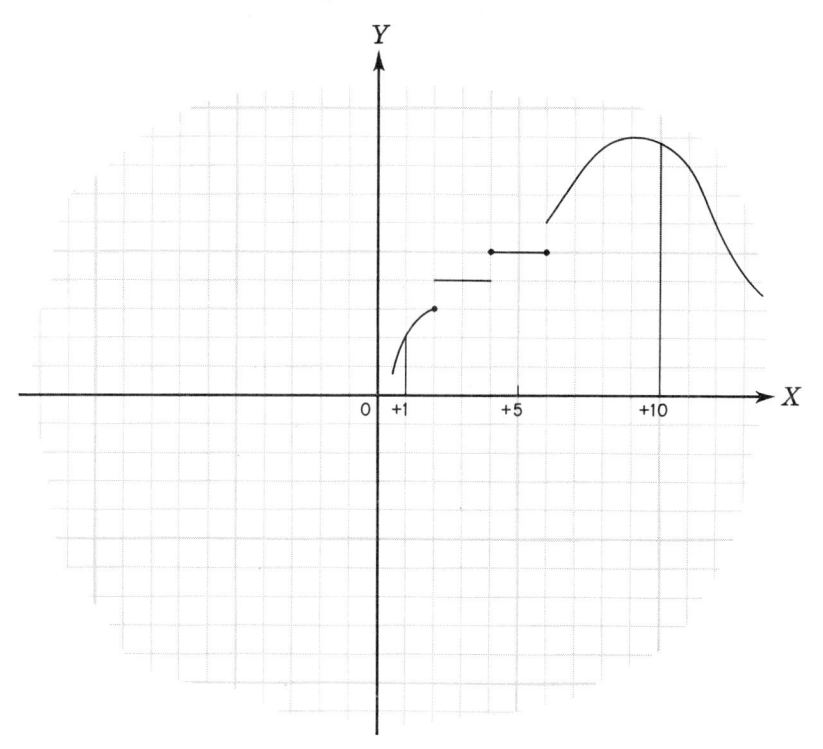

Figure 7-20

There are many other extensions of the Riemann integral, including those named for Stieltjes, Lesbegue, and Darboux. An interested student may learn about these in more advanced texts. The student is cautioned that there exist many integrals $\int_a^b f(x)\,dx$ which *cannot* be integrated in closed form [that is, no formula is known for a function $I(x)$ such that $I'(x) = f(x)$]. An important example is the so-called "elliptic integrals" which will be found tabulated in handbooks for various values of a and b.

7-11. Techniques of Integration

Before continuing with applications of integration, we shall gain a bit more proficiency in the techniques of integration. At this point

Integral Calculus

we are able to evaluate

$$\int_a^b kx^m \, dx = \frac{kx^{m+1}}{m+1}\Big|_a^b = \frac{k}{m+1}(b^{m+1} - a^{m+1})$$

for m a rational number *other than* -1 and suitable limits for a and b.

We are *unable* to integrate $\int_3^7 \frac{dx}{x}$ at this point although we know that the integral exists. (Why?) In Chap. 9 this important integral will be discussed.

Problem Set 7-11

1. $\int_1^3 x^3 \, dx$.   2. $\int_2^5 3y^2 \, dy$.   3. $\int_4^7 12t^3 \, dt$.

4. $\int_{-1}^1 8x^5 \, dx$. Does the result surprise you? Why?   5. $\int_{-1}^1 8x^6 \, dx$. Contrast with Prob. 4.

6. $\int_{-1}^1 8|x^5| \, dx$. Compare with Probs. 4 and 5.   7. $\int_0^4 (1 - 3x)^2 \, dx$.

8. $\int_0^{16} \sqrt{x} \, dx$.   9. $\int_1^4 (x^{-2} + 3x + \sqrt{x^3}) \, dx$.

10. $\int_1^8 (x^{-3} + 2x + \sqrt[3]{x}) \, dx$.

11. $\int_0^1 \sqrt{3x^2 + x} \, (6x + 1) \, dx$.   HINT: If $u = 3x^2 + x$, then $\frac{du}{dx} = 6x + 1$. The given integral is thus of the form $\int_{u=0}^{u=4} \left(u^{\frac{1}{2}} \frac{du}{dx} \right) dx$ and $I(u) = \frac{2}{3} u^{\frac{3}{2}}$ has the property that $\frac{dI}{dx} = u^{\frac{1}{2}} \frac{du}{dx}$. Thus, $\int_{u=0}^{u=4} \left(u^{\frac{1}{2}} \frac{du}{dx} \right) dx = \frac{2}{3} u^{\frac{3}{2}} \Big|_{u=0}^{u=4} = \cdots = \frac{16}{3}$.

12. $\int_0^1 \sqrt{3x+1} \, (3 \, dx)$.   HINT: If $u = 3x+1$, $\frac{du}{dx} = 3$ and the given integral is of the form $\int_{u=1}^{u=4} \left(u^{\frac{1}{2}} \frac{du}{dx} \right) dx$. Note changes of limits. Why?

13. $\int_1^{10} \sqrt{5x - 1} \, dx$.   HINT: Let $u = 5x - 1$.

$$\int_1^{10} \sqrt{5x - 1} \, dx = \tfrac{1}{5} \int_1^{10} \sqrt{5x - 1} \, (5 \, dx) = \tfrac{1}{5} \int_{x=1}^{x=10} u^{\frac{1}{2}} \frac{du}{dx} \, dx = \,?$$

Sec. 7-12. Self-test   213

14. $\int_1^5 (3x^2 + x)^{\frac{1}{2}}(x + 1)\, dx.$

15. $\int_1^4 (x + 2\sqrt{x})^{10}(1 + 1/\sqrt{x})\, dx.$

16. Evaluate the integrals of Probs. 1 to 15 as indefinite integrals (without limits).

7-12. Self-test

1. Evaluate $\sum_{x=3}^{10} (x^2 - 1).$

2. Give the exact meaning of the statement $\lim_{x \to \infty} f(x) = 27\pi$. Your statement should contain the phrase " . . . for each $\epsilon > 0$ there exists . . . ."

3. Obtain an upper and a lower bound for the area bounded by $y = x^3 + 1$, the x axis, and the lines $x = 1$ and $x = 4$. Use the rectangles with sides $x = 1$, $x = 2$, $x = 3$, and $x = 4$ to obtain your bounds.

4. Evaluate the $\lim_{n \to \infty} \sum_{i=1}^n (\xi^3 + 1)\dfrac{3}{n}$ involved in the definition of the area of Prob. 3. Check by using the fundamental theorem of Secs. 7-8 and 7-9.

5. Integrate, if possible.

(a) $\int_1^3 (x^3 - 4x + 3)\, dx.$

(b) $\int_1^4 (\sqrt{x} + x)\, dx.$

(c) $\int_1^{27} 4x^{\frac{1}{3}}\, dx.$

6. Which of the following integrals exist even though you may not be able to evaluate them, and which are meaningless? Justify your answers.

(a) $\int_1^3 \dfrac{dx}{x + 2}.$ (b) $\int_1^3 \dfrac{dx}{\sqrt{x - 2}}.$ (c) $\int_1^3 \dfrac{dx}{(x - 2)^3}.$

7. Evaluate $\int_2^{10} \sqrt{2x + 5}\, dx.$

8. Evaluate $\int_2^2 \sqrt{(4x^2 + 5)}\, (8x\, dx).$

8

Applications of the Integral

This is the chapter in which you will learn to think in terms of calculus. Study it well.

8-1. Motion

Since integration and differentiation are, in a sense, inverse operations,† it seems possible that integration can be employed to derive the functions used in the falling body problems of Sec. 6-3. This is indeed the case.

In rectilinear (straight-line) motion, if the distance s of a particle from the origin is expressed as a function of the time t, then the velocity is $v = \dfrac{ds}{dt} = s'(t)$, and the acceleration is $a = \dfrac{dv}{dt} = s''(t)$. Starting with either the acceleration function or the velocity function and certain observed facts, we may derive the distance function using integration. In elementary physics the well-known formula $s = -\tfrac{1}{2}gt^2 + v_0 t + s_0$ is used to obtain the distance above ground of a body falling under gravity if the velocity at time $t = 0$ is v_0, the distance above ground at time $t = 0$ is s_0, and the upward direction is taken as positive. *This formula is obtained by integration.*

† By now the reader is assumed to be sufficiently sophisticated to both ask and answer questions such as "What are inverse operations?" and "Why is the clause 'in a sense' inserted?" Not to ask and at least attempt to answer such questions suggests that a drastic change in study habits is needed.

Example

A ball is thrown *upward* from the top of a tower 400 ft high with an initial velocity of 120 ft/sec. Assuming that the acceleration due to gravity is 32 ft/sec/sec *downward* (a fact determined by experiment), determine the velocity with which the ball strikes the ground.

Let us take the upward direction as positive, the time of throwing as $t = 0$, and the origin at the earth's surface. Then

$$a(t) = -32 \text{ ft/sec/sec.}$$
$$v(t) = \int \left(\frac{dv}{dt}\right) dt = \int -32 \, dt = 32t + C.$$

The problem states $v(0) = 120$ ft/sec. Consequently, $120 = 0 + C$, and

$$v(t) = -32t + 120.$$

Now $h(t) = \int \left(\frac{dh}{dt}\right) dt = \int v(t) \, dt = \int (-32t + 120) \, dt$
$$= -16t^2 + 120t + k.$$

The domain of definition of each function is $0 \leq t \leq$ (value of t when the ball strikes the ground).

The problem states $h(0) = 400$. Hence,

$$400 = 0 + 0 + k \quad \text{and} \quad h(t) = -16t^2 + 120t + 400.$$

From here on, the problem is the same as Example 3, Sec. 6-3.

Problem Set 8-1

In all problems involving gravity, assume $g = 32$ ft/sec/sec directed downward.

1. A stone is thrown upward with an initial velocity of 32 ft/sec from the top of a building 560 ft tall. Starting with the assumption that the acceleration of the stone is -32 ft/sec/sec, derive the equation for the height $h(t)$ of the stone from the ground at any time t sec after it is thrown until it strikes the ground. From this, deduce the impact velocity with which the stone strikes the ground.

2. (a) Find the velocity of the stone of Prob. 1 as it passes a window 320 ft above the ground level.

(b) How high does the stone of Prob. 1 ascend?

3. A pellet is projected upward from ground level with an initial velocity of 96 ft/sec. Using the methods of this section, find an equation giving the distance of the pellet above the earth at any time t sec after it is projected. When will the pellet reach its highest point? How high will it go? When and with what velocity will it strike the ground?

4. If the pellet of Prob. 3 were projected from a point 80 ft above ground level with an initial velocity of 64 ft/sec, answer the same questions.

5. A ball is dropped from rest from a point 45 ft above the ground. Simultaneously, a second ball is thrown upward from a spot on the ground directly below the first ball. If the initial velocity of the second ball is 30 ft/sec, determine whether or not the two balls will meet while they are still in the air. If they do meet, find the speed, direction, and height of each ball at the moment of impact.

6. A ball is thrown upward with an initial speed of 128 ft/sec. How high above the ground is the ball 3 sec after it is thrown? In which direction is it moving? How high does the ball go? With what velocity does it strike the ground?

7. A bullet is shot upward with an initial velocity of 1,600 ft/sec. How high does it go? How long does it remain in the air? Will it have sufficient velocity when it strikes the ground to be dangerous?

8. David Delbert and Linda Small spent a day at the beach near Big Bugbite Falls. David notes that a piece of wood which was swept over the falls requires 4.5 sec to descend. How high are the falls?

9. David's little brother, who is interested in airplanes, guesses that the piece of wood of Prob. 8 must have been going more than 70 mph as it struck the water near the base of the falls. Is his estimate a reasonable one?

10. A hockey puck travels 216 ft before coming to rest. If the deceleration of the puck is 12 ft/sec/sec, find the initial velocity of the puck.

11. A package slides down a chute 60 ft long with an acceleration of 5 ft/sec/sec.

(a) Find the initial velocity of the package if it requires 4 sec to traverse the chute.

(b) How fast was the package moving when it was one-third of the way down the chute?

(c) How long did it take the package to get halfway down the chute?

(d) How far down the chute did the package go during the first half of the time of descent?

Sec. 8-1. Motion

12. What uniform acceleration is needed to increase the speed of an automobile from rest (0 mph) to 60 mph in a distance of 440 ft? HINT: First obtain the speeds in feet per second.

*__13.__ A bullet buries itself 9 in. into a tree in 0.01 sec. Assuming that the deceleration of the bullet was constant and that the bullet came to rest in the indicated time, find the speed of the bullet at the moment of impact.

14. With what approximate speed would a projectile need to be hurled to just reach the top of the Empire State Building, which is 1,250 ft high?

15. A long inclined plane is constructed in such a manner that objects slide down it with an acceleration of 12 ft/sec/sec. An object is thrown *up* the incline. It travels 3,750 feet up the incline before starting to slide back down. Find the initial velocity with which the object was thrown.

16. Work may be defined as force times distance if the force is constant. If the force is not constant but is a function of distance (as in a stretched spring, for example), then we define work as $W = \int_a^b F(x)\,dx$, where $F(x)$ is the force at distance x and W is the work done by $F(x)$ as x varies from a to b. If the force required to stretch a spring x in. from rest is $F(x) = (12x)$ lb, find the work required to stretch the spring from rest. (*a*) 3 in.; (*b*) 6 in.; (*c*) 12 in. In what units will the work be expressed? (*d*) How much work is done in stretching the spring from $x = 7$ to $x = 13$ in.? Set this up as an integral and integrate.

17. Work Prob. 16 if $F(x) = 4x$.

18. A balloon is rising at the rate of 15 ft/sec. A stone dropped from the balloon reached the ground in 8 sec. How high was the balloon? (The nature of the data does not merit an accuracy of more than the nearest ten feet, if that.)

19. A ball is thrown upward and reaches a height of 80 ft in 1 sec. How high will the ball go?

20. A stone is thrown upward with an initial speed of 32 ft/sec from the top of a building 100 ft above ground level. Determine the velocity of the stone as it passes a window 52 ft above ground level on its way down. Start with the assumption that $g = 32$, that is, $s''(t) = -32$ ft/sec/sec and derive all relationships used.

21. Consult a table of integrals (*Handbook of Chemistry and Physics*, for example) and note the variety of things yet to be learned. Find which integrals we have already studied in this course. Find three functions which are familiar but whose derivative and/or integral we have not studied.

22. Obtain the formula $s = -gt^2/2 + v_0 t + s_0$, mentioned earlier in this section. Start with the acceleration due to gravity as a constant g.

23. Imagine you are the navigator on an interplanetary rocket ship about to land on planet J5A. It is known that the rocket-braking power needed for a safe landing is proportional to the gravitational attraction of the planet. Your ship uses $\frac{1}{2}$ rocket-braking power on earth. If on the planet J5A a ball drops 100 ft from rest in 2 sec, can you safely land your ship on J5A?

24. A stone is thrown upward from the top of a building 180 ft high with an initial velocity of 8 ft/sec. Starting with 32 ft/sec/sec downward as the acceleration due to gravity, determine a function which expresses the height of the stone above the ground, as a function of the time t in seconds after the stone is thrown.

25. Determine the velocity of the stone in Prob. 1 as it passes a window 60 ft above ground level.

26. A curve has slope equal to $\frac{7}{2}$ times its abscissa (x value) and passes through $(2, -1)$. Determine the equation of the curve. HINT: If $y' = \frac{7}{2}x$, $y = \int \frac{7}{2}x \, dx = 7x^2/4 + C$. Since $(2, -1)$ lies on the curve $y = 7x^2/4 + C$, it is possible to determine C. Do so.

27. A curve has slope equal to four times its abscissa $[F'(x) = 4x]$ and passes through $(3, -6)$. Determine the equation of the curve.

28. The slope of a curve is six times the square of its abscissa. The curve has a y intercept of 5. Determine the equation of the curve.

29. Verify the theorem $\int x^n \, dx = \dfrac{x^{n+1}}{n+1} + C$ if $n \neq -1$ by differentiation. Be sure to consider the case in which $n = 0$.

30. The slope of a curve at a point is always four units less than twice the abscissa of that point. Find the equation of the family of curves having this property, and pick out the member of the family which passes through $(-2, 3)$.

31. If $\dfrac{dy}{dt} = 3t^2 - 7t + 6$ and $y = 30$ when $t = -2$, find an equation expressing y as a function of t. Graph this function.

32. If $\dfrac{dz}{dx} = 8x^7 - 10x^4 + 12x^3 - 7$ and $z = 40$ when $x = 1$, find an equation relating z and x. Make a rough sketch of this equation.

Sec. 8-2. On Setting up Problems

33. Prove, using integration, that if the derivative of a function is constant, then the function is a linear function (that is, its graph is a straight line).

34. $F(x)$ is a cubic polynomial such that $y = F(x)$ has a maximum at $(-3, 20)$ and a minimum at $(1, -12)$. Explain why the slope of $y = F(x)$ must be of the form $k(x-1)(x+3)$. Determine the equation of the curve.

35. Find the area bounded by the parabola $y + x^2 - 6x + 5 = 0$, and the x axis. Sketch the curve and shade the area found.

36. Find the area bounded by the x axis and the curve having equation $y = (x-2)(x-3)(x-5)$ between the ordinates $x = 1$ and $x = 6$. Sketch the curve and shade the area found.

37. Determine the area under $y = f(x)$ between $x = -2$ and $x = 3$ if

$$f(x) = \begin{cases} 2x + 5 & \text{for } x < 1. \\ 6x^2 + 1 & \text{for } x \geq 1. \end{cases}$$

8-2. On Setting up Problems

formulating problems

Now that we have the fundamental antiderivative theorem at our disposal for computing integrals, we might think that no further use would be made of the definition of the integral as a limit of a sum. On the contrary, the definition affords a powerful method of *formulating problems* in terms of the integral calculus. Once the problem is set up as an integral, we perform the calculations by means of the method of antidifferentiation. However, initially, it is often of tremendous advantage to *think* in terms of the sum and limit, basic in the definition, in order to arrive at the integral to be evaluated. Let us clarify these remarks with several examples.

Example 1

Find the area bounded by the curves $y = -x^3 + 12x^2 - 36x + 40$, $x = 3$, $x = 7$, $y = 0$. Note that in the region under consideration, $3 \leq x \leq 7$, y is always positive.

Before turning the page, the student should think a bit about the problem and make a rough sketch of the curve $y = -x^3 + 12x^2 - 36x + 40$ shading in the area to be determined.

Applications of the Integral

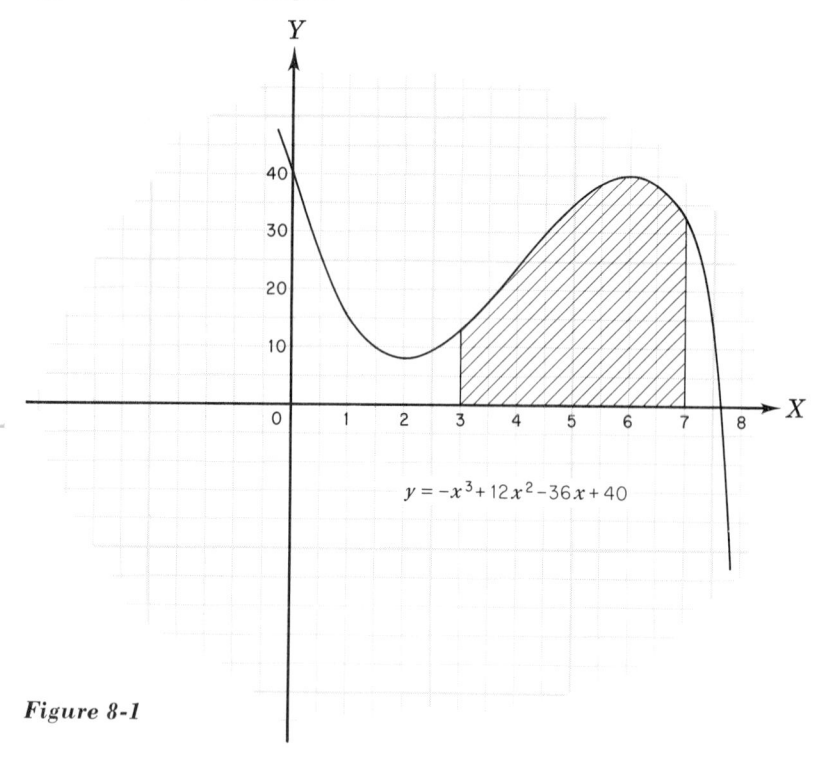

Figure 8-1

(graph showing $y = -x^3 + 12x^2 - 36x + 40$)

Solution

Using Fig. 8-1, we may *think* as follows: Divide the interval $3 \leq x \leq 7$ into n subintervals and let Δx_i be the width of the ith subinterval and let ξ_i be a point in this subinterval. From the sum $\sum_{i=1}^{n} f(\xi_i) \Delta x_i$, the area is

$$\lim_{\substack{n \to \infty \\ \text{all } \Delta x_i \to 0}} \sum_{i=1}^{n} f(\xi_i) \Delta x_i = \int_{3}^{7} f(x) \, dx$$

$$= \int_{3}^{7} (-x^3 + 12x^2 - 36x + 40) \, dx.$$

This integral is then evaluated, using the theorem of Secs. 7-7, 7-8, and 7-11.

To shorten the work we may *think* as follows (Fig. 8-2): One small rectangle used in the definition of area has area $f(\xi_i) \Delta x_i$.

Because of the end result we are seeking, we actually *write* this as $f(x) \, dx$ rather than $f(\xi_i) \Delta x_i$, although we *think* the latter. The total area will be the limiting value of the sum of all such small rectangles.

Sec. 8-2. On Setting up Problems 221

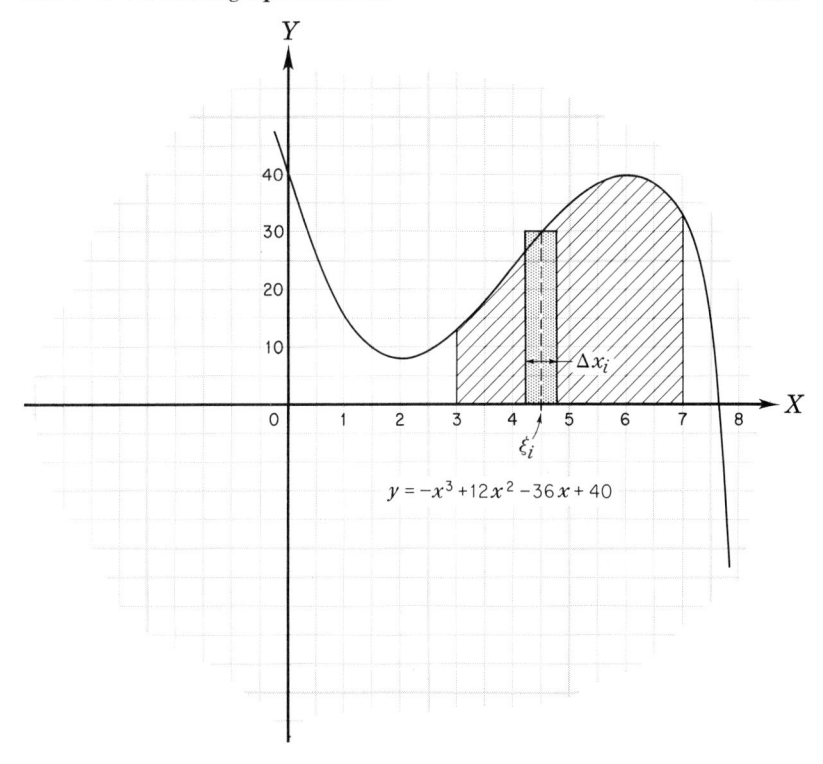

Figure 8-2

Hence we think:

$$A = \lim_{\substack{n \to \infty \\ \text{all } \Delta x_i \to 0}} \sum_{i=1}^{n} f(\xi_i)\, \Delta x_i.$$

However, we often write simply $A = \int_a^b f(x)\, dx$. Indeed, this is where the symbol ∫ originated. It is an elongated S, which may remind us of both *sum* and *limit*.

Study the wording of this example carefully.

Example 2

A 100-lb bag of sand is raised 5 ft, but steadily loses sand at the rate of 2 lb for each foot of elevation. Find the work done if at each instant the force is equal to the weight of the bag at that instant. Again the student is requested to study the problem before turning the page. Try to set up the desired integral yourself. This is how you learn to **do** mathematics.

Applications of the Integral

Solution

We *think:* Divide the interval (5 ft) into subintervals of width Δx_i (Fig. 8-3).

We *think:* The force required at $x = \xi_i$, namely, $f(\xi_i)$, is equal to the mass of the bag, namely, $(100 - 2\xi_i)$. The work is then, by definition,

$$\lim_{\substack{n \to \infty \\ \text{all } \Delta x_i \to 0}} \sum_{i=1}^{n} f(\xi_i) \, \Delta x_i$$

which is equal to

$$\lim_{\substack{n \to \infty \\ \text{all } \Delta x_i \to 0}} \sum_{i=1}^{n} (100 - 2\xi_i) \, \Delta x_i.$$

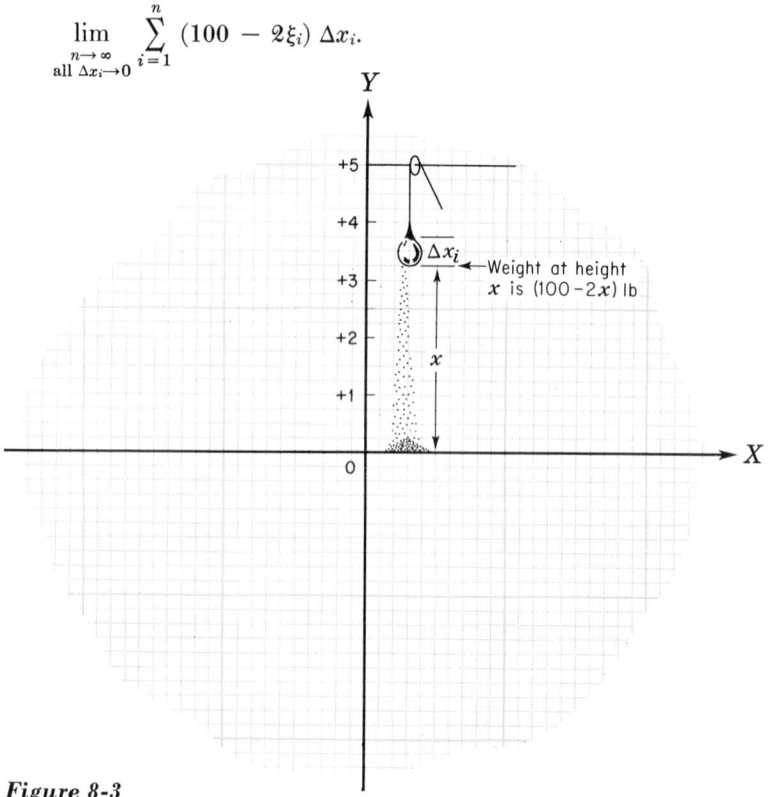

Figure 8-3

However, we *write:* At height x the bag weighs $(100 - 2x)$ lb. In lifting this weight through a distance dx, the work done is $(100 - 2x) \, dx$ and the total work is $\int_0^5 (100 - 2x) \, dx$. At this point we compute, by means of the fundamental theorem,

Sec. 8-2. On Setting up Problems

$$\text{Work} = \int_0^5 (100 - 2x)\, dx = 100x - x^2 \Big|_0^5$$
$$= 500 - 25 = 475 \text{ ft-lb}.$$

Example 3

A rectangular fish tank is 3 ft long, 2 ft wide, and 1 ft deep. It is filled with water weighing 62.4 lb/cu ft. Find the total force on one end of the tank (see Fig. 8-4).

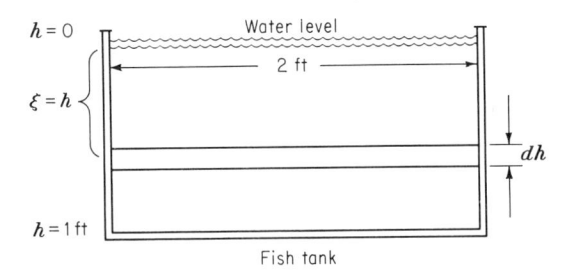

Figure 8-4

Solution

We *think:* Divide the interval corresponding to the depth of the tank into subintervals of height Δh_i. It is necessary to extend the definition of total force F due to constant pressure P over an area A (that is, $F = P \cdot A$) to include variable pressure [namely, $P(h) = 62.4 \cdot h$, where h is the distance below the surface] encountered in this problem. The extension is so similar to the extensions made in defining area and work in Chap. 7 that it is given immediately that the total force F resulting from the variable pressure $P(h)$ acting on the end of the tank (Fig. 8-4) is obtained as a limit of a sum of forces on narrow strips having approximately equal pressures throughout.

$$F = \lim_{\substack{n \to \infty \\ \text{all } \Delta A_i \to 0}} \sum_{i=1}^{n} P(\xi_i)\, \Delta A_i$$

where ξ_i is some point in the ith interval, ΔA_i. Since ΔA_i is a small strip with dimensions Δh_i by 2, it follows that

$$\Delta A_i = 2\Delta h_i \quad \text{and} \quad F = \lim_{\substack{n \to \infty \\ \text{all } h_i \to 0}} \sum_{i=1}^{n} (62.4h \cdot 2\Delta h_i).$$

However, to save time we often *say* that the force on the small strip at depth h is $62.4h\ 2dh$ and we write: The total force is

$$F = \int_0^1 62.4h\ 2dh = 62.4h^2 \Big|_0^1 = 62.4 \text{ lb.}$$

Example 4

Find the area bounded by the curves $x = y^2$ and $y = x - 2$ (Fig. 8-5).

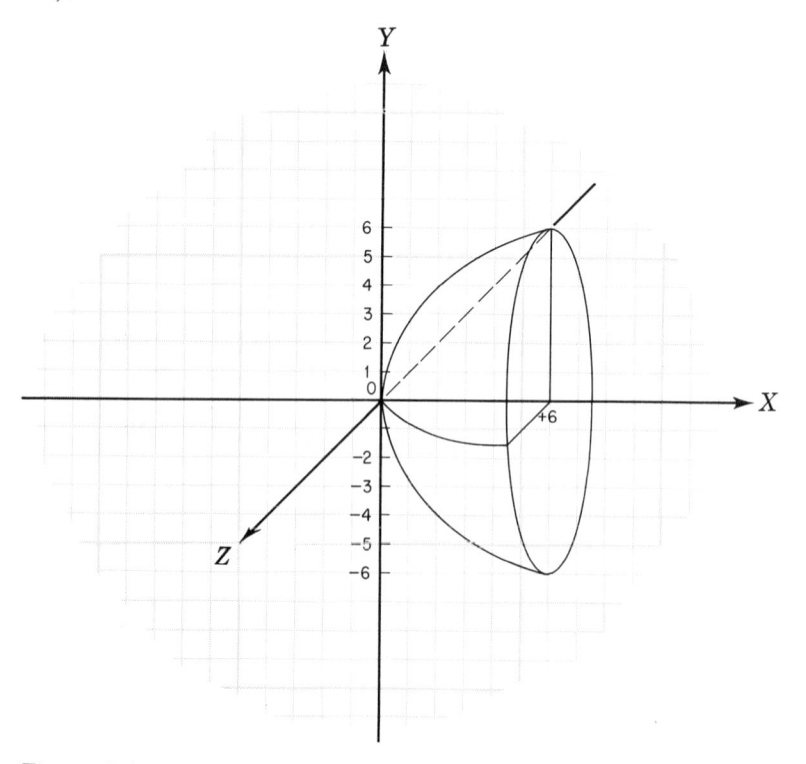

Figure 8-5

Solution

First we solve $x = y^2$ and $y = x - 2$ simultaneously to find the points of intersection. We get $x = (x - 2)^2$, or $x^2 - 5x + 4 = 0$ and the points of intersection are $(1, -1)$ and $(4, 2)$. Here a difficulty arises (Fig. 8-5). The area from $x = 0$ to $x = 1$ is bounded by $y = +\sqrt{x}$, $y = -\sqrt{x}$, and $x = 1$, while the area from $x = 1$ to $x = 4$ is bounded by $y = \sqrt{x}$, $y = x - 2$, $x = 1$, and $x = 4$. We

Sec. 8-2. On Setting up Problems

cannot set up the integral by dividing the x axis into subintervals, unless we treat the problem as two separate problems. To avoid this we make use of horizontal strips by dividing the y axis into subintervals (Fig. 8-6).

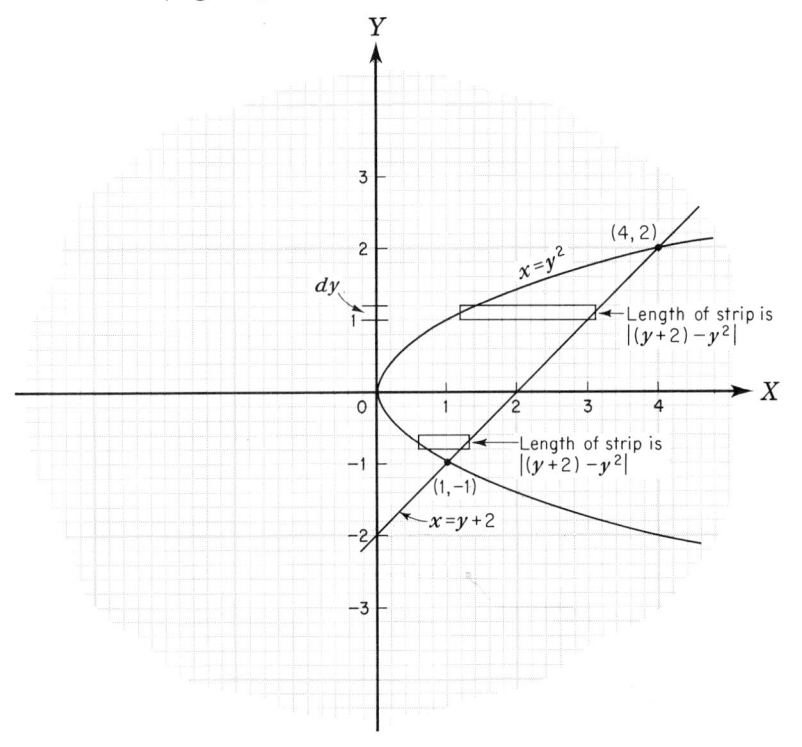

Figure 8-6

We *write:* The area of each small horizontal strip is

$$[(y+2) - y^2]\, dy$$

and the whole area is

$$\int_{-1}^{2} (y + 2 - y^2)\, dy = \left(\frac{y^2}{2} + 2y - \frac{y^3}{3}\right)\Big|_{-1}^{2}$$

$$= \left(\frac{4}{2} + 4 - \frac{8}{3}\right) - \left(\frac{1}{2} - 2 + \frac{1}{3}\right)$$

$$= \frac{20}{6} + \frac{7}{6}$$

$$= \frac{9}{2} \text{ square units.}$$

Problem Set 8-2

1. Show that the same answer is obtained in Example 4 by evaluating $\int_0^1 2\sqrt{x}\, dx + \int_1^4 [\sqrt{x} - (x-2)]\, dx$. Explain why this is so.
2. Find the area bounded by $y = 3x - 1$, $y = x + 1$, and $y = -1$.
3. Find the area bounded by $y = x^3 - x^2$ and $y = x^2$. *Ans.* $\frac{4}{3}$ square units.
4. Find the area bounded by $y = 4 - x^2$, $y = 2 - x$. *Ans.* $\frac{9}{2}$ square units.
5. Find the area bounded by $y = 2x^3$, $y = 2$, $y = 4$, $x = 0$. *Ans.* $\frac{3}{2}\sqrt[3]{16} - \frac{3}{2}$ square units.
6. Find the area bounded by the parabola $(y-3)^2 = 2(x+3)$ and the line $y = x + 2$. *Ans.* 18 square units.
7. A 300-lb bag of sand is raised 8 ft, but steadily loses sand at the rate of 4 lb/ft of elevation. Find the work done.
8. A trough is full of water. Its end is a rectangle 5 ft wide and 3 ft high. Find the total force on the end of the trough, assuming that water weighs 62.4 lb/cu ft.
9. Let the trough of Prob. 8 have a trapezoidal end as shown in Fig. 8-7. Show that the total force lies between 291 and 292 lb.

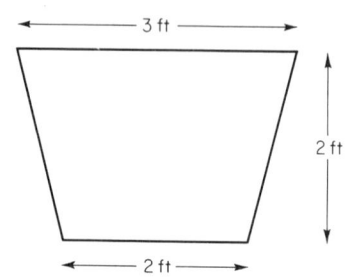

Figure 8-7

In Probs. 10 to 17, set up an integral which is equal to the area bounded by the given curves, but *do not attempt to perform* the integration.

10. Circle of radius r.
11. Ellipse $(x^2/a^2) + (y^2/b^2) = 1$.
12. $y = \tan x$, $x = 0$, $x = \pi/2 - |\epsilon|$, $y = 0$.
13. $y = \tan x$, $x = 0$, $x = \pi/2$, $y = 0$. HINT: See Sec. 7-10.
14. $y = \log_a x$, $x = 1$, $x = 2$, $y = 0$.

Sec. 8-3. Further Applications of Integration

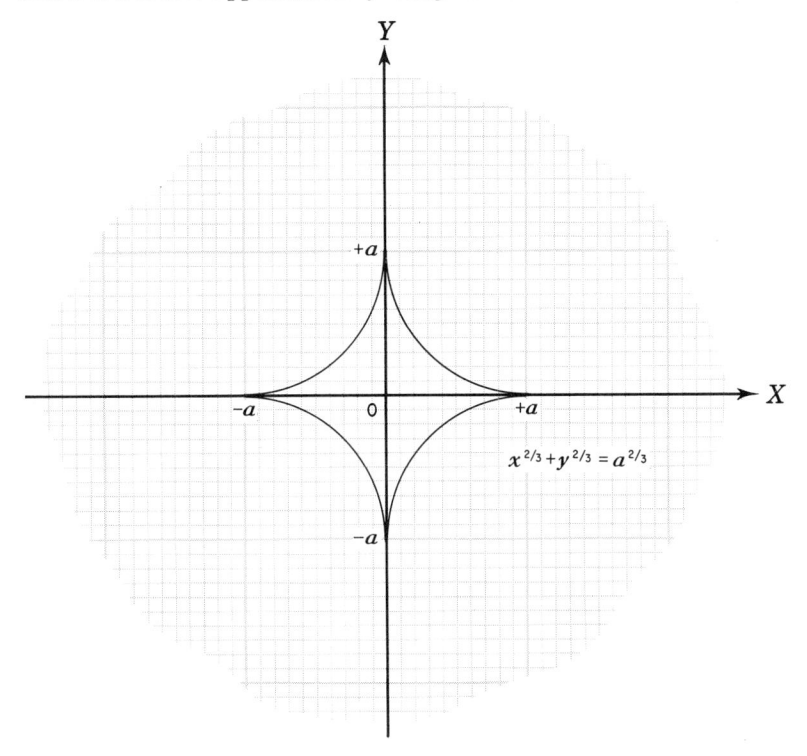

Figure 8-8

15. $x^{\frac{2}{3}} + y^{\frac{2}{3}} = a^{\frac{2}{3}}$ (see Fig. 8-8).
16. To the right of the parabola $y^2 = x$ and inside the circle $x^2 + y^2 = 1$.
17. Inside $x^2 + y^2 = 4$ and to the right of $y^2 = x$.
18. The end of a cylindrical tank is a circle with a diameter of 10 ft. The axis of the cylinder is horizontal. If the tank is filled with oil weighing 50 lb/cu ft, set up an integral which, if evaluated, would give the total force on the circular end of the tank.
19. Set up Prob. 18 if the tank is only half full.

8-3. Further Applications of Integration

Example 1

A cylindrical cistern is 4 ft in diameter and 15 ft deep (Fig. 8-9). It is two-thirds full of water. Find the work done in emptying the cistern by using a pump which discharges the water at a point 3 ft above the top of the cistern.

The work done, W, in lifting a weight P, in pounds, through a vertical distance h feet is $W = hp$ ft-lb.

To ensure that the reader keeps in mind the nature of the definite integral, we return briefly to the methods of Secs. 7-3 and 7-4.

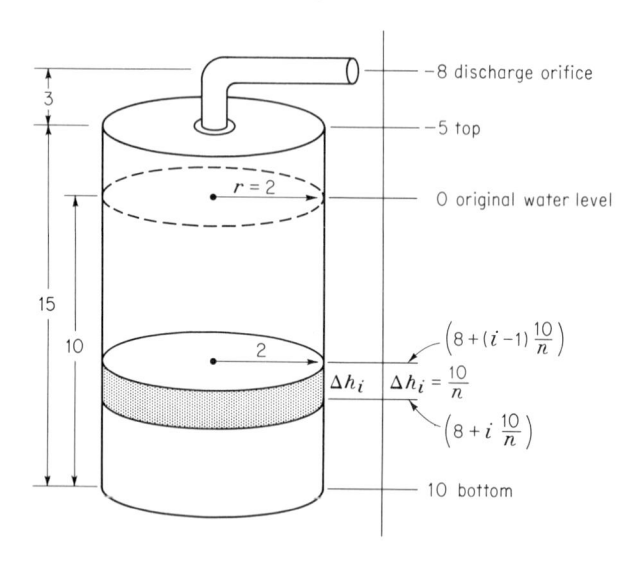

Figure 8-9

Select an h axis parallel to the axis of the cylinder with $h = 0$ at the original water level and *positive direction downward*. The water lies between $h = 0$ and $h = 10$. Divide the axis between $h = 0$ and $h = 10$ into n equal parts, each of width $\Delta h_i = 10/n$. The distance from the pump orifice at $h = 8$ to the top of the ith interval is $[8 + (i-1)10/n]$ while the distance from $h = -8$ to the bottom of the ith interval is $(8 + i10/n)$ (Why?) If W_i is the work done in raising the small disk (radius 2, thickness Δh_i) opposite the ith interval to the orifice level, then

$$\left[8 + (i-1)\frac{10}{n}\right] 62.4\pi(2)^2 \Delta h_i \leq W_i \leq \left(8 + i\frac{10}{n}\right) 62.4\pi(2)^2 \Delta h_i.$$

The weight of the disk of water is 62.4 (volume of disk) $= 62.4\pi(2)^2 \Delta h_i$. Thus, $W = hp =$ (distance raised) $\times$ (weight).

Upon summing up all the small disks between $h = 0$ and $h = 10$,

Sec. 8-3. Further Applications of Integration 229

we obtain, since $W = \Sigma W_i$,

$$\sum_{i=1}^{n} \left[8 + (i-1)\frac{10}{n} \right] 62.4\pi(2)^2 \, \Delta h_i \leq W$$

$$\leq \sum_{i=1}^{n} \left(8 + i\frac{10}{n} \right) 62.4\pi(2)^2 \, \Delta h_i.$$

Upon taking the limit, as n increases without bound (and $\Delta h_i \to 0$), we obtain

$$\lim_{\substack{n \to \infty \\ \Delta h_i \to 0}} \sum_{i=1}^{n} \left[8 + (i-1)\frac{10}{n} \right] 62.4\pi(2)^2 \, \Delta h_i \leq W$$

$$\leq \lim_{\substack{n \to \infty \\ \Delta h_i \to 0}} \sum_{i=1}^{n} \left(8 + i\frac{10}{n} \right) 62.4\pi(2)^2 \, \Delta h_i.$$

Since $f(h) = (8 + h)62.4\pi(2)^2$ is a continuous function (a straight line, to be precise), it follows from the theorem of Sec. 7-6 that the outer limits are equal and we may write

$$W = \lim_{n \to \infty} \sum_{i=1}^{n} (8 + \xi_i) 62.4\pi(2)^2 \, \Delta h_i$$

where ξ_i is any point of the ith interval. This is precisely the definition (Sec. 7-6) of

$$W = \int_0^{10} (8 + h) 62.4\pi(2)^2 \, dh.$$

We have again gone through the rather laborious process by which we *first discover* a generalized relationship.

Now, however, to "rediscover" it, using Fig. 8-10, we *think:* A typical disk of water having radius 2 and thickness Δh_i will have volume $= \pi(2)^2 \, \Delta h_i$ and weight $= 62.4\pi(2)^2 \, \Delta h_i$. If this is raised through a distance $(8 + h_i)$, the work done will be

$$W_i = (8 + h_i) 62.4\pi(2)^2 \, \Delta h_i$$

and the total work will be

$$\lim_{n \to \infty} \sum_{i=1}^{n} (8 + h_i) 62.4\pi(2)^2 \, \Delta h_i.$$

230 Applications of the Integral

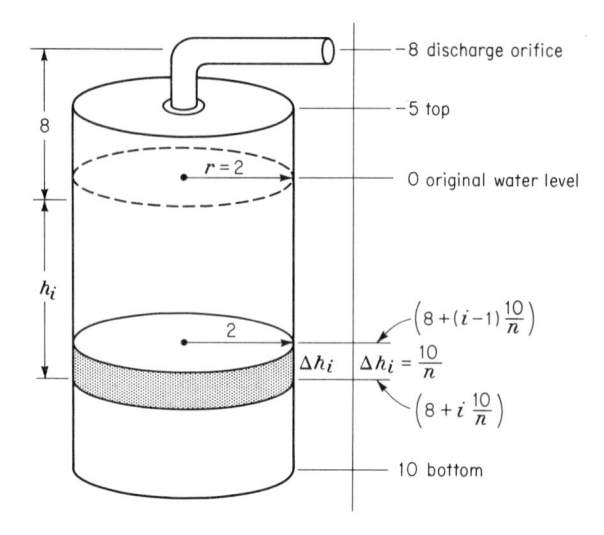

Figure 8-10

However, since the ultimate limit process is always in our mind, we *write*

$$W_i = (8 + h_i) 62.4\pi (2)^2 \, dh \quad \text{and}$$

$$W = \int_0^{10} (8 + h) 62.4\pi (2)^2 \, dh.$$

This is easily evaluated, using $\int_a^b kf(x) \, dx = k \int_a^b f(x) \, dx$.

$$\begin{aligned}
W &= \int_0^{10} (8 + h) 62.4\pi (2)^2 \, dh \\
&= 62.4\pi (2)^2 \int_0^{10} (8 + h) \, dh \\
&= 62.4\pi (2)^2 \left[8h + \frac{h^2}{2} \right]_0^{10} \\
&= 62.4\pi (2)^2 (80 + 50 - 0 - 0) \\
&= 1.02 \times 10^5 \text{ ft-lb.}
\end{aligned}$$

It is easier to work with integrals than with the "limit of a sum," but if you wish to be able to formulate new problems, it is essential that when you *write* $\int_a^b f(x) \, dx$, you *think*, "This is a limit of a *sum* of the form $\sum_{i=1}^n f(\xi_i) \, \Delta x_i$." However, there is no need to write this out each time it occurs.

Sec. 8-3. Further Applications of Integration

Example 2

Use integration to determine the volume of the cone obtained by rotating about the x axis that portion of the line $y = \frac{1}{2}x$ lying between $x = 0$ and $x = 10$ (Fig. 8-11).

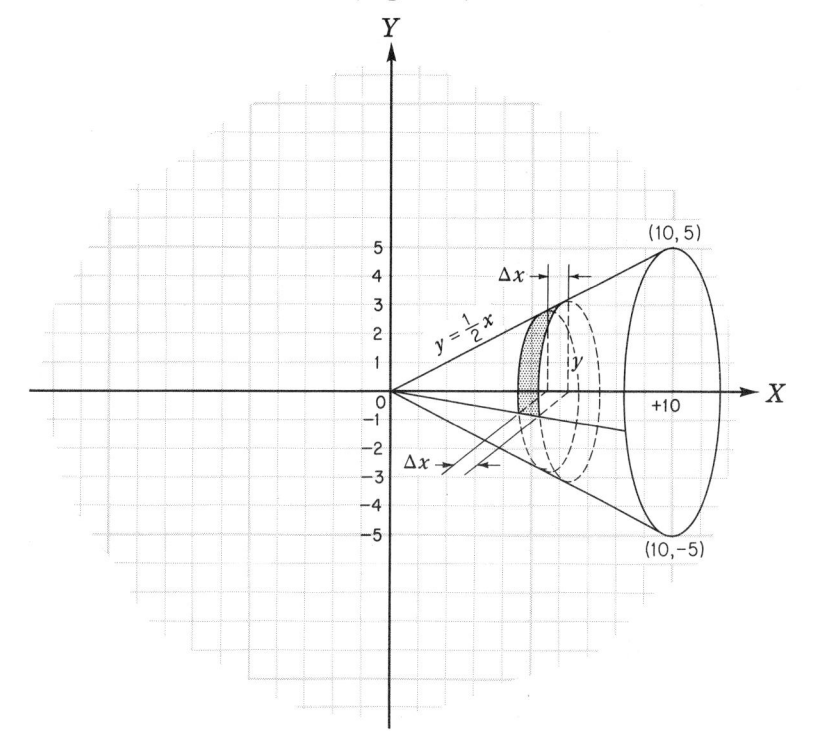

Figure 8-11

We *think:* A typical disk has thickness Δx and radius y, hence has volume $\pi y^2 \, \Delta x_i = \pi(x/2)^2 \, \Delta x$. The desired volume is

$$V = \lim_{\substack{n \to \infty \\ \text{all } \Delta x_i \to 0}} \sum_{i=1}^{n} \pi(x/2)^2 \, \Delta x.$$

We *write*

$$V = \int_0^{10} \pi y^2 \, dx = \int_0^{10} \pi \left(\frac{x}{2}\right)^2 dx = \frac{\pi}{4} \int_0^{10} x^2 \, dx = \frac{\pi}{4} \frac{x^3}{3} \bigg|_0^{10}$$

$$= \frac{\pi}{4}\left(\frac{1{,}000}{3} - \frac{0}{3}\right) = \frac{250\pi}{3} \text{ cubic units} = \text{?}$$

Applications of the Integral

Example 3

Find the volume of the "bullet nose" obtained by revolving the parabola $y^2 = 4x$ about the x axis and cutting it with the plane $x = 4$ (Fig. 8-12).

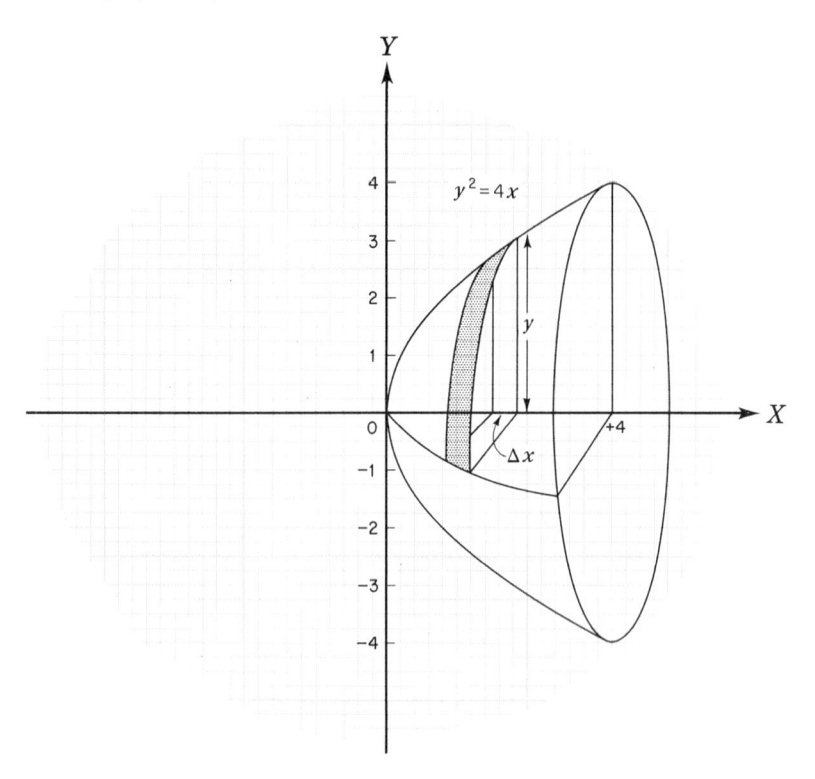

Figure 8-12

The desired volume is the limit of a sum of small disks having volume $\Delta V_i = \pi y^2 \Delta x$; that is,

$$V = \int_0^4 \pi y^2 \, dx = \int_0^4 \pi(4x) \, dx = 4\pi \int_0^4 x \, dx = 2\pi x^2 \Big|_0^4 = 32\pi - 0$$
$$= 32\pi \text{ cubic units} = 100.5 \text{ cubic units}.$$

Example 4

A hemispherical bowl is 10 in. in diameter. It is filled with wine to a depth of 3 in. Find the volume of the wine (Fig. 8-13).

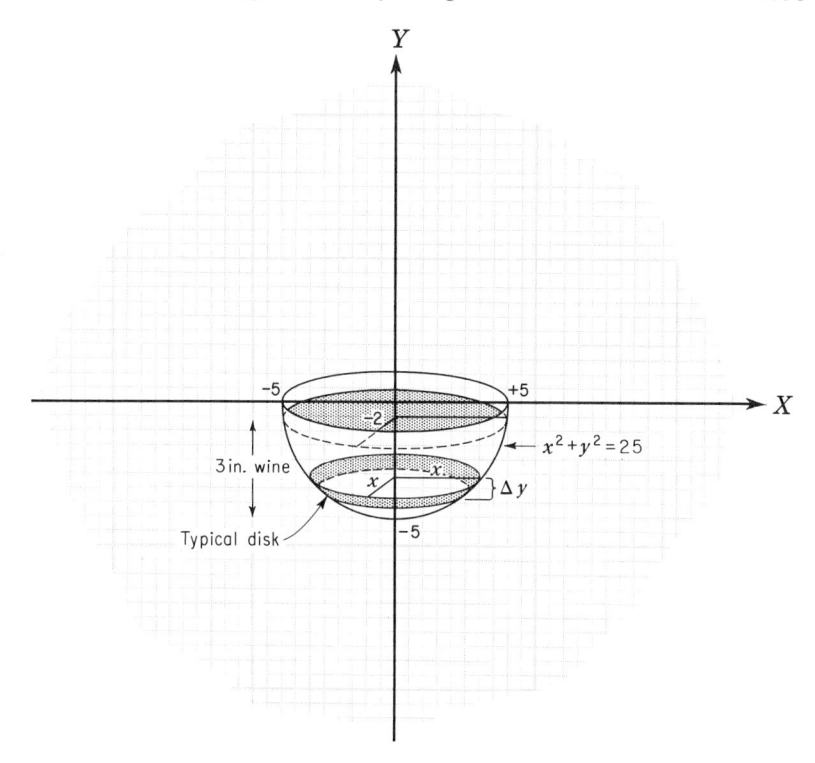

Sec. 8-3. Further Applications of Integration 233

Figure 8-13

A typical disk of wine has volume $\Delta V_i = \pi x^2 \, \Delta y$. Hence we *think*

$$V = \lim_{n \to \infty} \sum_{i=1}^{n} \Delta V_i = \lim_{n \to \infty} \sum_{i=1}^{n} \pi x^2 \, \Delta y.$$

The cross-sectional circle has as its equation $x^2 + y^2 = 25$.
We *write:*

$$V = \int_{-5}^{-2} \pi x^2 \, dy = \int_{-5}^{-2} \pi (25 - y^2) \, dy = \pi \int_{-5}^{-2} (25 - y^2) \, dy$$

$$= \pi \left[\left(25y - \frac{y^3}{3} \right) \Big|_{-5}^{-2} \right] = \pi \left[\left(-50 + \frac{8}{3} \right) - \left(-125 + \frac{125}{3} \right) \right]$$

$$= 36\pi \text{ cu in., slightly less than } \tfrac{1}{2} \text{ gal of wine.}$$

Note that the limits are taken from the smaller (-5) value to the larger (-2) value to obtain the Δy_i as positive quantities and the resulting volume as positive.

Applications of the Integral

Example 5

A long straw is placed in the bowl of Prob. 4, in such a manner that its top is 6 in. above the surface of the liquid (Fig. 8-14). If the wine weighs 64 lb/cu ft, determine the work done in sucking all the wine through the straw.

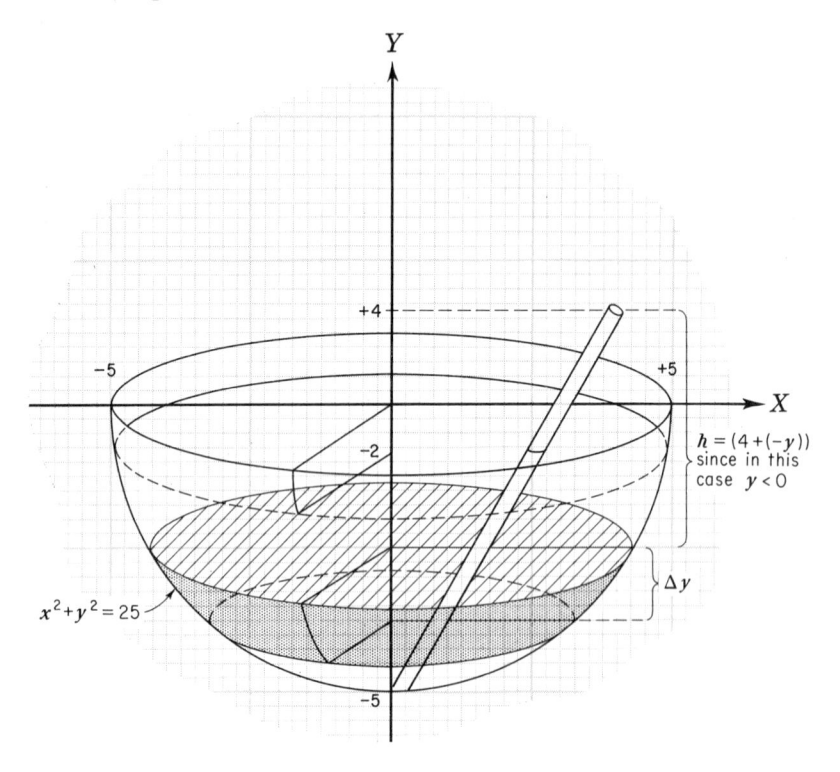

Figure 8-14

Since the disk is below the x axis, its y *coordinate is negative* and the distance which the disk must be raised is given by the expression $h = (4 + |y|) = (4 - y)$. The volume of the disk is $\pi x^2 \, \Delta y$ and its weight is $\Delta W_i = \left(\dfrac{64\pi}{1{,}728} x^2 \, \Delta y\right)$. We *think:*

$$W = \lim_{n \to \infty} \sum_{i=1}^{n} h_i \, \Delta W_i = \lim_{n \to \infty} \sum_{i=1}^{n} (4 - y)\left(\dfrac{64\pi x^2}{1{,}728} \Delta y\right).$$

Sec. 8-3. Further Applications of Integration

We write:
$$W = \int_{-5}^{-2} (4-y)\left(\frac{64\pi x^2}{1{,}728} dy\right)$$
$$= \int_{-5}^{-2} (4-y)\left[\frac{64\pi}{1{,}728}(25-y^2)\,dy\right]$$
$$= \frac{64\pi}{1{,}728}\int_{-5}^{-2}(100 - 25y - 4y^2 + y^3)\,dy.$$

This integration is left as an exercise for the student.

Problem Set 8-3

1. Complete Example 5. Be sure you indicate the units.

2. Determine the volume of the cone obtained by revolving the portion of line $y = 2x$ between $x = 0$ and $x = 6$ about the x axis.

3. The parabola $y^2 = 10x$ is revolved about the x axis.
(a) Determine the volume of the bullet nose cut off at $x = 10$.
(b) Determine the volume of the frustrum between $x = 4$ and $x = 7$.

4. The ellipse $3x^2 + 4y^2 = 108$ is revolved about the x axis. Determine the volume of the prolate spheroid so generated.

5. If the ellipse of Prob. 4 is revolved about the y axis, an oblate spheroid is generated. Determine its volume.

6. A sphere of radius 6 in. is cut by two parallel planes, one of which passes 2 in. from the center of the sphere and the other of which is 5 in. from the center of the sphere and on the same side of the center as the first plane. Find the volume of that portion of the sphere which lies between the two planes.

7. Determine the volume of the small "cap" which a plane 6 in. from the center of the sphere of radius 7 cuts from the sphere.

8. A hemispherical bowl with radius 6 in. contains 6 in. of water (weight 62.4 lb/cu ft). If the level of the water is lowered to 3 in., find
(a) the volume of the water removed.
(b) the work done if the water is pumped to an orifice 4 in. above the top of the bowl.

9. Repeat Prob. 8, but completely empty the bowl this time, starting from the original volume.

10. A cylindrical cistern is 10 ft deep and 4 ft in diameter. Determine the work done in lowering the water level from 8 ft to 4 ft if the discharge pump is located at ground level.

Applications of the Integral

11. How much work is done in pumping the remaining half of the water out of the cistern of Prob. 10?

*12. If David and Peter each agree to do one-half the work in pumping the water out of the cistern of Prob. 10, and if Peter agrees to pump first, how deep should the water be when David takes over?

13. A spring has a normal length of 4 ft. A 6-lb weight will stretch it to 7 ft (that is, stretch it $s = 3$ ft; hence $F = ks$ yields $6 = k3$ or $k = 2$, and in general $F = 2s$, where F is the force and s is the distance stretched). Determine the work done in stretching the spring's length

(a) from 4 ft to 10 ft.
(b) from 6 ft to 12 ft.
(c) from 100 ft to 101 ft.
(d) from 101 ft to ——— oops, the spring broke!

14. A trough 6 ft long has its end in the shape of the parabola $y = 6x^2$. The trough is 2 ft wide at the top (see Fig. 8-15).

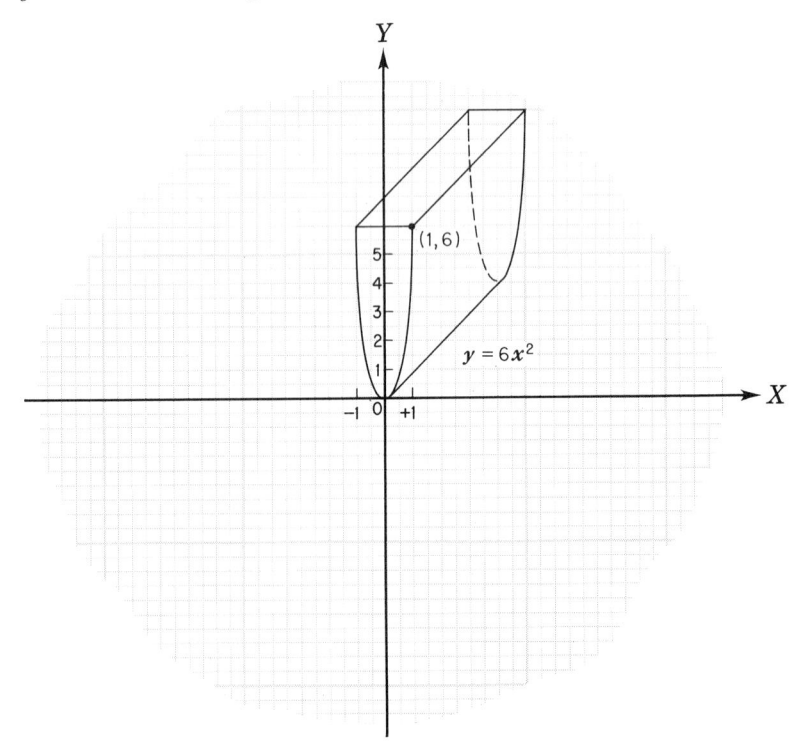

Figure 8-15

(a) How deep is the trough?
(b) What is the volume of the trough?
(c) What is the total force on the end of the trough due to water pressure if the trough is full?

15. Shirley and Bill were exploring an old shack near Robbers' Cave. Shirley found a cylindrical drum partly filled with a deep red liquid. The drum was lying on its side (axis horizontal) and the liquid came one-third of the way up the diameter of the cylinder. If the cylinder was 24 in. in diameter and 36 in. long, how many gallons of liquid did it contain?

8-4. Self-test

1. A ball is thrown upward with an initial velocity of 128 ft/sec.
(a) How high above the ground is the ball 2.5 sec after it is thrown?
(b) In which direction is the ball going?
(c) How high does it go?
(d) With what velocity does it strike the ground?
Derive all equations, starting with $V(t) = S''$, $t = -32$.

2. What uniform deceleration is needed to stop a car going 60 mph in a distance of 440 ft?

3. A spring obeys the law $F = 10s$, where F is force in pounds and s is stretch in feet from an original unstretched length. Determine the work done in stretching the spring from $s = 1$ to $s = 3$ ft and from $s = 3$ to $s = 5$ ft.

4. Determine the total force on the end of a rectangular tank 5 ft high, 4 ft wide, and 30 ft long which is filled with a plating solution which weighs 66 lb/cu ft.

5. Determine the finite area bounded by $y = 4 - x^2$ and $y + x = 2$.

6. The parabola $y^2 = 6x$ is revolved about the x axis. Determine the finite volume generated by cutting the figure so generated with the plane, $x = 6$.

7. Water is pumped out of a cylindrical cistern through a pipe which discharges 4 ft above the top of the cistern. The cistern is 10 ft deep and 4 ft in diameter. It contains water to a depth of 6 ft when the pumping begins. Determine the work done in pumping out half of the water in the cistern.

8. Which is larger, $\int_{30}^{31} 10^x \, dx$ or $\int_{1}^{3} 10^x \, dx$?

9

Logarithmic and Exponential Functions

9-1. $\int \frac{dx}{x}$, ln x, and e

We have shown that $\int x^n \, dx = \frac{x^{n+1}}{n+1} + C$ if $n \neq -1$. The integral $\int \frac{dx}{x}$ has not been discussed. This important integral will be investigated next.

Consider $\int_1^x \frac{dt}{t}$, which is meaningful for $x > 0$. The curve $y = 1/t$, $t > 0$, is one branch of a hyperbola (Fig. 9-1). The

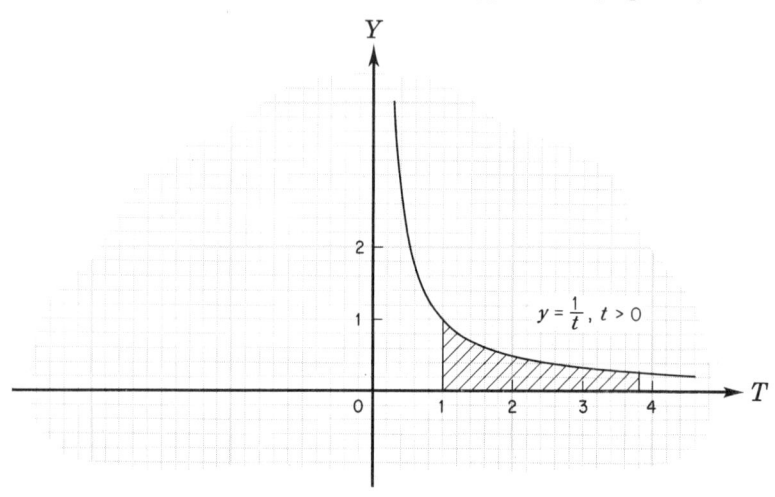

Figure 9-1

Sec. 9-1. $\int \frac{dx}{x}$, $\ln x$, and e

integral $\int_1^x \frac{dt}{t}$ is the area under $y = 1/t$ between $t = 1$ and $t = x$ if $x > 1$. If $0 < x < 1$, then $\int_1^x \frac{dt}{t} = -\int_x^1 \frac{dt}{t}$. If $x = 1$, $\int_1^1 \frac{dt}{t} = 0$. The integral is then a function of x, L, for $x > 0$.

$$L(x) = \int_1^x \frac{dt}{t} \quad \text{if } x > 0.$$

$L(x)$ is monotone increasing; that is, as x increases, $L(x)$ also increases. (Why?)

We turn our attention away from the curve $y = 1/t$ to concentrate on properties of $L(x)$, the desired integral. The curve $y = L(x)$ passes through the point $(1,0)$ since $L(1) = 0$. Furthermore, $\frac{dL(x)}{dx} = 1/x$. (Why?) Thus, at $(1,0)$ $y = L(x)$ has slope 1.

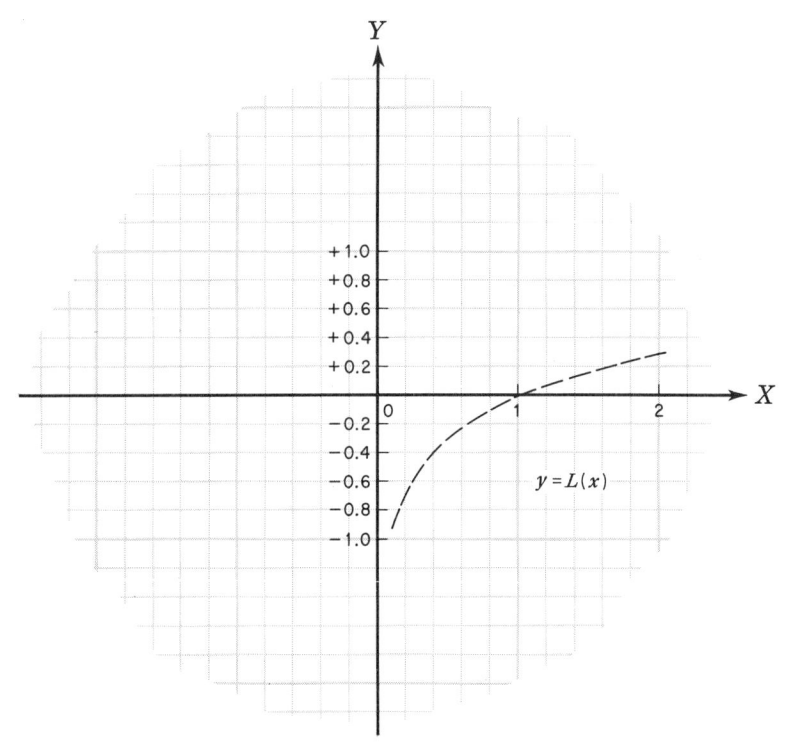

Figure 9-2

Logarithmic and Exponential Functions

The slope of the tangent line at any point on the curve $[x_0, L(x_0)]$ is $1/x_0$. (Why?) Since x_0 is positive, then the slope is also positive. $L(x) = \int_1^x \frac{dt}{t}$ is without meaning for $x \leq 0$, since $y = 1/t$ has a discontinuity at $x = 0$.

As x increases, $L(x)$ also increases. However, as x increases, the slope of $y = L(x)$, namely, $L'(x) = 1/x$, approaches zero. Thus the curve is monotone increasing, but relatively flat (almost horizontal) for x large. For $0 < x < 1$, we find that $1/x$ is still positive, but as $x \to 0^+$, the slope $1/x$ becomes large without bound. We are, then, led to believe that the graph of $y = L(x) = \int_1^x \frac{dt}{t}$ has the general shape indicated in Fig. 9-2.

$(1,0)$ is on the graph of $y = L(x) = \int_1^x \frac{dt}{t}$. Slope for $y = L(x)$ at $x = x_0$ is $1/x_0$. Slope of $y = L(x)$ is 1 at $(1,0)$. Slope is always positive.

These conclusions were deduced from two facts,

Fact 1 $\qquad \dfrac{dL(x)}{dx} = \dfrac{1}{x} \qquad x > 0.$

Fact 2 $\qquad L(1) = 0.$

From these same two facts, we now derive the important fact

$$L(ax) = L(a) + L(x).$$

PROOF: Both a and x are restricted to be positive by the definition of the function L.

$$\frac{dL(ax)}{dx} = \frac{1}{ax} \cdot \frac{d(ax)}{dx} = \frac{1}{ax} \cdot a = \frac{1}{x}$$

or $\qquad \dfrac{dL(ax)}{dx} = \dfrac{dL(x)}{dx}.$

Hence $L(ax) = L(x) + C$ for some constant C. (Why? See the middle of page 196.) Letting $x = 1$, we obtain $L(a) = 0 + C$, since $L(1) = 0$, and it follows that

funda-mental law

$$L(ax) = L(x) + L(a).$$

From this *fundamental law* it follows, by putting $x = b$, that

Sec. 9-1. $\int \frac{dx}{x}$, ln x, and e

Rule I

$$L(a \cdot b) = L(a) + L(b).$$

By putting $x = b/a$ in the fundamental law, it follows that

$$L(b) = L\left(\frac{b}{a}\right) + L(a)$$

which yields

Rule II

$$L\left(\frac{b}{a}\right) = L(b) - L(a).$$

Some students are already familiar with a function obeying Rules I and II. Can you think of one?

Rule III

$$L(a^n) = n \cdot L(a).$$

This is established noting that since $\dfrac{dL(u)}{dx} = \dfrac{1}{u} \cdot \dfrac{du}{dx}$,

$$\frac{dL(x^n)}{dx} = \frac{1}{x^n} \cdot \frac{d(x^n)}{dx} = \frac{nx^{n-1}}{x^n} = \frac{n}{x}$$

and also that

$$\frac{dn \cdot L(x)}{dx} = n \cdot \frac{1}{x} = \frac{n}{x}.$$

Since $L(x^n)$ and $n \cdot L(x)$ have the same derivative, they differ, at most, by a constant. (Why?) Thus

$$L(x^n) = n \cdot L(x) + C.$$

The constant C remains unchanged, no matter what positive value is substituted for x. Hence, substituting $x = 1$ shows that $C = 0$. (Why?) Thus, $L(x^n) = n \cdot L(x)$, establishing Rule III.

The fact is that $L(1) = 0$, and Rules I, II, III are fundamental properties of a logarithm. It is usual to use the abbreviation ln x for the function $L(x)$. The function ln x is a logarithm but ln x is *not* the *common* or *base* 10 *logarithm*. We have now established

242   *Logarithmic and Exponential Functions*

$$\int_1^x \frac{dt}{t} \equiv \ln x.$$

$\ln (1) = 0.$

rules of logarithms

$\ln (a \cdot b) = \ln a + \ln b.$ I

$\ln (a/b) = \ln a - \ln b.$ II

$\ln (a^n) = n \cdot \ln a.$ III

Furthermore, if ln (*x*) has a derivative, as we assumed here, then it must be a *continuous* function for $x > 0$. (Why? Consult Prob. 25, Set 4-2, if in doubt.)

Therefore, the equation $\ln x = 1$ has a real solution. Let the letter *e* denote this solution, $\ln e = 1$. Then

$$(\ln x) \cdot (1) = \ln x$$
$$(\ln x) \cdot (\ln e) = \ln x.$$

Using III, $\ln (e^{\ln x}) = \ln x.$

If ln *x* is a single-valued function, then

$$e^{\ln x} = x.$$

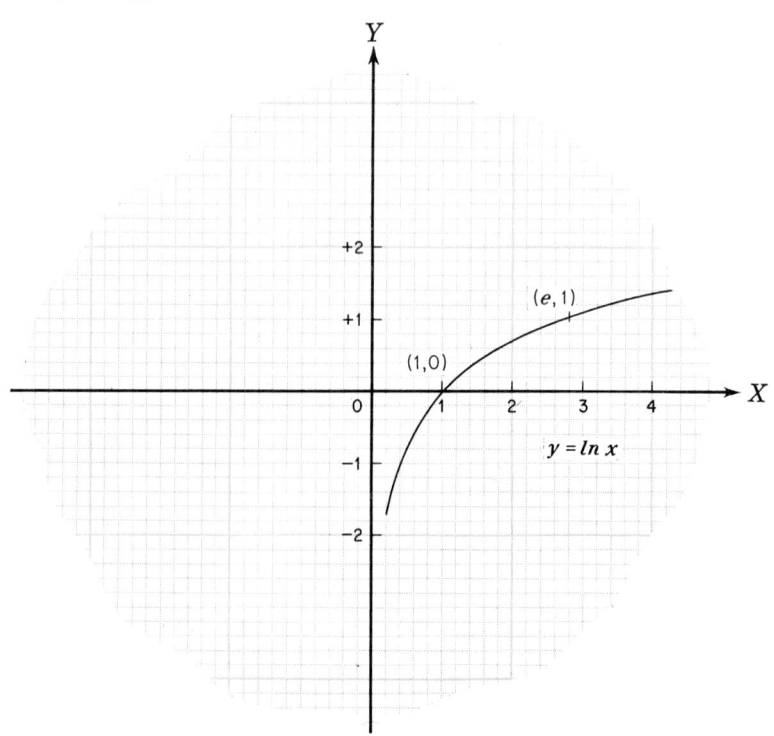

Figure 9-3

Sec. 9-1. $\int \frac{dx}{x}$, ln x, and e

This corresponds to the well-known definition: $b^{\log_b x} = x$, where $1 \neq b > 0$; or, as it is often stated, "If $b^y = x$, then $\log_b x = y$."

The function ln x (or $\log_e x$) is known as the *natural logarithm* of x. It is "natural" in the sense that it arises naturally as the solution of the differential equation $\frac{dy}{dx} = \frac{1}{x}$. This equation occurs often in applications involving the natural, physical, and social sciences. It can be shown that the number e having the property ln $e = 1$ lies between 2 and 3.

$$2 < e < 3.$$

More exactly, $2.718281828 < e < 2.718281829$.

The number e is irrational; in fact, e is transcendental, that is, not even a solution of a polynomial equation having rational coefficients. The approximation $e = 2.718282$ is accurate enough for most physical computations. However, within recent years, certain statistical work used in nuclear and atomic physics has made it necessary to find a very large number of places for e. The journal *Mathematical Tables and Other Aids to Computation* gives the most recent data, which is now above *27,000 decimal place accuracy*.

We have established

$$\int_1^x \frac{dt}{t} = \ln x \quad \text{for } x > 0.$$

$$\frac{d(\ln x)}{dx} = \frac{1}{x} \quad \text{for } x > 0.$$

The function ln x is undefined for $x \leq 0$, but $\int_{-b}^{-a} \frac{dt}{t}$ exists if *both* limits are negative (see Fig. 9-4). To see this, consider the area bounded by $t = -b$, $t = -a$, the t axis and the curve $y = 1/t$. From geometric symmetry, we find this area to be the same as $\int_a^b \frac{dt}{t}$, although the region is below the axis. Hence,

$$\int_{-b}^{-a} \frac{dt}{t} = -\int_a^b \frac{dt}{t} = -(\ln b - \ln a) = \ln a - \ln b$$
$$= \ln |-a| - \ln |-b|.$$

Thus, in both cases we have

$$\boxed{\int_a^b \frac{dt}{t} = \ln |b| - \ln |a|}$$

providing zero does not lie between a and b.

Logarithmic and Exponential Functions

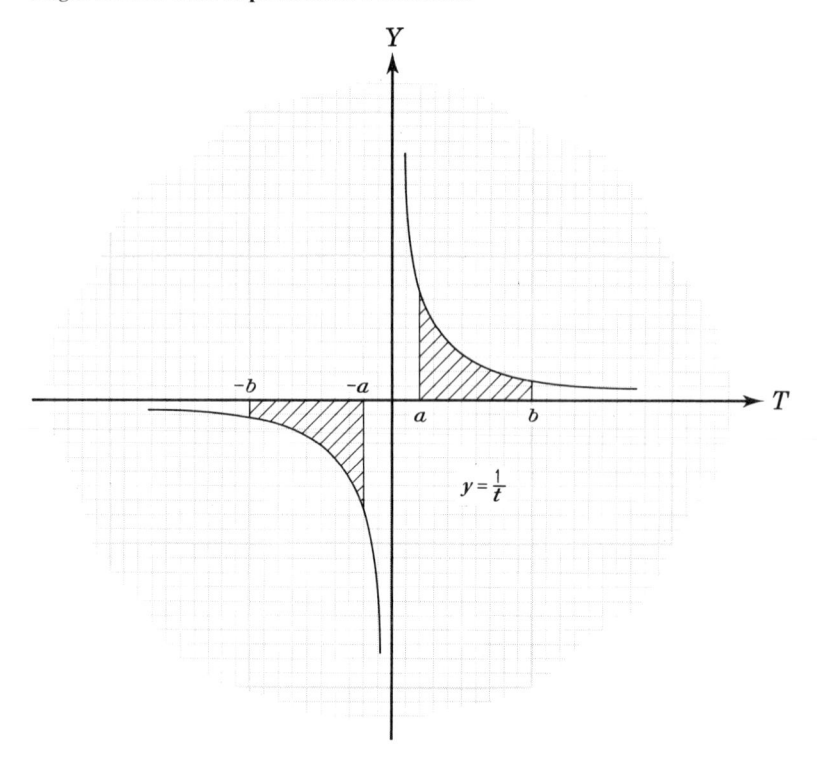

Figure 9-4

Since

$$\frac{d[\ln u(x)]}{dx} = \frac{1}{u(x)} \cdot \frac{du(x)}{dx}$$

for $u(x)$ a positive function, it follows from the fundamental theorem of the integral calculus (Sec. 7-8) that

$$\int_a^x \frac{1}{u(t)} \cdot \frac{du(t)}{dt} \, dt = \ln u(t) \Big|_a^x = \ln u(x) - \ln u(a).$$

Or, if you prefer

$$\int \frac{1}{u(x)} \cdot \frac{du(x)}{dx} \, dx = \ln u(x) + C \qquad \text{using Sec. 7-8.}$$

The argument for negative $u(x)$ is similar to that given above for $\int_{-b}^{-a} 1/t \, dt$. In general then, if $u = u(x)$ is continuous and never zero

Sec. 9-1. $\int \frac{dx}{x}$, **ln** x, **and** e 245

in the region of integration, then

$$\int \frac{1}{u}\frac{du}{dx}\,dx = \ln |u| + C.$$

Supply the missing steps in Prob. 1 of Set 9-1.

In more advanced work ln z will be defined for negative and even for complex values of z. This is needed in electrical theory, for example. Such logarithms are many-valued complex-numbered relations which will not be needed in this text. *The logarithm of zero is never defined.*

Example 1

Find y' where $y = \ln (x^3 - 5)^{\frac{2}{3}}$.

Using Rule III, we obtain

$$y = \frac{2}{3} \ln (x^3 - 5) = \frac{2}{3} \ln u.$$

Then $y' = \dfrac{2}{3} \dfrac{1}{u} \cdot \dfrac{du}{dx} = \dfrac{2 \cdot 3x^2}{3(x^3 - 5)} = \dfrac{2x^2}{x^3 - 5}.$

Example 2

Find $\displaystyle\int \frac{2x\,dx}{x^2 + 7}.$

$$\int \frac{2x\,dx}{x^2 + 7} = \int \frac{1}{x^2 + 7} \cdot 2x \cdot dx = \int \frac{1}{u}\frac{du}{dx}\,dx = \ln |u| + C$$

where $u = x^2 + 7$. (Why?) Thus

$$\int \frac{2x\,dx}{x^2 + 7} = \ln |x^2 + 7| + C.$$

Example 3

Find $\displaystyle\int \frac{(x - 1)\,dx}{x^2 - 2x + 3}.$

The form suggests $\displaystyle\int \frac{1}{u}\frac{du}{dx}\,dx$ with $u = x^2 - 2x + 3$. However, in this case we would need $du/dx = 2x - 2$, not $x - 1$. However, since $\int \frac{1}{2} f(x)\, dx = \frac{1}{2} \int f(x)\, dx$,

$$\int \frac{(x-1)\, dx}{x^2 - 2x + 3} = \int \frac{\frac{1}{2}(2x - 2)\, dx}{x^2 - 2x + 3} = \frac{1}{2} \int \frac{(2x - 2)\, dx}{x^2 - 2x + 3}$$
$$= \frac{1}{2} \int \frac{1}{u} \frac{du}{dx}\, dx = \frac{1}{2} \ln |x^2 - 2x + 3| + C.$$

Problem Set 9-1

1. Prove that if $u(x)$ is a negative-valued function of x, then
$$\frac{d(\ln |u|)}{dx} = \frac{1}{u} \cdot \frac{du}{dx} \quad \text{and} \quad \int \frac{1}{u} \frac{du}{dx}\, dx = \ln |u| + C.$$

2. (a) Use the relationship $x = e^{\ln x}$ to show that
$$\log_{10} x = (\log_{10} e) \cdot (\ln x).$$

(b) From this, deduce that $\dfrac{d \log_{10} u}{dx} = \dfrac{(\log_{10} e)}{u} \cdot \dfrac{du}{dx}.$

In Probs. 3 through 15, find $\dfrac{dy}{dx}$.

3. $y = 3 \ln x^5 + 5.$
4. $y = 3 \ln (x^5 + 5).$ Note the contrast with Prob. 3.
5. $y = \ln (1 - x^3).$ **6.** $y = \ln (3x^4 - 2x + 5).$
7. $y = \ln [(x^2 - 5x + 3)(2x - 7)].$
HINT: $\ln A \cdot B = \ln A + \ln B.$
8. $y = \ln \sqrt[5]{4x^3 - 3x - 7}.$ HINT: The relationship $\ln a^n = n \ln a$ provides a useful short cut.
9. Work Probs. 7 and 8 without using the hints, then using the hints, and compare the work involved.
10. $y = \ln [(x^2 - 5x + 2)^3 (x^3 - 17\pi)^5].$ HINT: See Probs. 7 and 8.
11. $y = \ln \dfrac{3x - 4}{2x + 5}.$ HINT: Use Rule II before taking the derivative.
12. (a) $y = (x^2 - 3) \ln x.$ (b) $y = \ln x^{(x^2 - 3)}.$
13. (a) $y = (\ln x)^3.$ This is often written $\ln^3 x.$
(b) $y = \ln (x^3).$ This is often written $\ln x^3.$
14. $y = \ln \dfrac{x^3}{x^2 + 5}.$ **15.** $y = \dfrac{7x}{\ln x}.$

Sec. 9-2. Use of Tables of ln x

In Probs. 16 through 25, perform the indicated integrations.

16. $\displaystyle\int \frac{dx}{x+3}.$ **17.** $\displaystyle\int 2\,\frac{x\,dx}{x^2+5}.$ **18.** $\displaystyle\int \frac{dx}{4-x}.$

19. $\displaystyle\int \frac{x\,dx}{x^2-5}.$ NOTE: This is not of the form $\displaystyle\int \frac{1}{u}\frac{du}{dx}\,dx.$ Why not?

20. $\displaystyle\int \frac{dx}{x\ln x}.$ HINT: If $u = \ln x$, what is $\dfrac{du}{dx}$?

21. $\displaystyle\int \frac{dy}{y^{\frac{1}{3}}(y^{\frac{2}{3}}-3)}.$ HINT: If $u = y^{\frac{2}{3}} - 3$, what is $\dfrac{du}{dx}$?

22. $\displaystyle\int \frac{3x-1}{x^2}\,dx.$

HINT: $\displaystyle\int \frac{3x-1}{x^2}\,dx = \int\left(\frac{3x}{x^2} - \frac{1}{x^2}\right)dx = 3\int\frac{dx}{x} - \int\frac{dx}{x^2} = ?$

23. $\displaystyle\int \frac{7z^2-5}{z}\,dz.$ HINT: $\dfrac{7z^2-5}{z} = \dfrac{7z^2}{z} - \dfrac{5}{z} = 7z - \dfrac{5}{z}.$

24. $\displaystyle\int \frac{x^{-\frac{1}{2}} - 6x^2}{x^{\frac{1}{2}} - x^3 + 4}\,dx.$ **25.** $\displaystyle\int \frac{(t - \frac{1}{2})^2\,dt}{t}.$

26. Find the vertices of the triangle of largest area which may be formed by the positive axes and a tangent to the curve $xy = 1$ in the first quadrant.

9-2. Use of Tables of ln x

Taking logarithms to the base 10 of both members of the identity

$$x = e^{\ln x}$$
$$\log_{10} x = \log_{10}(e^{\ln x})$$
$$= \ln x \cdot \log_{10} e.$$

Hence, $\ln x = \dfrac{\log_{10} x}{\log_{10} e} = \dfrac{\log_{10} x}{0.43429448}.$

This permits the calculation of ln x from a table of common (base 10) logarithms. Actually, tables of natural or naperian† logarithms (base e) are available. The student who plans to enter applied

---

† Named for John Napier, a Scot, who first published logarithmic tables in 1614, at which time exponents were unknown.

248 *Logarithmic and Exponential Functions*

mathematics or one of the sciences should become proficient in their use. The main difference in technique between the use of ln tables and $\log_{10}$ tables is that the *characteristic* in ln x is *not* merely the exponent of the power of 10.

In base 10	*In base e*
log 346 = log (3.46 × 10²)	ln 346 = ln (3.46 × 10²)
= log 3.46 + log 10²	= ln 3.46 + ln 10²
log 3.46 = 0.5391	ln 3.46 = 1.2413
log 10² = 2.0000	ln 10² = 4.6052
log 346 = 2.5391	ln 346 = 5.8465

Since $\ln 2 = \int_1^2 \frac{1}{t} \, dt$, we may set up the integral as the limit of

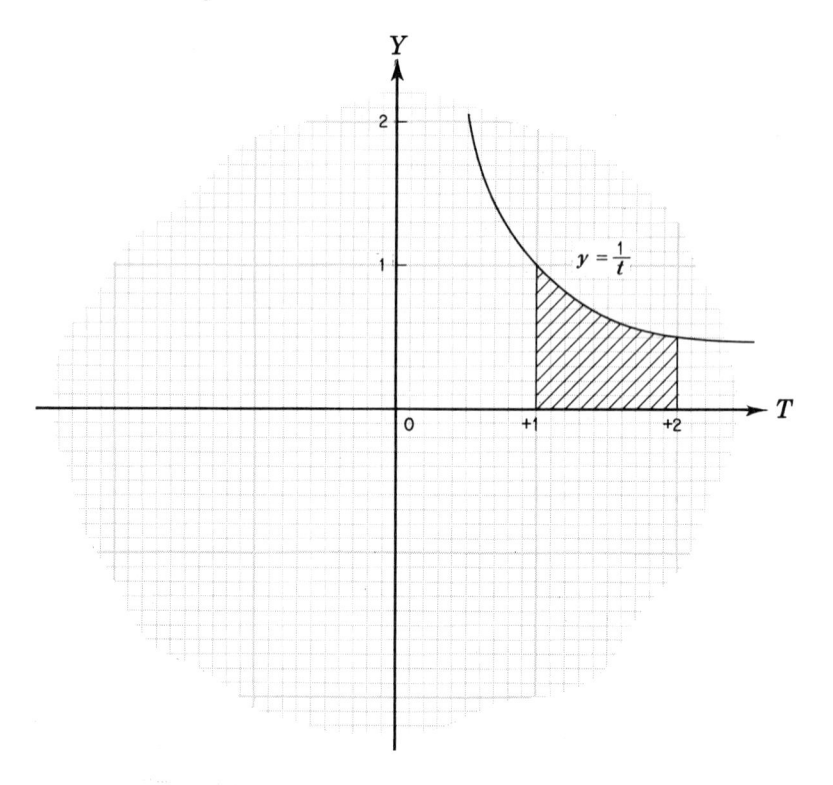

Figure 9-5

sums, and so obtain an expression for ln 2. From the definition of the definite integral (Sec. 7-6) we have

Sec. 9-2. Use of Tables of ln x

$$\ln 2 = \int_1^2 \frac{1}{t} dt = \lim_{n \to \infty} \sum_{i=0}^{n-1} \left[\left(\frac{1}{1 + (i/n)}\right) \cdot \left(\frac{1}{n}\right)\right]$$

$$= \lim_{n \to \infty} \sum_{i=0}^{n-1} \left[\left(\frac{n}{n + i}\right) \cdot \left(\frac{1}{n}\right)\right]$$

$$= \lim_{n \to \infty} \sum_{i=0}^{n-1} \left(\frac{1}{n + i}\right).$$

Thus $\ln 2 = \lim_{n \to \infty} \left(\frac{1}{n} + \frac{1}{n+1} + \frac{1}{n+2} + \cdots + \frac{1}{n+n-1}\right).$

In a similar fashion, it is possible to find $\ln k$ as the limit of a sum of fractions for any $k > 0$. The actual exact arithmetic computation of $\ln 2$ is impossible (just as is the computation of $\log_{10} 2$, or of $\sqrt{3}$, or of π since none of these expressions is rational). Approximations, accurate to five or seven (or even several hundred) decimal places, may be obtained using series methods. The series obtained above, $\frac{1}{n} + \frac{1}{n+1} + \cdots + \frac{1}{n+n}$, is not particularly convenient for actual computation, but it illustrates the reasoning more easily than advanced techniques.

Example

$$\int_2^4 \frac{dx}{x+1} = \ln(x+1)\Big|_2^4 = \ln 5 - \ln 3$$
$$= 1.60944 - 1.09861 = 0.51083.$$

Problem Set 9-2

In Probs. 1 through 8, evaluate the indicated integral by using previous theorems and appropriate numerical tables.

1. $\int_2^3 \frac{dx}{x}.$ *Ans.* 0.40547.

2. $\int_5^{30} \frac{dx}{x+5}.$ *Ans.* $\ln \frac{35}{10} = \ln 3.5 = 1.25276.$

3. $\int_0^1 \frac{t\,dt}{t^2+1}.$

4. $\int_e^5 \frac{dx}{x \log_e x}.$ HINT: If $u = \log_e x = \ln x$, then $\frac{du}{dx} = \frac{1}{x}.$

5. $\int_1^3 \frac{(x+2)\,dx}{x^2+4x+3}.$

250 *Logarithmic and Exponential Functions*

6. (a) $\displaystyle\int_1^2 \frac{dz}{5z-18}.$

   (b) $\displaystyle\int_3^4 \frac{dz}{5z-18}.$ HINT: Part (b) is a booby trap.

7. $\displaystyle\int_5^{10} \frac{dt}{t-1}.$

8. $\displaystyle\int_2^4 \frac{3x-1}{x^2}\, dx.$ HINT: See Prob. 22, Set 9-1.

9. A piston encloses gas in a cylinder. The force exerted on the piston by the gas is $F = 4/x$, where x is the distance between the end of the piston and the end of the cylinder. The work done in compressing the gas to one-half its volume is given by $W = k\displaystyle\int_1^2 F\, dx$ where k depends upon the units used. Compute W.

10. Find the area bounded by the lines $x = 5$, $x = 7$, $y = 0$ and the curve $xy = 4$.

11. Find the area bounded by $x + y = 10$ and $xy = 4$.

12. A culture contains $N(t)$ bacteria at time t ($0 \leq t \leq 100$ hr). It is known that $\dfrac{dN}{dt} = \dfrac{N}{4}$ and that $N(0) = 1{,}000$. Find $N(t)$.

9-3. The Inverse Function of $y = \ln x$, Namely, $y = e^x$

The relation $x = e^{\ln x}$ establishes that $E(x) = e^x$ and $L(x) = \ln x$ are inverse functions. That is, since for any $x > 0$, $e^{\ln x} = x$ and $\ln(e^x) = x \cdot \ln e = x \cdot 1 = x$, it follows that $E[L(x)] = x$ and $L[E(x)] = x$. The graphs $y = e^x$ and $y = \ln x$ are reflections (mirror images) of one another about the line $y = x$, as is usual with inverse functions (see Fig. 9-6).

We now proceed to find de^u/dx and $\int e^t\, dt$.

Let $E(x) = e^u$

then $\ln[E(x)] = \ln e^u = u$

$$\frac{d\{\ln[E(x)]\}}{dx} = \frac{1}{E(x)} \cdot \frac{d[E(x)]}{dx} = du$$

$$\frac{1}{E(x)} \frac{dE(x)}{dx} = \frac{du}{dx}$$

$$\frac{dE(x)}{dx} = E(x)\frac{du}{dx}$$

or

$$\frac{de^u}{dx} = e^u \frac{du}{dx}.$$

Sec. 9-3. The Inverse Function of $y = \ln x$, Namely, $y = e^x$ 251

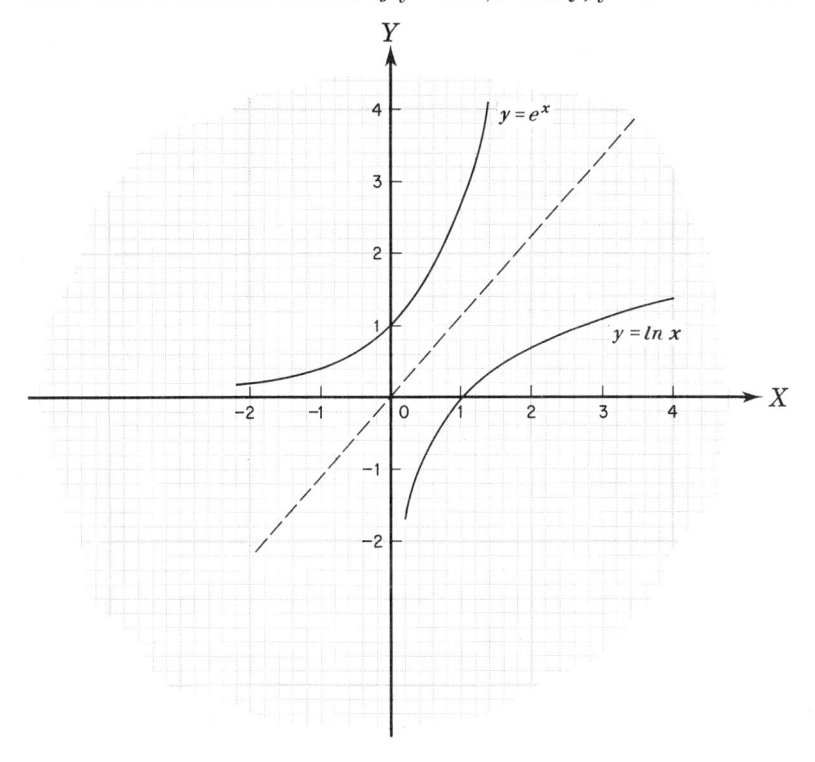

Figure 9-6

Then by the fundamental theorem, we have

$$\int e^u \frac{du}{dx} dx = e^u + C.$$

This rather remarkable property of the function, $a \cdot e^x$, of being its own derivative, namely,

$$\frac{d(a \cdot e^x)}{dx} = a \cdot e^x,$$

is unique† among continuous functions.

Example

$$\frac{d[x^2 \cdot e^{(4x^2-7x+3)}]}{dx} = \frac{d(u \cdot v)}{dx}$$

† The astute reader may well point out that the zero function, $f(x) = 0$, has this property. However, this function is included in the family $a \cdot e^x$ by taking $a = 0$.

with $u = x^2$ and $v = e^{(4x^2-7x+3)}$. Hence

$$\frac{d[x^2 \cdot e^{(4x^2-7x+3)}]}{dx} = \frac{d(u \cdot v)}{dx} = u\frac{dv}{dx} + v\frac{du}{dx}$$

$$= x^2 \cdot e^{(4x^2-7x+3)} \cdot (8x - 7) + e^{(4x^2-7x+3)} \cdot (2x).$$

Problem Set 9-3

1. If $f(x) = 3e^{7x}$, find $f'(x)$. 2. If $y = e^{-x^2}$, find $\frac{dy}{dx}$.

3. Find y' if $y = (x^4 - 3x)e^{x^3-5}$. 4. Find y' if $y = e^{-3x} \log 5x$.

5. Find the slope of $y = e^x$ at $x = 5$. Use tables.

6. Find $\phi'(t)$ if $\phi(t) = \dfrac{t^2 + 3t}{e^{4t}}$. 7. $\displaystyle\int \frac{dt}{e^t} = -\int e^{-t}(-dt) = ?$

8. $\displaystyle\int (e^4 + x)\,dx$. Ans. $xe^4 + \tfrac{1}{2}x^2 + C$.

9. In radioactive decay and in many other natural phenomena, it has been found that the *rate of change of the amount present* is proportional to the amount present. That is, if x represents the amount present, then $\dfrac{dx}{dt} = kx$. If $k < 0$, then decay takes place. (Why?) Find x as a function of time t by solving $\displaystyle\int \frac{dx}{x} = \int k\,dt$. The constant of integration C and the proportionality constant k must be determined by "boundary value conditions." For example, if it is known that when $t = 0$, $x = 2$ and that when $t = 1{,}500$, $x = 1$, it is possible to evaluate these constants.

10. Since radium has a half life of 1,500 years, the function derived in Prob. 9 may be used to tell how much radium will be present t years after a given 2-gram sample was observed.
  (a) Find the amount present after 1,000 years.
  (b) In how many years will only 0.5 gram remain?

11. Find the vertices of the triangle of largest area which may be formed by the positive coordinate axes and a line tangent to $y = e^{-x}$.

12. Sketch the functions f and g given by:
  (a) $f(x) = \dfrac{e^x + e^{-x}}{2}$. This function is often called the hyperbolic cosine of x, abbreviated cosh x.
  (b) $g(x) = \dfrac{e^x - e^{-x}}{2}$. This function is often called the hyperbolic sine of x, abbreviated sinh x.

Sec. 9-4. Self-test

13. Prove, using the definitions of Prob. 12, that

$$\frac{d}{dx} \cosh x = \sinh x,$$

$$\frac{d}{dx} \sinh x = \cosh x.$$

14. Let $H(x) = (\cosh x)^2 - (\sinh x)^2$. Form $H'(x)$, and from this deduce that $H(x) = \cosh^2 x - \sinh^2 x = $ a constant. Find $H(0)$ to evaluate this constant, thus deriving a fundamental identity of hyperbolic functions.

15. Define $\tanh x = \dfrac{\sinh x}{\cosh x}$ and $\operatorname{sech} x = \dfrac{1}{\cosh x}$. From these definitions and the results of Probs. 12 and 13, show that

$$\frac{d \tanh x}{dx} = \operatorname{sech}^2 x.$$

16. If $y = (x^2 - 3)^4 \sqrt[3]{(3x - 11)^5}$, find $\dfrac{dy}{dx}$. HINT: A neat short cut makes use of the fact that

$$\log y = 4 \log (x^2 - 3) + \frac{5}{3} \log (3x - 11).$$

Then $\dfrac{\dfrac{dy}{dx}}{y} = 4 \cdot \dfrac{2x}{x^2 - 3} + \dfrac{5}{3} \cdot \dfrac{3}{3x - 11}.$

17. Find $f'(x)$, if $f(x) = (x^2 - 5)^5 \cdot (4x + 9)^{13} \sqrt[3]{12x + 7}$. HINT: See hint of Prob. 16.

9-4. Self-test

1. Sketch the following graphs on the same set of axes and identify each. Show the intercept and the coordinates of the point at which each curve crosses the line $x = 2$.

(a) $y = \ln x$.   (b) $y = e^x$.   (c) $y = 10^x$.   (d) $y = 2e^x$.

2. Determine the equation of the line tangent to the curve $y = \ln x$ at the point $(e, 1)$ which lies on the curve. Give an approximation of the slope and y intercept of this line which is accurate to the nearest tenth of a unit.

3. Integrate: (a) $\displaystyle\int 5e^{3x}\, dx.$ (b) $\displaystyle\int \frac{2x\, dx}{x^2 + 1}.$ (c) $\displaystyle\int \frac{4x^2 + x - 3}{x^2}.$

4. Differentiate, using logarithmic techniques:
(a) $y = (4x - 3)^3 \sqrt[5]{(2x + 7)^2}$.
(b) $y = \ln\left(\dfrac{4x - 3}{x^2 + x + 1}\right)$.
(c) $y = 4e^{3x^2} + 2$.

5. Use a table of natural logarithms to determine

(a) $\ln 436$. \qquad (b) $\ln 0.00273$. \qquad (c) $\ln (1.62 \times 10^{30})$.

6. Sketch $y = 1/t$ and shade an area which is equal to $\ln 4$ square units.

7. Prove that $(\cosh x)^2 - (\sinh x)^2 = 1$ either directly from the definition, using algebra, or more subtly using the derivative of $H(x) = (\cosh x)^2 - (\sinh x)^2$.

8. Which is larger, e^π or π^e? Don't guess.*

---

* The astute reader may recognize this as a trap at once. If not, it may be advisable for you to reread the bottom half of page 150. Your instructor may or may not wish to discuss the intricacies which arise when exponents are generalized to include the possibility of non-rational exponents such as 2^π or even $2^{\sqrt{3}}$. Actually exponents have been generalized to include complex numbers as exponents, and the theory is indeed fruitful, but you are *not* ready to discuss 2^{i+3} yet.

10

Trigonometric Functions

10-1. Trigonometric Definitions

Trigonometric functions are defined, not for angles, but *for numbers.* Electrical theory makes use of the "phase angle." However, phase angle is not an angle at all, but a displacement, that is, a distance or a number. It is possible to associate phase angle with an actual angle by considering a rotating vector field. However, this is unnecessary and often undesirable. In chemistry the rate of ion exchange may be proportional to tan t, where t is *time* (a number), not an angle. In small vibration theory (flutter analysis in aeronautical engineering, for example) it is even necessary to take the sine, cosine, and tangent of a matrix. Trigonometric functions of complex numbers, such as sin $(3 - 5i)$, are in regular use in electronics.

The trigonometric functions of a real number t are defined by constructing a circle with center $(0,0)$ and radius 1, as in Fig. 10-1a. The point $(1,0)$ is called the starting point. Measure along the circumference of the circle beginning at the starting point $(1,0)$ an arc of length t. If t is a positive number, measure in the counterclockwise direction. If t is negative, measure in the clockwise direction. Let $P(C_t, S_t)$ be the end point of this arc.

In doing this, we have set up a correspondence between the points on a number axis and the points on the circumference of a unit circle by placing the origin of the number axis at the point $(0,1)$ on the circle and then wrapping the number line around the circle.

255

256   *Trigonometric Functions*

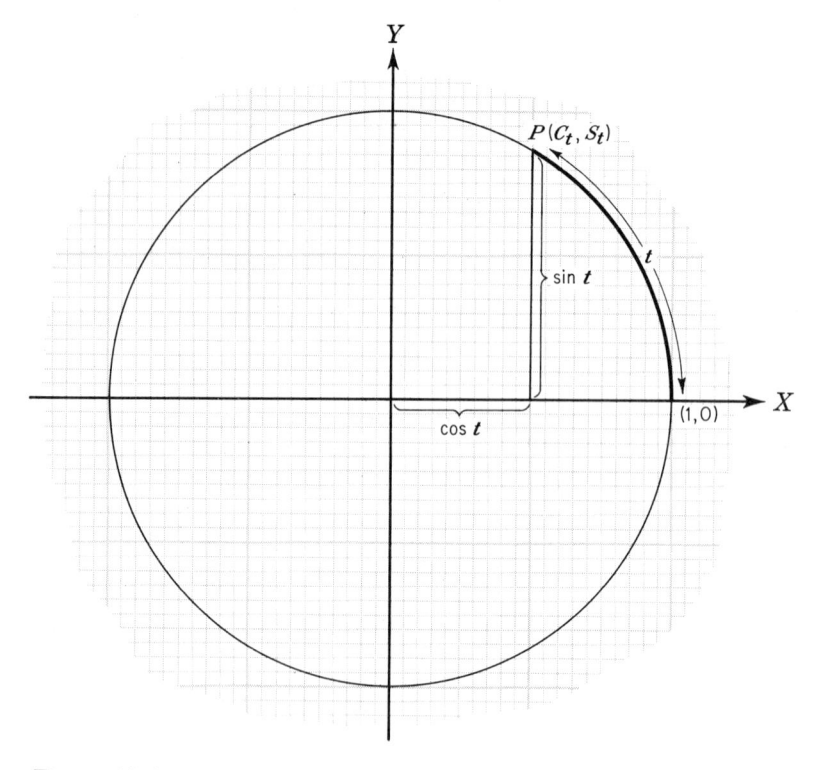

Figure 10-1a

For each point t on the number line, there corresponds a unique point $P(C_t, S_t)$ on the circle. However, each point on the circle is the correspondent of infinitely many points on the number axis since the line is wrapped around the circle many times. Such a correspondence satisfies the definition of a function since for each real number t (that is, to each point on the number axis) there corresponds a unique point $P(C_t, S_t)$ on the circle. We use the coordinates of this point to define the trigonometric functions of the number t. (See Fig. 10-1b.)

The cosine of the real number t is defined to be the x coordinate of $P(C_t, S_t)$. The y coordinate of $P(C_t, S_t)$ is sine t (sin t). Furthermore:

$$\tan t = \frac{\sin t}{\cos t}, \quad \cot t = \frac{\cos t}{\sin t},$$

$$\sec t = \frac{1}{\cos t}, \quad \csc t = \frac{1}{\sin t},$$

Sec. 10-1. Trigonometric Definitions 257

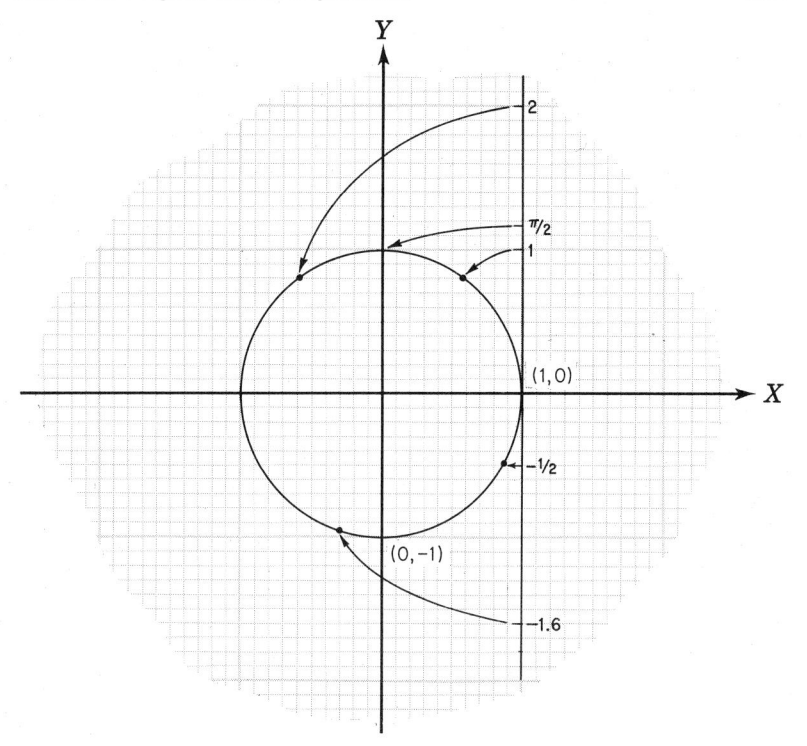

Figure 10-1b

when these fractions have meaning. In case a denominator is zero, we say the corresponding function is *undefined* for that number t. These definitions are consistent with the angular definitions often given in trigonometry, and for many purposes they are much more desirable. When trigonometric tables are computed, they are computed for the trigonometric functions of a number, not of an angle. It is possible to convert them into tables of trigonometric functions of a degree, but the actual computation is done for trigonometric

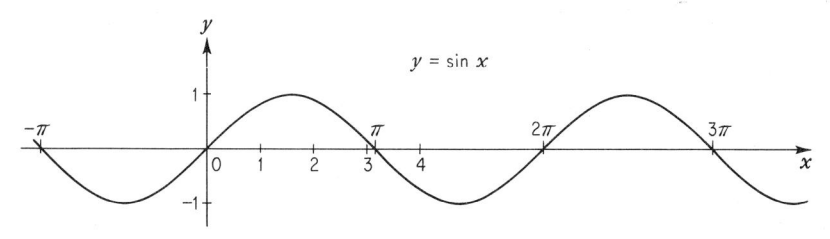

Figure 10-2

258   *Trigonometric Functions*

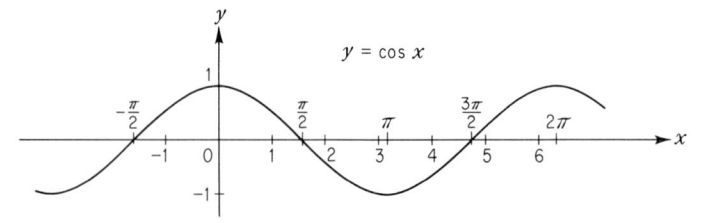

Figure 10-3

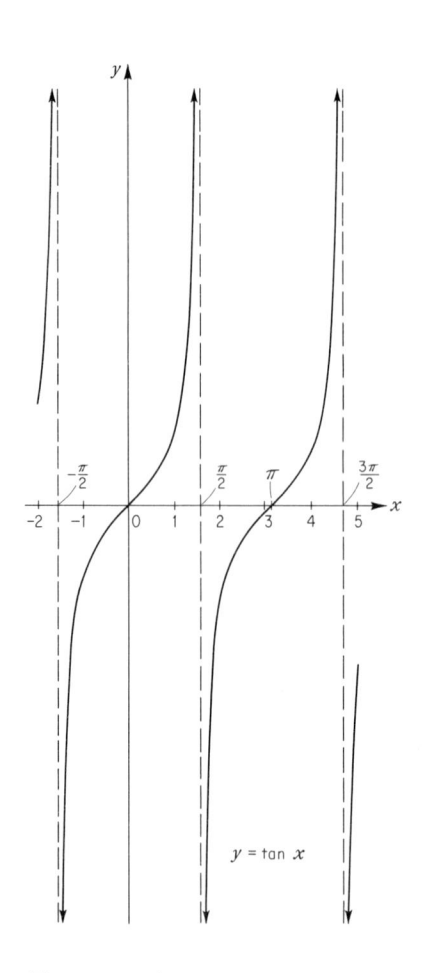

Figure 10-4

Sec. 10-1. Trigonometric Definitions

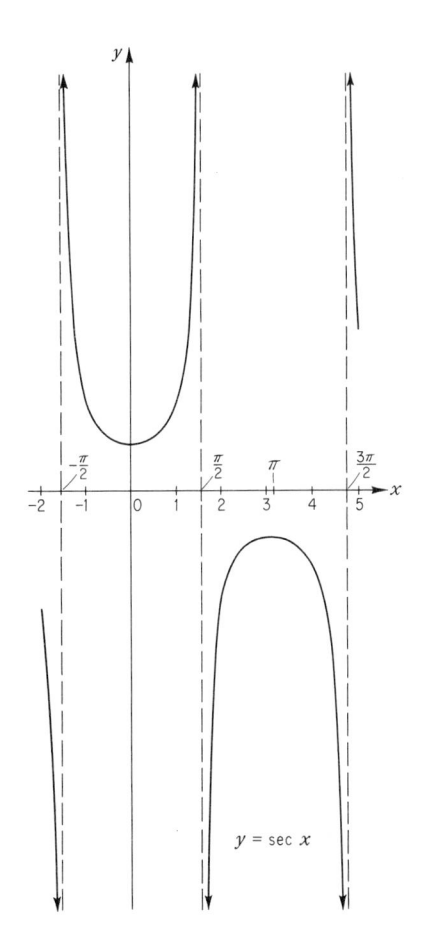

Figure 10-5

functions of numbers. Differentiation and integration of trigonometric functions of numbers are neater than for functions of angles in degrees, just as the derivative of ln x is neater than the derivative of $\log_{10} x$.

Graphs of the functions $y = \sin x$, $y = \cos x$, $y = \tan x$, $y = \sec x$, followed by the graphs of $y = a \sin bx$ for several values of a and b are presented in Figs. 10-2 to 10-7. The reader is expected to verify that these graphs are consistent with the definitions just given.

260   *Trigonometric Functions*

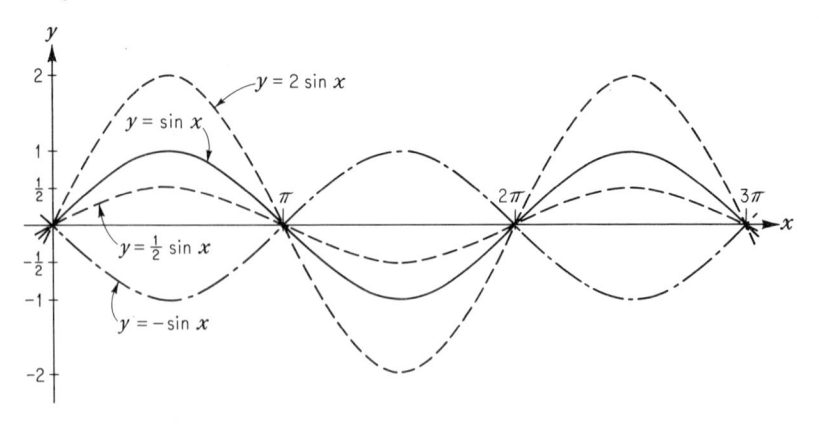

Figure 10-6

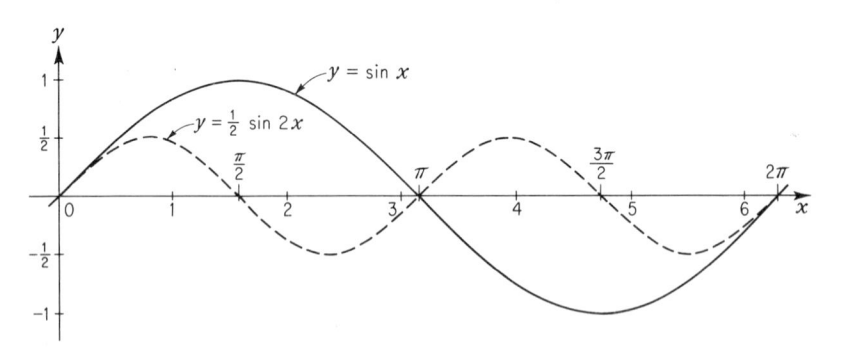

Figure 10-7

10-2. Limits of Trigonometric Functions

Let t be a real number such that $-\pi/2 < t < \pi/2$. The wrapping function of Sec. 10-1 maps t onto the right half of the unit circle. If a line is drawn through the origin and through the end point P of the arc of length t on the circle used in the definitions of the trigonometric functions, and if the tangent to the circle at the point $S(1,0)$ is drawn, then the distance from $S(1,0)$ to the intersection Q of the tangent line and the line OP is equal in length to the tangent of the number t. The student is asked to prove this geometrically in the next set of exercises. He may also show that the secant of t is equal to the hypotenuse of the right triangle OSQ. This is presented in Fig. 10-8. The given diagram is valid for $-\pi/2 < t < \pi/2$.

Sec. 10-2. Limits of Trigonometric Functions

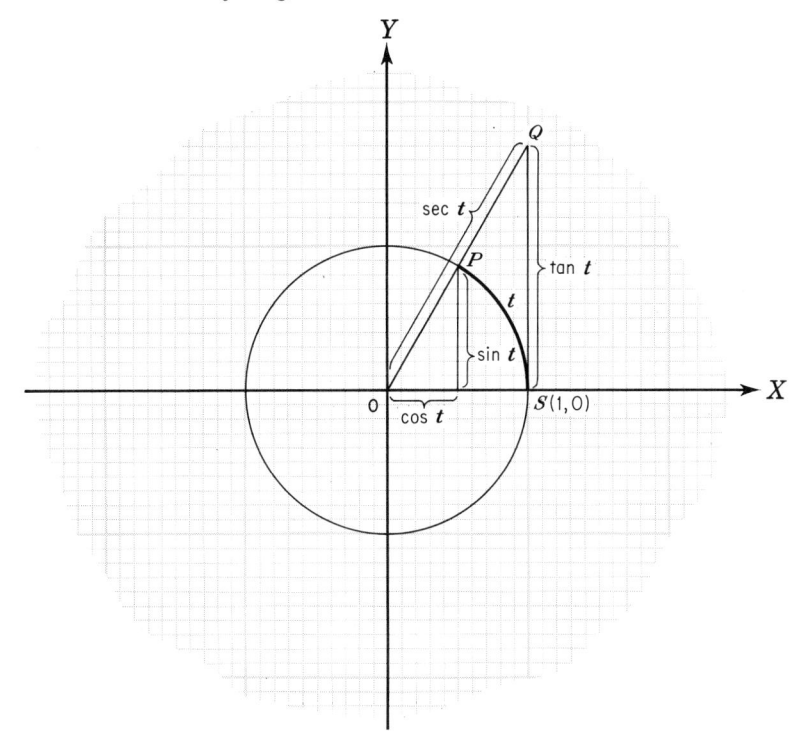

Figure 10-8

Problem Set 10-2

1. Sketch the curve $y = 3 \cos 5x$ for x between $-\pi$ and $3\pi/2$.
2. Find the limit of $\sin s$ as s approaches $\pi/4$.
3. Find the limit of $\tan s$ as s approaches 0.
4. Find the limit of $\tan s$ as s approaches $\pi/2$.
5. Find the limit of $\sec s$ as s approaches π.
6. Find the limit of $\sec s$ as s approaches $\pi/2$.
7. Find the limit of $\sin 7s$ as s approaches $\pi/3$.
8. Find the limit of $\sec 3s$ as s approaches $\pi/6$.
9. Use the unit circle diagram given in Fig. 10-8 and the areas of certain triangles and a sector of a circle to show the validity of $\sin t \cdot \cos t < t < \tan t$ if $0 < t < \pi/2$.
10. Use the unit circle diagram given in Fig. 10-8 and the areas of certain triangles and a sector to show that if $-\pi/2 < t < 0$, then $\sin t \cdot \cos t > t > \tan t$.

11. Using the results of Probs. 9 and 10, show the validity of the inequality $\cos t < \dfrac{t}{\sin t} < \dfrac{1}{\cos t}$ if $-\dfrac{\pi}{2} < t < \dfrac{\pi}{2}$ and $t \neq 0$.

12. Show that $\lim\limits_{t \to 0} \dfrac{t}{\sin t} = 1$. Show also that $\lim\limits_{t \to 0} \dfrac{\sin t}{t} = 1$. Be sure to use the definition of t as a number, as given in Sec. 10-1.

13. Sketch the function $f(x) = \dfrac{\sin x}{x}$. What value must be given to $f(0)$ in order to make $f(x)$ a continuous function? Note that $f(x)$ is not defined for $x = 0$ as given.

14. Show that the segment of the line tangent to the unit circle at the point $(1,0)$ between the points $(1,0)$ and the line drawn through the center of the circle and the end point of the arc t is equal in length to $\tan t$ (see Fig. 10-8).

15. Show that the hypotenuse of the triangle formed by the x axis, the line tangent to the unit circle at $(1,0)$, and the line through the origin and the end point of the arc t is equal in length to $\sec t$ (see Fig. 10-8).

10-3. Derivatives of $\cos u$ and $\sin u$

By noting the areas of triangles OAP and OSQ and the area of the sector of the circle OSP in Fig. 10-9, we see that

Area of triangle OAP < area of sector OSP < area of triangle OSQ.

Recall that OA is $\cos t$ and AP is $\sin t$, OS is 1 and SQ is $\tan t$. Since the area of a sector of a unit circle whose periphery is t units is $t/2$ units, we have

$$0 < \tfrac{1}{2} \sin t \cdot \cos t < \tfrac{t}{2} < \tfrac{1}{2} \tan t \quad \text{if} \quad 0 < t < \tfrac{\pi}{2}.$$

Upon dividing by $\dfrac{\sin t}{2} > 0$,

$$0 < \cos t < \dfrac{t}{\sin t} < \dfrac{\tan t}{\sin t}, \quad \text{if } 0 < t < \dfrac{\pi}{2}.$$

If $-\dfrac{\pi}{2} < t < 0$, $\quad 0 > \tfrac{1}{2} \sin t \cos t > \tfrac{t}{2} > \tfrac{1}{2} \tan t$.

Since $\dfrac{\sin t}{2} < 0$, the inequalities must be reversed upon dividing

Sec. 10-3. *Derivatives of cos u and sin u* 263

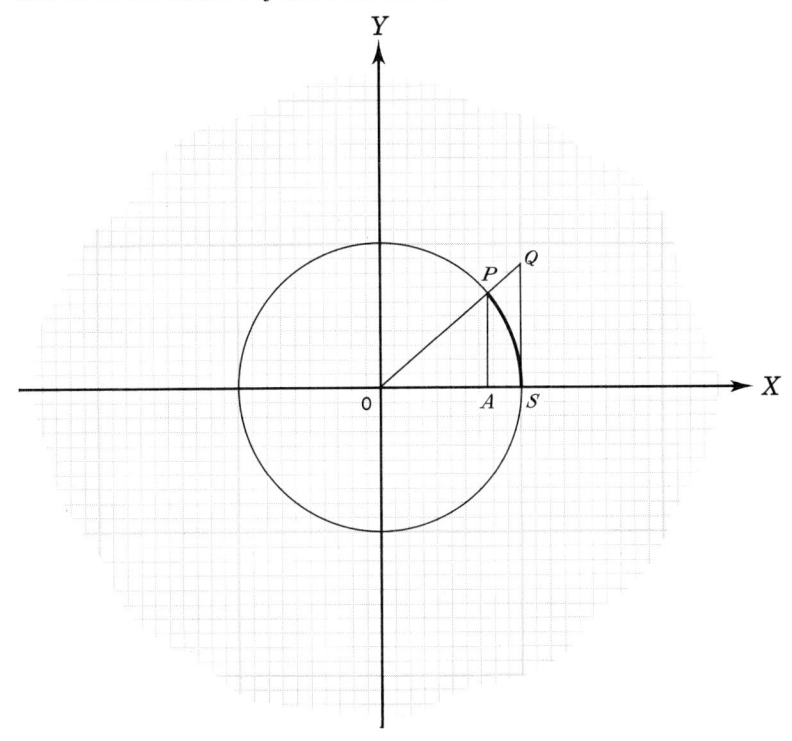

Figure 10-9

by $\frac{\sin t}{2}$ to obtain

$$0 < \cos t < \frac{t}{\sin t} < \frac{\tan t}{\sin t} \quad \text{if } \frac{-\pi}{2} < t < 0.$$

If $t \neq 0$, then $\frac{\tan t}{\sin t} = \frac{1}{\cos t}$.

Thus $0 < \cos t < \frac{t}{\sin t} < \frac{1}{\cos t}$, if $\frac{-\pi}{2} < t < \frac{\pi}{2}, t \neq 0$.

Inverting and changing the order of the inequalities, we obtain

$$\frac{1}{\cos t} > \frac{\sin t}{t} > \cos t \quad \text{if } t \neq 0 \text{ and } \frac{-\pi}{2} < t < \frac{\pi}{2}.$$

Since $\lim_{t \to 0}$ implies that t does not equal 0, the given relationship may

Trigonometric Functions

be used to show

$$\lim_{t \to 0} \frac{1}{\cos t} \geq \lim_{t \to 0} \frac{\sin t}{t} \geq \lim_{t \to 0} \cos t$$

$$1 \geq \lim_{t \to 0} \frac{\sin t}{t} \geq 1.$$

Hence, $\lim_{t \to 0} \frac{\sin t}{t} = 1$.

Upon writing $\frac{\Delta x}{2}$ for t, since $\Delta x \to 0$ implies $\frac{\Delta x}{2} \to 0$, we have

$$\lim_{\Delta x \to 0} \frac{\sin \frac{\Delta x}{2}}{\frac{\Delta x}{2}} = 1.$$

This limit will be made the basis of the calculus of trigonometric and inverse trigonometric functions.

To compute $\frac{d \sin x}{dx}$, recall (or consult any trigonometry text or handbook) that

$$\sin a - \sin b = 2 \cos \left(\frac{a+b}{2} \right) \cdot \sin \left(\frac{a-b}{2} \right)$$

and make use of the basic definition of derivative as

$$\lim_{\Delta x \to 0} \frac{f(x + \Delta x) - f(x)}{\Delta x} \quad \text{from Sec. 4-2.}$$

Let $f(x) = \sin x$. Then

$$f(x + \Delta x) - f(x) = \sin (x + \Delta x) - \sin x.$$

Taking $a = x + \Delta x$ and $b = x$ in the trigonometric identity above, we obtain

$$f(x + \Delta x) - f(x) = 2 \cos \left(\frac{x + \Delta x + x}{2} \right) \cdot \sin \left(\frac{x + \Delta x - x}{2} \right)$$

$$= 2 \cos \left(x + \frac{\Delta x}{2} \right) \cdot \sin \left(\frac{\Delta x}{2} \right).$$

Therefore, $\frac{f(x + \Delta x) - f(x)}{\Delta x} = 2 \cos \left(x + \frac{\Delta x}{2} \right) \cdot \frac{\sin \frac{\Delta x}{2}}{\Delta x}$

$$= \cos \left(x + \frac{\Delta x}{2} \right) \cdot \frac{\sin \frac{\Delta x}{2}}{\frac{\Delta x}{2}}.$$

Sec. 10-3. Derivatives of cos u and sin u

Now, by passing to the limit as $\Delta x \to 0$, we obtain

$$\frac{d \sin x}{dx} = \lim_{\Delta x \to 0} \frac{f(x + \Delta x) - f(x)}{\Delta x}$$

$$= \lim_{\Delta x \to 0} \left[\cos\left(x + \frac{\Delta x}{2}\right) \cdot \frac{\sin \frac{\Delta x}{2}}{\frac{\Delta x}{2}} \right]$$

$$= \lim_{\Delta x \to 0} \left[\cos\left(x + \frac{\Delta x}{2}\right) \right] \cdot \lim_{\Delta x \to 0} \frac{\sin \frac{\Delta x}{2}}{\frac{\Delta x}{2}}$$

$$= \lim_{\Delta x \to 0} \left[\cos\left(x + \frac{\Delta x}{2}\right) \right] \cdot 1 \quad \text{(Why?)}$$

$$= \cos x.$$

Applying the theorem of Sec. 4-10 produces the more general result

$$\boxed{\frac{d \sin u}{dx} = \cos u \frac{du}{dx}}$$

where $u = u(x)$ is a differentiable function of x.

It is possible to determine $\dfrac{d \cos u}{dx}$ in a similar fashion. However, instead, we shall employ the identity $\cos u = \sin(\pi/2 - u)$ to determine $\dfrac{d \cos u}{dx}$.

$$\frac{d(\cos u)}{dx} = \frac{d}{dx}\left[\sin\left(\frac{\pi}{2} - u\right)\right]$$

$$= \cos\left(\frac{\pi}{2} - u\right) \cdot \frac{d}{dx}\left(\frac{\pi}{2} - u\right)$$

$$= \cos\left(\frac{\pi}{2} - u\right) \cdot \left(-\frac{du}{dx}\right)$$

$$= -\sin(u) \frac{du}{dx}.$$

Hence, we have shown

$$\frac{d(\sin u)}{dx} = \cos u \frac{du}{dx}$$

$$\frac{d(\cos u)}{dx} = -\sin u \frac{du}{dx}.$$

Trigonometric Functions

From these we may obtain derivatives of the other trigonometric functions by using the formulas

$$\frac{d(u \cdot v)}{dx} = u\frac{dv}{dx} + v\frac{du}{dx}$$

and $$\frac{d(u/v)}{dx} = \frac{v\frac{du}{dx} - u\frac{dv}{dx}}{v^2}.$$

Example 1

$$\frac{d\, 5\sin(x^2 - 4x)}{dx} = 5[\cos(x^2 - 4x)](2x - 4).$$

Example 2

$$\frac{d\,(e^{\sin 3x})}{dx} = e^{\sin 3x}(\cos 3x) \cdot 3 = 3\cos 3x \cdot e^{\sin 3x}.$$

Problem Set 10-3

In Probs. 1 through 14, find the derivative of the function given.

1. $f(x) = 3\cos(2x - 5)$.
2. $g(t) = e^{3\sin 2t}$.
3. $f(x) = e^{x^2} - 5\sin x$.
4. $F(x) = \sin(1/x)$.
5. $G(y) = \cos(\sin y)$.
6. $B(t) = 3\sin^2 t + 3\cos^2 t$.
7. $f(x) = x\sin x$.
8. $g(t) = \cos\sqrt{t}$.
9. $R(t) = \sqrt{1 + \sin t}$.
10. $L(x) = \ln(\sin x)$.
11. $K(x) = 9\cos(x^2)\sin(x^3)$.
12. (a) $\Gamma_1(x) = \ln(1/\sin x)$.
    (b) $\Gamma_2(x) = \ln(\cos x)$.
13. $p(x) = e^{-x}\sin x + 9$.
14. $c(x) = x + \dfrac{\sin 2x}{2}$.

15. Find the equation of the line tangent to $y = \sin x$ at $(0,0)$.

16. Find $\dfrac{d\,\tan x}{dx}$. HINT: $\tan x = \dfrac{\sin x}{\cos x} = \dfrac{u}{v}$. Now use the relation

$$\frac{d\left(\dfrac{u}{v}\right)}{dx} = \frac{v\dfrac{du}{dx} - u\dfrac{dv}{dx}}{v^2}.$$

17. Find $\dfrac{d\,\cot x}{dx}$.
18. Find $\dfrac{d\,\sec x}{dx}$.
19. Find $\dfrac{d\,\csc x}{dx}$.

20. (a) The following curves pass through the origin. Determine

Sec. 10-3. Derivatives of cos u and sin u

the relative steepness of each curve *at the origin;* that is, rank the curves in order of steepness at (0,0).

$$y = e^x - 1 \qquad y = \tan x$$
$$y = \sin x \qquad y = x$$

(b) Rank the curves of part (a) in steepness at $x = 0.01$.
(c) Rank them at $x = 0.1$.
(d) Consider y'' at $x = 0$ and $x = .01$ for each curve listed above. $y'' = \dfrac{d(y')}{dx}$ indicates the *rate of change of the slope* of $y = f(x)$. (Why?) Are the results of (d) compatible with the results you obtained in parts (a), (b), and (c) above? Explain.

21. Show that $\dfrac{d}{dx} \tan(x/2) = \tfrac{1}{2} \sec^2(x/2)$.

22. Show that $\dfrac{d}{dx} \cos^2(1/x) = \dfrac{1}{x^2} \sin(2/x)$.

23. Show that $\dfrac{d}{dx}(\tan x - x) = \tan^2 x$.

24. Differentiate (a) $x \sin(1/x)$. (b) $x^2 \sin(1/x)$.

25. If $y = [x]e^x$, where $[x]$ = the greatest integer $\leq x$, find $\dfrac{dy}{dx}$. Be sure of your answer. Does y exist for $x = 3$? Does y'?

26. (a) Integrate $\int \tan x \, dx$. HINT:

$$\int \tan x \, dx = \int \frac{\sin x \, dx}{\cos x} = -\int \frac{-\sin x \, dx}{\cos x} = -\int \frac{1}{u} \frac{du}{dx} dx$$

where $u = \cos x$.
(b) What restrictions are placed on the limits of integration a and b if $\displaystyle\int_a^b \tan x \, dx$ is meaningful?

27. (a) Integrate $\int \cot x \, dx$.
(b) What restrictions are placed on the limits of integration a and b if $\displaystyle\int_a^b \cot x \, dx$ is meaningful?

28. Consider the results of one of the Probs. 17, 18, or 19. From this deduce the integral of some function $f(x)$ which has not yet been considered in this text. Are there any restrictions on the limits of integration a, b, if $\displaystyle\int_a^b f(x) \, dx$ is to have meaning for the function f you have chosen?

268   *Trigonometric Functions*

29. One table lists $\int \tan x \, dx = -\ln \cos x$ while another table lists $\int \tan x \, dx = \ln \sec x$. Which is correct? Explain.

10-4. Derivatives of Other Trigonometric Functions

In the last problem set, we computed the derivative of $\tan x$. A little trigonometric juggling† brings the result into the form

$$\frac{d \tan u}{dx} = \sec^2 u \, \frac{du}{dx}.$$

In a similar fashion:

$$\frac{d \cot u}{dx} = -\csc^2 u \, \frac{du}{dx}.$$

$$\frac{d \sec u}{dx} = \sec u \cdot \tan u \, \frac{du}{dx}.$$

$$\frac{d \csc u}{dx} = -\csc u \cdot \cot u \, \frac{du}{dx}.$$

10-5. Integration of Trigonometric Functions

The graph of the curve $y = \sin x$ crosses the x axis at $x = 0$ and $x = \pi$ (see Fig. 10-10). The curve is continuous at every point. Hence, $\int_0^\pi \sin x \, dx$ is meaningful. It is a *number*, actually, the area bounded by the x axis and one arch of the sine curve, since

$$\int_0^\pi \sin x \, dx = \lim_{n \to \infty} \sum_{i=0}^{n-1} \left[\left(\sin \frac{\pi i}{n} \right) \cdot \left(\frac{\pi}{n} \right) \right]$$

by definition.

† $\dfrac{d \tan u}{dx} = \dfrac{d \dfrac{\sin u}{\cos u}}{dx}$

$= \dfrac{\cos u \cdot \cos u \dfrac{du}{dx} - \sin u \, (-\sin u) \dfrac{du}{dx}}{\cos^2 u}$

$= \dfrac{1}{\cos^2 u} \cdot \dfrac{du}{dx}$

$= \sec^2 u \, \dfrac{du}{dx}.$

Sec. 10-5. Integration of Trigonometric Functions 269

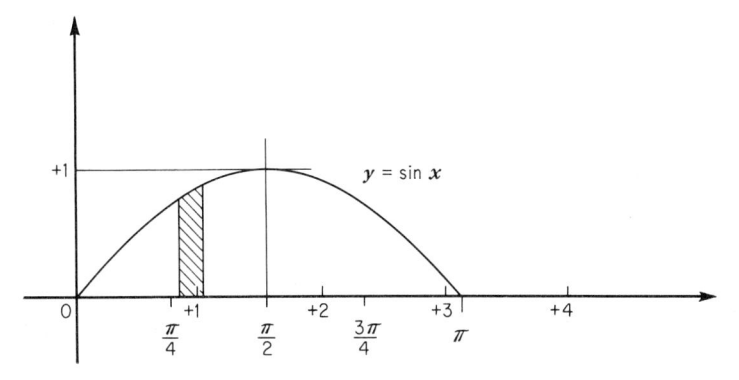

Figure 10-10

Before evaluating $\int_0^\pi \sin x\, dx$ by use of the fundamental theorem, try to approximate it by estimating the area represented.

Actually, since

$$\frac{d(-\cos x)}{dx} = +\sin x,$$

$$\int_0^\pi \sin x\, dx = -\cos x \Big|_0^\pi = -\cos \pi + \cos 0$$

$$\doteq -(-1) + 1 = 2.$$

Hence, the area under one arch of $y = \sin x$ is 2 units.

At this state, the reader is probably ready to adopt the usual abbreviation of writing du in place of $\frac{du}{dx} dx$. Thus,

$$\int \frac{1}{u} \frac{du}{dx} dx = \ln |u| + C$$

condenses to $\int \frac{du}{u} = \ln |u| + C.$

$\int \frac{1}{u} \frac{du}{dx} dx$
$= \int \frac{du}{u}$

This is a convenient shorthand, which is universally used. It has not been used until now in this volume but will be used hereafter, with an occasional return to the more complete notation for emphasis.

The following integration formulas are obtained by applying the fundamental (antiderivative) theorem to the results of Sec. 10-4.

In general, application of the fundamental theorem using the results of the last section yields

Trigonometric Functions

useful trigonometric integrals

$$\int \sin u \, du = -\cos u + C$$
$$\int \cos u \, du = \sin u + C$$
$$\int \sec^2 u \, du = \tan u + C$$
$$\int \csc^2 u \, du = -\cot u + C$$
$$\int \sec u \tan u \, du = \sec u + C$$
$$\int \csc u \cot u \, du = -\csc u + C$$

where du is used as an abbreviation for $\dfrac{du}{dx} dx$.

It requires a bit more ingenuity to integrate $\tan u$.

$$\int \tan u \, du = \int \frac{\sin u}{\cos u} du$$
$$= -\int \frac{1}{\cos u}(-\sin u) \, du = \int \frac{1}{v}\left(\frac{dv}{du}\right) du$$

with $v = \cos u$. (Why?)

Thus $\int \tan u \, du = -\int \left(\dfrac{1}{v}\dfrac{dv}{du}\right) du = -\ln |v| + C$
$$= -\ln |\cos u| + C = \ln |\sec u| + C$$

∫ tan u du that is, $\int \tan u \, du = \ln |\sec u| + C$.

Show in a similar fashion that

$$\int \cot u \, du = \ln |\sin u| + C.$$

The integration of $\sec u$ requires more trigonometric adeptness than the integration of $\tan u$, but the technique is essentially similar. If $\sec u$ is multiplied by $\dfrac{\sec u + \tan u}{\tan u + \sec u}$, the resulting

$$\int \sec u \, du = \int \left(\frac{\sec^2 u + \sec u \tan u}{\tan u + \sec u}\right) du$$

is in the form

$$\int \frac{1}{v}\frac{dv}{dx} dx = \ln |v| + C \quad \text{with } v = (\tan u + \sec u).$$

∫ sec u du Hence $\int \sec u \, du = \ln |\tan u + \sec u| + C$.

In a similar fashion we obtain

∫ csc u du $\int \csc u \, du = \ln |\csc u - \cot u| + C.$

In each case, du is an abbreviation for $\dfrac{du}{dx} dx$.

The more familiarity with trigonometric identities we have, the easier these manipulations become. This manipulation, along with

Sec. 10-5. Integration of Trigonometric Functions 271

the graphing of trigonometric functions of the type $y = x + 3 \sin x$ or $y = 4t - \sin t + 3 \cos 2t$ are the important parts of trigonometry. In modern trigonometry the "solution of triangles" is a relatively unimportant sidelight which can be learned in an hour or two if and when the need arises.

Some of the most useful of these identities are included here for easy reference. A handbook will contain others.

useful trigonometric identities

$$\csc \theta = \frac{1}{\sin \theta}$$

$$\sec \theta = \frac{1}{\cos \theta}$$

$$\tan \theta = \frac{\sin \theta}{\cos \theta}$$

$$\cot \theta = \frac{1}{\tan \theta} = \frac{\cos \theta}{\sin \theta}$$

$\sin^2 \theta + \cos^2 \theta = 1$ (This is "your best friend.")
$\tan^2 \theta + 1 = \sec^2 \theta$ (Divide "your best friend" by $\cos^2 \theta$.)
$1 + \cot^2 \theta = \csc^2 \theta$ (Divide "your best friend" by $\sin^2 \theta$.)

$$\cos^2 A = \frac{1 + \cos 2A}{2}$$

$$\sin^2 A = \frac{1 - \cos 2A}{2}$$

(You won't confuse these two if you ask yourself "What happens when $A = 0$?")

$\sin 2A = 2 \sin A \cos A$
$\cos 2A = \cos^2 A - \sin^2 A$

(Same comment as above.)

$ = 2 \cos^2 A - 1$
$ = 1 - 2 \sin^2 A$

(Learn the first one and ask "your best friend" to recall the others.)

Know that 180° corresponds to the number π, since a circle of radius 1 has semiperimeter π. It follows that trigonometric functions of $k180°$ correspond to the trigonometric functions of the number $k\pi$.

Angular degree notation	Corresponding real number notation	
0°	0	
30°	$\pi/6$	($\frac{1}{6}$ **of semicircle**)
45°	$\pi/4$	($\frac{1}{4}$ **of semicircle**)
60°	$\pi/3$	($\frac{1}{3}$ **of semicircle**)
90°	$\pi/2$	($\frac{1}{2}$ **of semicircle**)

Trigonometric Functions

Example 1

Find $\int \sin^2 y \, dy$.

This, again, requires trigonometric dexterity. Using the formula $\sin \frac{\theta}{2} = \pm \sqrt{\frac{1 - \cos \theta}{2}}$ in the form $\sin^2 y = \frac{1 - \cos 2y}{2}$, we obtain

$$\int \sin^2 y \, dy = \int \frac{1 - \cos 2y}{2} \, dy = \int \frac{1}{2} \, dy - \frac{1}{4} \int \cos 2y \, (2dy)$$

$$= \frac{y}{2} - \frac{1}{4} \sin 2y + C.$$

Example 2

Let $u = \sin 5x$. Then, $\frac{du}{dx} = [(\cos 5x)5]$.

$$\int \sin^3 5x \cos 5x \, dx = \frac{1}{5} \int \sin^3 5x (\cos 5x \, 5dx)$$

$$= \frac{1}{5} \int u^3 \, du = \frac{1}{5} \cdot \frac{1}{4} \sin^4 5x + C$$

$$= \frac{1}{20} \sin^4 5x + C.$$

Problem Set 10-5

In Probs. 1 through 8, prove the formulas stated.

1. $\int \csc x \, dx = \ln |\csc x - \cot x| + C.$

2. $\int \frac{\sin x \, dx}{\cos^2 x} = \sec x + C.$

3. $\int \cos^2 x \, dx = \frac{1}{2}x + \frac{1}{4} \sin 2x + C.$

4. $\int \cos^3 t \, dt = \frac{1}{3} \sin t (\cos^2 t + 2) + C.$

HINT: $\int \cos^3 t = \int (1 - \sin^2 t) \cos t = \int (1 - u^2) \, du.$

5. $\int \frac{dt}{\cos^2 t} = \tan t + C.$

6. $\int \frac{d\theta}{\sin^2 2\theta} = -\frac{1}{2} \cot 2\theta + C.$

7. $\int \cot^2 \theta \, d\theta = -\cot \theta - \theta + C.$

Sec. 10-5. Integration of Trigonometric Functions

8. $\int \dfrac{dx}{\sin x \cos^2 x} = \dfrac{1}{\cos x} + \ln\left|\tan \dfrac{x}{2}\right| + C.$

9. Find the area under the curve $y = \cos x$ between $x = -\pi/2$ and $x = \pi/2$. *Ans.* 2 square units.

10. The curves $y = \sin x$ and $y = \cos x$ intersect at $x = \pi/4$ and $x = 3\pi/4$. Find the area enclosed by them between these points of intersection.

11. Compute, accurate to the nearest 0.1 square unit, the area bounded by the curves $y = \tan x$ and $y = 2x$, $0 \leq x \leq \pi/4$.

12. The line $y = x$ intersects the curve $y = \tan(\pi x/4)$ at $(0,0)$ and at $(1,1)$. Compute to the nearest 0.01 square unit the area between these curves in this region.

13. One table lists $\int \sec x \, dx$ as $\ln|\sec x + \tan x| + C$ while another lists $\ln|\tan(\pi/4 + x/2)| + C$. Which is correct?

14. Another table lists $\int \sec x \, dx = \tfrac{1}{2} \ln\left|\dfrac{1 + \sin x}{1 - \sin x}\right| + C$. Reconcile this with the results of Prob. 13.

15. Does $\ln|\csc u - \cot u| = \ln|\tan u/2|$? Could either formula represent $\int \csc u \, du$?

16. Integrate $\int \tan^4 x \, dx$. HINT: $\tan^2 x = \sec^2 x - 1$.

17. Integrate $\int \sec^6 t \, dt$. HINT: $\sec^2 t = 1 + \tan^2 t$.

18. $\int \sqrt{\sin^3 x} \cos x \, dx$.

In Probs. 19 to 27, compute the given definite integrals accurate to the nearest 0.01 square unit.

19. $\int_{\pi/3}^{\pi/2} \sin 3x \, dx$. *Ans.* $-\tfrac{1}{3}$.
20. $\int_0^{\pi/3} 5 \tan y \, dy$.

21. $\int_{\pi/12}^{\pi/6} \sec^2 2t \, dt$. *Ans.* $1/\sqrt{3}$.
22. $\int_{\pi/4}^{\pi/2} \cos^3 t \sin t \, dt$.

23. $\int_0^{\pi/3} e^{3t} \, dt$. *Ans.* $\tfrac{1}{3}(e^\pi - 1)$.
24. $\int_{-\pi/4}^{\pi/3} \tan y \, dy$.

25. $\int_{-\pi/4}^{\pi/3} \cot x \, dx$. (Booby trap).
26. $\int_{-1}^{5} \tan 3x \, dx$.

27. $\int_0^7 \sin(x/3) \, dx$. *Ans.* 5.073.

28. (a) $\int \sec^7 \tan x \, dx$.

HINT: $\int \sec^7 x \tan x \, dx = \sec^6 x (\sec x \tan x \, dx) = \int \left(u^6 \dfrac{du}{dx}\right) dx.$

(b) $\int \dfrac{\sin x \, dx}{\cos^8 x}$. HINT: $\int \left(u^{-8} \dfrac{du}{dx}\right) dx.$

274   *Trigonometric Functions*

29. Integrate $\int \dfrac{\sqrt{16-x^2}}{x}\,dx$.

HINT: Make the substitution $x/4 = \sin\theta$ suggested by Fig. 10-11. Note that in this case $1 = \dfrac{dx}{dx} = 4\cos\theta\,\dfrac{d\theta}{dx}$ and $\sqrt{16-x^2} = 4\cos\theta$.

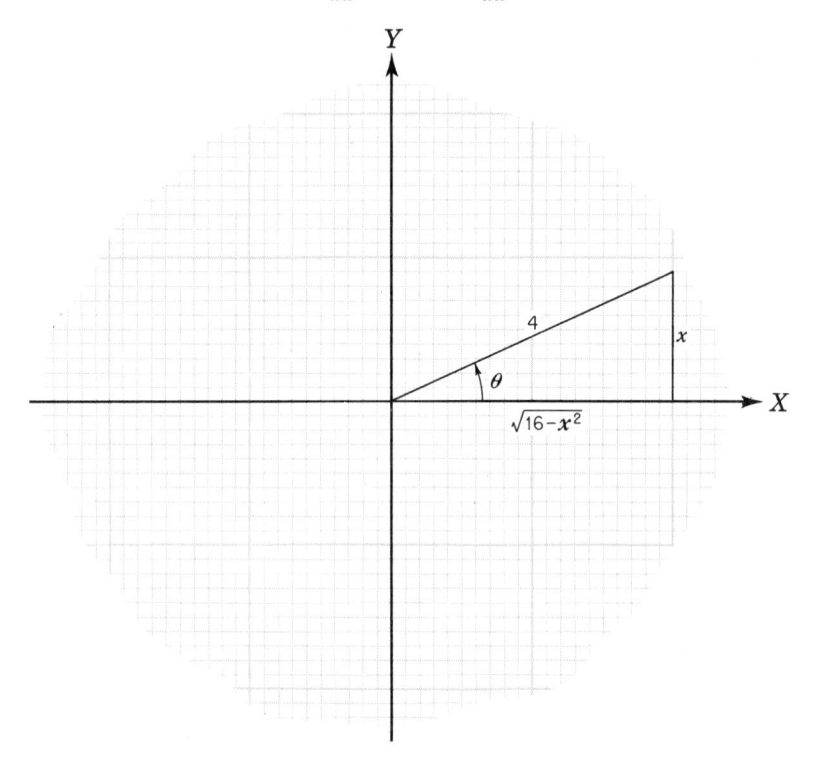

Figure 10-11

10-6. Self-test

1. Make use of the wrapping function to define $\cos x$ for the real number x.

2. Let $0 < x < 3$. Which is larger, $\cos x$ or $\cos(x+6)$? Justify your statement.

3. $\int_a^b \sin x\,dx$ exists for any choice of a and b. However, $\int_a^b \tan x\,dx$ may not exist for certain values of a and b. Give three different

sets of limits $\int_a^b$ such that $\int_a^b \tan x \, dx$ does not exist. Explain why the integral fails to exist in each case.

4. The function $f(x) = \dfrac{\sin 3x}{x}$ is defined and continuous (see Sec. 3-3) for every value of x except $x = 0$. Define $f(0)$ so that $f(x)$ is continuous at $x = 0$ or show that this is impossible.

5. (a) Write the equation of the line tangent to $y = \sin x$ at the point where $y = \sin x$ crosses the line $x = 1$. Where does the tangent line cross the x axis?

6. Derive a formula for $\int \tan x \, dx$.

7. Integrate: (a) $\int \cos^2 x \, dx$. (b) $\int 5 \sin 3x \, dx$. (c) $\int \csc x \, dx$.

8. Show that the integral $\int_2^5 x \sin x \, dx$ exists even though you cannot integrate the given function easily. Obtain a crude estimate of the size of this integral (that is, correct to within 10 square units).

11

Techniques of Integration

11-1. Techniques

Much time in the traditional course in calculus is spent in learning certain techniques of integration. It is not the purpose of this course to emphasize the details of such techniques but instead *to provide an understanding of the important basic concepts*. Many of the techniques can be circumvented by consulting a table of integrals. However, two particularly important ones provide excellent examples of techniques and are also highly important in their own right.

11-2. Integration by Parts

Recall from Sec. 4-8 the formula for the derivative of a product

$$\frac{d(u \cdot v)}{dx} = u\frac{dv}{dx} + v\frac{du}{dx}$$

where $u = u(x)$ and $v = v(x)$. Upon integrating this relation with respect to x, we obtain

Sec. 11-2. Integration by Parts

$$\int \frac{d(u \cdot v)}{dx} dx = \int \left(u \frac{dv}{dx} + v \frac{du}{dx} \right) dx$$

$$u \cdot v = \int \left(u \frac{dv}{dx} \right) dx + \int \left(v \frac{du}{dx} \right) dx.$$

Solving for $\int [u(dv/dx)]\, dx$,

integration by parts

$$\int \left(u \frac{dv}{dx} \right) dx = u \cdot v - \int \left(v \frac{du}{dx} \right) dx.\dagger$$

This important technique solves many integration problems and is also an important tool in statistics and more advanced mathematics courses.

Example 1

$\int x \sin x\, dx.$

This defies our usual method of integration. However after letting $u = x$ and $dv/dx = \sin x$, we compute $du/dx = 1$ and $v = -\cos x$. (The constant of integration is taken as zero here, but need not be.) Then, using the fundamental formula,

$$\int \left(u \frac{dv}{dx} \right) dx = u \cdot v - \int \left(v \cdot \frac{du}{dx} \right) dx$$

$$\int (x \sin x)\, dx = x(-\cos x) - \int [-\cos x(1)]\, dx$$

$$= -x \cos x + \int \cos x\, dx$$

$$= -x \cos x + \sin x + C.$$

Check that $\dfrac{df}{dx} = x \sin x$ if $f = -x \cos x + \sin x + C.$

---

† The student who reads other texts may find that the definition $du = \dfrac{du}{dx} dx$ has been adopted and the resulting identity is stated as $\int u\, dv = u \cdot v - \int v\, du$. This is a valid and convenient thing to do but tends to mask the true nature of the terms under discussion. We shall not do it here even though it is conventional. In a similar manner we may write $\int \sin u\, du$ in place of $\int \left(\sin u \dfrac{du}{dx} \right) dx$, etc.

Techniques of Integration

Example 2

$\int x^2 e^{3x}\, dx.$

Again the previous methods are of no avail and we try integration by parts. Let

$$u = x^2 \qquad \frac{dv}{dx} = e^{3x}$$

then $\dfrac{du}{dx} = 2x \qquad v = \dfrac{1}{3} e^{3x}.$ (Why?)

$$\int \left(u \frac{dv}{dx} \right) dx = u \cdot v - \int \left(v \frac{du}{dx} \right) dx$$

$$\int (x^2 e^{3x})\, dx = x^2 \left(\frac{1}{3} e^{3x} \right) - \int \left(\frac{1}{3} e^{3x} \cdot 2x \right) dx$$

$$= \frac{x^2 e^{3x}}{3} - \frac{2}{3} \int x e^{3x}\, dx.$$

Since we cannot integrate $\int x e^{3x}\, dx$, our effort seems fruitless. Actually, the exponent on x^2 has been reduced from 2 to 1. It could be further reduced from 1 to 0; then $x^0 \cdot e^{3x} = e^{3x}$ could be integrated.

Let us now try *another* integration by parts on the same problem.

$$\int x^2 e^{3x}\, dx = \frac{x^2 e^{3x}}{3} - \left(\frac{2}{3} \right) \int x e^{3x}\, dx.$$

In the last integral let

$$u = x \qquad \frac{dv}{dx} = e^{3x}$$

then $\dfrac{du}{dx} = 1 \qquad v = \dfrac{1}{3} e^{3x}.$

Hence $\displaystyle \int x^2 e^{3x}\, dx = \frac{x^2 e^{3x}}{3} - \frac{2}{3} \left\{ \left[x \frac{1}{3} e^{3x} \right] - \int \frac{1}{3} e^{3x}(1)\, dx \right\}$

$$= \frac{x^2 e^{3x}}{3} - \frac{2x e^{3x}}{9} + \left(\frac{2}{9} \right) \int e^{3x}\, dx$$

$$= \frac{x^2 e^{3x}}{3} - \frac{2x e^{3x}}{9} + \left(\frac{2}{9} \right) \left(\frac{1}{3} e^{3x} \right) + C.$$

The desired integration is complete. In some problems it may be necessary to integrate by parts eight, ten, or even a hundred times to reach the desired reduction of exponent. In others it may not help at all.

Sec. 11-2. Integration by Parts

Example 3 (A clever one)

$\int e^x \cos x \, dx.$

Let $\quad u = e^x \qquad \dfrac{dv}{dx} = \cos x.$

Then $\dfrac{du}{dx} = e^x \qquad v = \sin x.$

Using the identity

$$\int u \frac{dv}{dx} \, dx = u \cdot v - \int v \frac{du}{dx} \, dx.$$

$$\int e^x \cos x \, dx = e^x \sin x - \int (\sin x) e^x \, dx.$$

Since we cannot integrate the result, let us try to use integration by parts again. This time let

$u = e^x \qquad \dfrac{dv}{dx} = \sin x$

$\dfrac{du}{dx} = e^x \qquad v = -\cos x.$

giving

$\int e^x \cos x \, dx = e^x \sin x - [e^x(-\cos x) - \int(-\cos x)e^x \, dx]$
$\int e^x \cos x \, dx = e^x \sin x + e^x \cos x - \int e^x \cos x \, dx.$

The last integral is the same as the integral we were first attempting to integrate. If we continue with our integration by parts program, we shall chase our tail around a circle until dawn. *However, all is not lost!* Note that if

$Y = [\sim\sim\sim] - Y$
then $2Y = [\sim\sim\sim]$
and $\quad Y = \dfrac{[\sim\sim\sim]}{2}$ no matter what Y and $[\sim\sim\sim]$ are.

In the example,

$$\int e^x \cos x \, dx = (e^x \sin x + e^x \cos x) - \int e^x \cos x \, dx$$

$$2 \int e^x \cos x \, dx = (e^x \sin x + e^x \cos x)$$

$$\int e^x \cos x \, dx = \frac{(e^x \sin x + e^x \cos x)}{2} + C. \quad \text{(Why annex the } +C?\text{)}$$

280   *Techniques of Integration*

The desired integral has been found. This clever technique has been used for many years, but still finds new applications in today's mathematics.

Integration by parts has applications in probability and statistics as well as in integrating numerous functions. The integrals $\int_x^n g(x)\, dx$ where $g(x)$ is some integrable function of x, say, $\sin bx$, $\cos bx$, e^{ax}, $\sqrt{f(x)}$, or even $\log x$; and $\int_e^{ax} \sin bx\, dx$, $\int_e^{ax} \cos bx\, dx$, and other forms all yield to this powerful tool.

In general, *integration by parts is a method of swapping one integration problem for another.* If you are judicious in your choice of swaps, you may be able to obtain a simpler integration than you had before. If not, don't swap.

11-3. Trigonometric Substitution

Problem 29, Set 10-5 suggests another important technique of integration, namely, trigonometric substitution. Many expressions involving the sum or the difference of two squares yield to this method.

Example 1

$\int \sqrt{9 - x^2}\, dx.$

If we think of a right triangle having $\sqrt{9 - x^2}$ as one side, the following arrangement suggests itself. (See Fig. 11-1.) Let θ be the acute angle opposite side x. Then: $x/3 = \sin \theta$ or $x = 3 \sin \theta$, and $\dfrac{\sqrt{9 - x^2}}{3} = \cos \theta$, or $\sqrt{9 - x^2} = 3 \cos \theta$.

Thus

$\int \sqrt{9 - x^2}\, dx = \int 3 \cos \theta\, dx.$
                          ↑   ↑

However, we cannot integrate this expression since the $\cos \theta$ requires a $\dfrac{d\theta}{dx}$ factor. We obtain this by differentiating $x = 3 \sin \theta$ with respect to x,

Sec. 11-3. Trigonometric Substitution

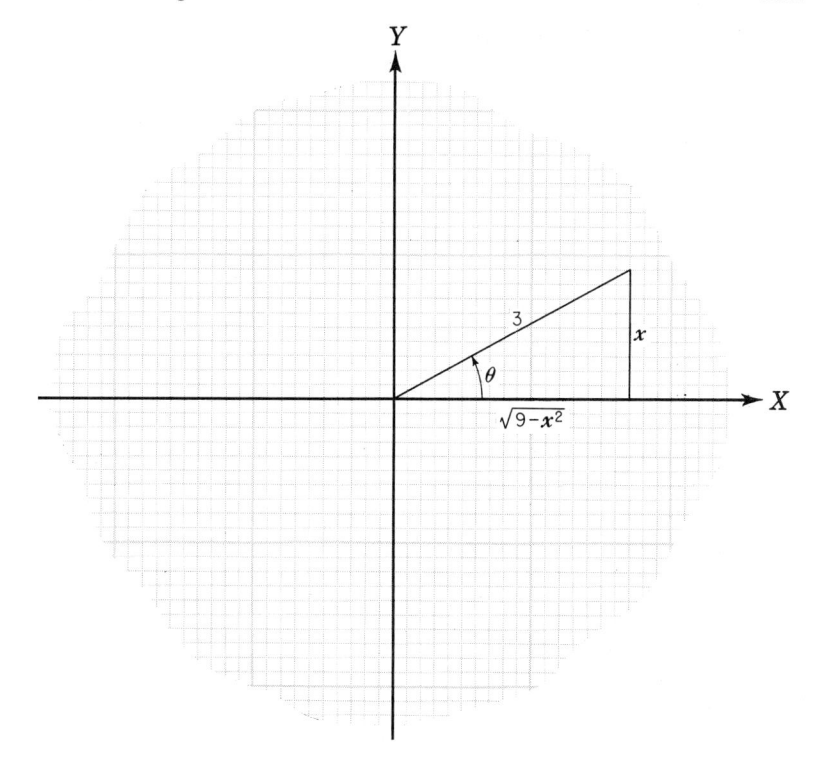

Figure 11-1

$$x = 3 \sin \theta$$
$$\frac{dx}{dx} = 3 \cos \theta \frac{d\theta}{dx}$$
$$1 = 3 \cos \theta \frac{d\theta}{dx} \cdot †$$

Since $\dfrac{dx}{dx} = 1$ (Why?), we substitute this equivalent of 1 into

---

† Again, other texts which have adopted the convention that $du = \dfrac{du}{dx} dx$ would write $dx = 3 \cos \theta \, d\theta$ here, which tends to mask he point involved. The convention is desirable, but not at this stage of development.

Techniques of Integration

the integral

$$\int \sqrt{9-x^2}\, dx = \int 3\cos\theta (1)\, dx$$
$$= \int 3\cos\theta \left(3\cos\theta \frac{d\theta}{dx}\right) dx$$
$$= 9\int \cos^2\theta \frac{d\theta}{dx}\, dx.$$

This may be integrated by the methods of Chap. 10, using

$$\cos^2\theta = \frac{1+\cos 2\theta}{2}$$

to obtain $9\int \cos^2\theta \left(\frac{d\theta}{dx}\right) dx$

$$= \frac{9}{2}\int (1+\cos 2\theta)\left(\frac{d\theta}{dx}\right) dx$$
$$= \frac{9}{2}\int \left(\frac{d\theta}{dx}\right) dx + \frac{9}{4}\int \cos 2\theta \left(2\frac{d\theta}{dx}\right) dx$$
$$= \frac{9}{2}\theta + \frac{9}{4}(\sin 2\theta) + C.$$

Since $\sin 2\theta = 2\sin\theta\cos\theta$, we have Fig. 11-2

$$= \frac{9}{2}\theta + \frac{9}{2}\sin\theta\cos\theta + C$$
$$= \frac{9}{2}\left(\text{the angle whose sine is } \frac{x}{3}\right) + \frac{9}{2}\left(\frac{x}{3}\right)\left(\frac{\sqrt{9-x^2}}{3}\right) + C.$$

Example 2

$$\int \frac{1}{\sqrt{4+x^2}}\, dx.$$

If $\sqrt{4+x^2}$ is to be one of the sides of a triangle, the diagram of Fig. 11-3 suggests itself. From the diagram,

$$\sqrt{4+x^2} = 2\sec\theta$$
$$x = 2\tan\theta$$
$$1 = \frac{dx}{dx} = 2\sec^2\theta \frac{d\theta}{dx}.$$

Sec. 11-3. Trigonometric Substitution

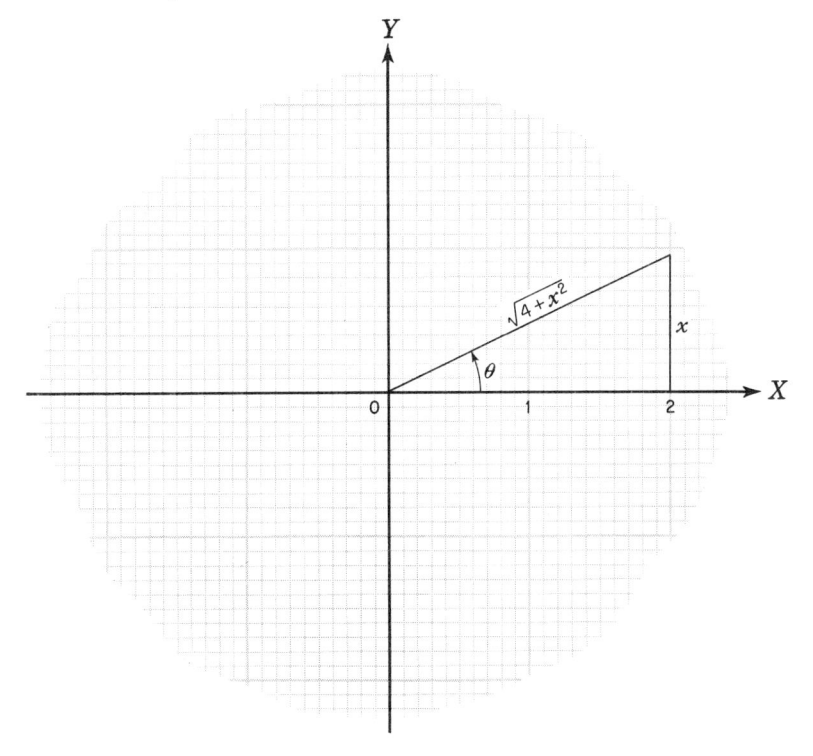

Figure 11-3

$$\int \frac{1}{\sqrt{4+x^2}}\, dx = \int \frac{1}{2\sec\theta}\, dx$$
$$= \int \frac{1}{2\sec\theta} \cdot \left(2\sec^2\theta \frac{d\theta}{dx}\right) dx$$
$$= \int \sec\theta \frac{d\theta}{dx}\, dx$$
$$= \ln|\sec\theta + \tan\theta| + C$$
$$= \ln\left|\frac{\sqrt{4+x^2}}{2} + \frac{x}{2}\right| + C$$
$$= \ln\left|\frac{\sqrt{4+x^2} + x}{2}\right| + C.$$

It may be desirable in applications to write this in the form

$$= \ln|\sqrt{4+x^2} + x| - \ln 2 + C$$
$$= \ln|\sqrt{4+x^2} + x| + C_1$$

since $-\ln 2 + C$ is merely another constant, C_1.

Techniques of Integration

The three triangles shown in Figs. 11-4 to 11-6 are particularly useful in making trigonometric substitutions. They *need not* be memorized since they readily suggest themselves.

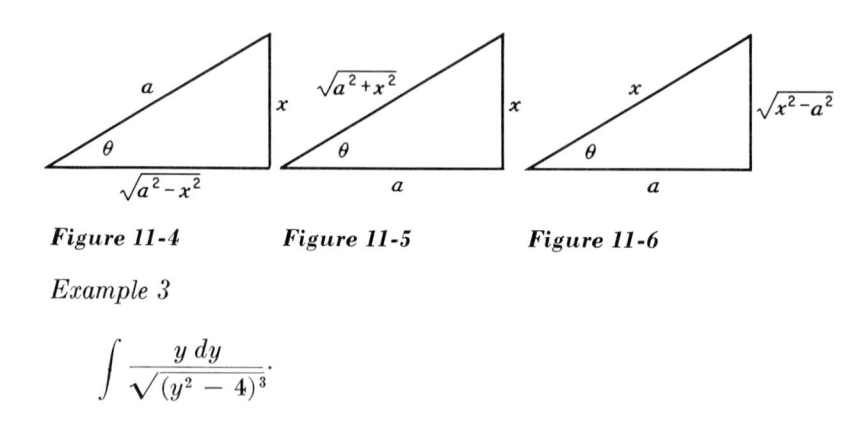

Figure 11-4 *Figure 11-5* *Figure 11-6*

Example 3

$$\int \frac{y \, dy}{\sqrt{(y^2 - 4)^3}}.$$

The observation that $\sqrt{(y^2 - 4)^3}$ appears rather than $\sqrt{(y^2 - 4)^1}$ need not alarm us or alter our approach. Neither does the extra y in the numerator necessarily imply additional difficulty. Of the several methods of integration available, let us choose trigonometric substitution for illustrative purposes.

A triangle having $\sqrt{y^2 - 4}$ as one of its sides is shown in Fig. 11-7.

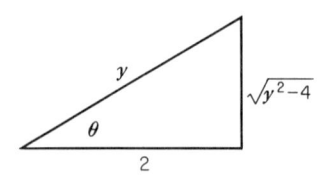

Figure 11-7

Sec. 11-3. Trigonometric Substitution

Let $\quad y = 2\sec\theta$

$$1 = \frac{dy}{dy} = 2\sec\theta\tan\theta\frac{d\theta}{dy}$$

$$\sqrt{y^2 - 4} = 2\tan\theta.$$

Substituting into the original integral shows

$$\int \frac{y \cdot 1 \cdot dy}{\sqrt{(y^2-4)^3}} = \int \frac{2\sec\theta \cdot 2\sec\theta\tan\theta\frac{d\theta}{dy}\cdot dy}{8\tan^3\theta}$$

$$= \frac{1}{2}\int \tan^{-2}\theta\left(\sec^2\theta\frac{d\theta}{dy}dy\right)$$

$$= \frac{1}{2}\frac{1}{\tan\theta} + C = \frac{-1}{\sqrt{y^2-4}} + C.$$

This may be checked by differentiation. The astute reader will have noticed by now that the original integral is of the form $\int u^k\, du$ $\frac{1}{2}\int u^{-\frac{3}{2}}\frac{du}{dy}dy = -u^{-\frac{1}{2}} + C$. Thus, the integration could have been performed at once, without trigonometric substitution. This would not have been the case, of course, if the y in the numerator had been replaced by y^3 or y^2 in the original problem. Nevertheless, it is often propitious to check on simpler forms before jumping into a substitution method.

Trigonometric substitution illustrates one reason that the study of trigonometry is important today. The "solving of triangles," which 20 years ago often occupied as much as one-third of the time spent in a trigonometry course, is a relatively *un*important application today.

Additional work on techniques of integration may be found in any standard calculus text. The two examples given, integration by parts and trigonometric substitution, are fairly typical of the techniques, and are two of the most useful ones. If you understand the basic principles of calculus and can formulate new problems in terms of derivatives and integrals, it will always be possible to find

Techniques of Integration

other persons actually to carry out the integrations. It is hoped that the social scientists of the current generation will formulate problems in terms of calculus for nonengineering fields as carefully as did the physicists and physical chemists of the past decade.

Problem Set 11-3

Integrate:

1. $\int x \cos x \, dx.$
2. $\int x^2 e^x \, dx.$
3. $\int e^x \sin x \, dx.$
4. $\int x \ln x \, dx.$
5. $\int x^2 e^{3x} \, dx.$
6. $\int \sqrt{1 - x^2} \, dx.$
7. $\int \frac{1}{\sqrt{x^2 + 1}} \, dx.$
8. $\int \sqrt{x^2 + 4} \, dx.$
9. $\int \frac{dx}{x\sqrt{9 - x^2}}.$
10. $\int \frac{dx}{\sqrt{(x^2 + 1)^3}}.$

11. Find by integration the area under $y = +\sqrt{4 - x^2}$ between $x = 0$ and $x = 2$.

12. Determine the area of a circle of radius a by integration.

13. Rework Example 1, using the substitution suggested by the triangle of Fig. 11-8. Is an equivalent result obtained?

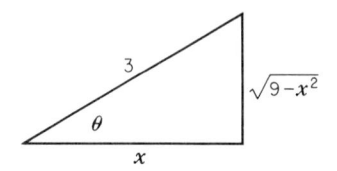

Figure 11-8

14. $\int \frac{x^3 \, dx}{\sqrt{(x^2 - 4)^3}}.$

15. $\int y^3 \sqrt{16 + y^2} \, dy.$

16. $\int x \sqrt{x^2 - 49} \, dx.$

11-4. Self-test

1. Use integration by parts to integrate $\int x \cos x \, dx.$

2. Integrate $\int \frac{dx}{\sqrt{16 + x^2}}.$

Sec. 11-4. Self-test

3. Integrate $\int e^x \sin x \, dx$.

4. Integrate $\int \sqrt{16 - x^2} \, dx$.

5. Integrate $\int \sqrt{(x^2 - 25)^3} \, dx$.

6. Determine the area bounded by the x axis and that portion of the curve $y = +\sqrt{25 + x^2}$ between $x = 3$ and $x = 7$. Obtain your answer correct to the nearest tenth of a unit.

7. Integrate $\int \log x \, dx$. HINT: Let $u = \log x$ and $dv = dx$ and integrate by parts.

8. Show that $\int_{3.2}^{4.7} \frac{\sin x}{x} \, dx$ exists even though you are unable to evaluate the integral. Give a rough approximation of the value of this integral.

12

Epilogue

12-1. A Look Backward

Let us turn our attention now to what you have accomplished in this book and to what you should logically study next to further your own interests.

Chapter 1 was devoted to a review of basic algebraic theory. Very possibly, if you are like the majority of our readers, about half of Chap. 1 was already familiar to you and the other half required some refresher. Chapter 2 was a similar review of geometric theory, including analytic geometry. Chapter 3 began a development of the basic ideas of calculus. In a standard undergraduate calculus book designed primarily for engineering students, a good deal of time is spent learning to differentiate and integrate certain complicated functions. In this text, on the other hand, the emphasis has been on the procedures used to set up new problems. It may still be necessary for you to consult a mathematician—or even an undergraduate student of mathematics—to actually perform some of the integrations which you are now capable of setting up, but this situation is not unusual. In any applied mathematical laboratory the majority of the functions set up do not have easy closed-form integrals. Often a computer must be used in order to obtain a reasonable approximation of such integrals. Later chapters have developed important special functions, such as the logarithmic and exponential functions as well as trigonometric functions. Finally, Chap. 11 developed two of the many important techniques of integration.

This book is not designed to make you an expert in the techniques of calculus. It is intended to refresh your knowledge of the theories of calculus, or if you have never studied calculus before, to give you a fundamental understanding of what calculus is all about so that you will be capable of setting up problems of your own—problems which perhaps have never before been expressed through the medium of calculus.

12-2. Where Next?

The question of what, if any, mathematics you should take next will depend largely upon your own interests. You probably should not plan to take a more advanced course in calculus right away. If you have the time and inclination to take some additional mathematics, then the next step should be a study of finite mathematics. This study includes matrices, along with some work on groups, fields, and rings. A one-semester survey course in modern abstract algebra would be ideal, or you may even find a special course in finite mathematical models. Look at the index of the proposed text. If *Markov Chains* are not mentioned, then quite possibly this is not the course you need. This does not mean that a first course need include Markov Chains; it should not. The course should, however, contain the basic material on matrices needed to study Markov Chains, which would then be mentioned in passing. The concept of a *characteristic root* or *eigenvalue* is also of considerable importance and will be mentioned in most texts which are suitable for your next course of study.

Then, if you have not already studied computer programming, you should arrange to take such a course. Modern research demands an understanding of basic computer programming. *Even though you may not be programming the computer yourself* in your research, you *must* understand the basic principles to make efficient use of modern computers. You need not be an expert computer programmer, any more than you are an expert in calculus; but it is essential, no matter what your special field, that you know enough about computer programming to be able to express your problems in language compatible with the computer's ability.

A third course which you should certainly consider is one in basic mathematical statistics. You now have enough calculus

Epilogue

to understand the fundamental ideas of the probability density function and of sample space, without which work in statistics has little foundation today.

Some of the most important work that remains to be done will not be found in courses at all but in reading and study on your own. Appended to this book in an extensive, annotated reading list, which contains a variety of books. Some were chosen merely for pleasure; others are fairly deep and difficult mathematical treatises. Although all of them have been successfully read and understood by high school students in the Oklahoma area, most of these books contain ideas which will challenge a graduate student in mathematics. Hence, we feel justified in recommending them to you.

Do not let your mathematical training lapse. Continue it, even if only by reading some recreational mathematics each week. You will be surprised what even twenty minutes' work, done regularly each week, can do to help you improve. If you should have further questions, please feel free to write directly to the author, Dr. Richard V. Andree, Chairman, Department of Mathematics, The University of Oklahoma, Norman, Oklahoma. He will be glad to give individual counseling if he can be of any assistance either to you personally or to some of your students or colleagues.

Reading List

Each book in this list contains material deemed suitable for high school students. Most of the books also contain portions of interest to college math majors. We hope they will also interest you. Mathematics is an absorbing subject in its own right; it also provides powerful tools for other disciplines. Mathematics is the analysis of the basic structure of almost anything—a geometry, an economy, a language, or some physical or social situation. When you study vital, basic interrelations in structure, you are doing mathematics whether you realize it or not. If you spend twenty minutes a week reading mathematics, you can keep your newly found knowledge fresh and growing.

Abbott, *Flatland*, Dover Publications, New York.
> A science-fiction classic of life in a two-dimensional world that is also a first-rate introduction to some aspects of modern science and hyperspace.

Adler, *Thinking Machines*, The John Day Company, Inc., New York, 1961.
> An explanation of the theory of electronic computers, made intelligible to anyone who has studied high school mathematics.

Adler, *The Magic House of Numbers*, New American Library of World Literature, Inc., 1961.
> Explains the basic whys and hows of our number system in a clear, entertaining way. Packed with curiosities, riddles, tricks, games.

Adler, *The New Mathematics*, The John Day Company, Inc., New York, 1958.
> With elementary algebra and plane geometry as tools, Adler skillfully builds up many interesting concepts of modern mathematics.

Allendoerfer and Oakley, *Fundamentals of Freshman Math*, McGraw-Hill Book Company, Inc., New York, 1959.
> An excellent college freshman text which introduces calculus in a framework of college algebra and analytic geometry.

Altshiller-Court, *College Geometry*, Barnes & Noble, Inc., New York, 2d rev. and enlarged edition, 1952.
> Probably the best known of all advanced plane geometry texts. Has even been translated into Chinese. Requires no background beyond the usual high school course.

Anderson, Raymond W., *Romping Through Mathematics*, Alfred A. Knopf, Inc., New York, 1961.
> The story of how the necessary mental tools for counting and measuring have been collected and arranged over a period of four thousand years.

Andree, *Selections from Modern Abstract Algebra*, Holt, Rinehart, and Winston, New York, 1958.
> A college level book which includes congruences (modular algebra), Boolean algebra (set theory), groups, fields, matrices, etc. Has been used successfully as a text for advanced high school students, as well as in dozens of summer institutes for teachers.

Andree, R. V., *Programming the IBM 650 Magnetic Drum Computer and Data Processing Machine*, Holt, Rinehart, and Winston, New York, 1958.
> Basic programming for modern computers.

Ball and Coxeter, *Mathematical Recreations and Essays*, The Macmillan Company, New York, 1939.
> A classic book on recreational mathematics which should be on every library shelf. Very enjoyable.

Barnett, I. A., *Some Ideas About Number Theory*, NCTM, Washington, D.C., 1961.
> An informal account of some of the more elementary results of number theory. Provides many direct applications for the classroom.

Reading List

Beckenback and Bellman, *An Introduction to Inequalities*, Random House, Inc., New York, 1961.
> This book provides an introduction to the fascinating world of quantities which can be ordered according to their size.

Begle, *Introductory Calculus with Analytic Geometry*, Holt, Rinehart, and Winston, New York, 1954.
> Covers topics usually treated in a first course in calculus; includes analytic geometry through the conics. Carefully written and authoritative.

Bell, E. T., *Men of Mathematics*, Simon and Schuster, Inc., New York.
> A book which brings the student to a realization of the great men in mathematics and gives him insight into their lives.

Boehm, G. A. W., *New World of Mathematics*, The Dial Press, Inc., New York, 1959.
> An excellent, inexpensive book derived from three articles in *Fortune*. Highly recommended.

Bowers, *Arithmetical Excursions*, Dover Publications, New York, 1961.
> Lighthearted collection of facts and entertainments for anyone who enjoys manipulating numbers or solving arithmetical puzzles. 529 numbered problems and diversions, all with answers.

Brixey and Andree, *Fundamentals of College Mathematics*, Holt, Rinehart, and Winston, New York, 1961.
> This lively book carefully integrates introductory calculus and statistical inference with college algebra, trigonometry, and analytic geometry.

Carroll, *Pillow Problems and a Tangled Tale*, Dover Publications, New York, 1958.
> "Pillow Problems" (1893) contains 72 mathematical puzzles, all typically ingenious. The problems in "A Tangled Tale" (1895) are in story form. Carroll not only gives the solutions, but uses answers sent in by readers to discuss wrong approaches and misleading paths, and grades them for insight.

Courant and Robbins, *What Is Mathematics?*, Oxford University Press, New York, 1941.
> Not an easy book, but well worth the effort. Contains excellent work on basic mathematical analysis. Portions can be read and enjoyed by all, but some parts require mature cogitation.

Court, *Mathematics in Fun and Earnest*, The Dial Press, Inc., New York, 1958.
    An entertaining book which illustrates Dr. Court's thesis that mathematics in earnest should be fun; mathematics in fun may be in earnest. Requires little background, but will exercise your reasoning power.

Coxeter, *Introduction to Geometry*, John Wiley & Sons, Inc., New York, 1961.
    A survey of all important areas of modern geometry, with relationships to modern algebra stressed.

Dadourián, *How to Study; How to Solve*, Addison-Wesley Publishing Company, Reading, Mass., 1958.
    This book is designed to help students who are having trouble with mathematics. An excellent volume, and well worth the modest price.

Dantzig, *Number, The Language of Science*, The Macmillan Company, New York, 1954.
    Subtitled "A Critical Survey Written for the Cultured Non-Mathematician," this is a historical treatment of the number concept and its importance in modern life. Well written.

Davis, *The Lore of Large Numbers*, Random House, Inc., New York, 1961.
    Arithmetic and the uses of large numbers are explained to introduce the reader to the horizons of modern mathematics.

Degrazia, Joseph, *Math Is Fun*, Emerson Books, Inc., New York, 1961.
    A book of wit-piquing and brain-teasing problems and puzzles—from trifles to cryptograms, and from clock and speed puzzles to problems of arrangement. The author has included answers at the back of the book.

Dubisch, *The Nature of Number; An Approach to Basic Ideas of Modern Mathematics*, The Ronald Press Company, New York, 1952.
    A direct and understandable way to gain an over-all view of what mathematics is about, and an insight into the nature of its theory.

Dudeny, *Amusements in Mathematics*, Dover Publications, New York, 1958.

A selection of over 400 mathematical puzzles. Everyone will enjoy this book.

Eves and Newsom, *An Introduction to the Foundations and Fundamental Concepts of Mathematics,* Holt, Rinehart, and Winston, New York, 1958.

A sound book which good high school students will enjoy reading.

Fadiman, Clifton, *Fantasia Mathematica,* Simon and Schuster, Inc., New York.

A fantastic collection of pseudo-mathematical materials. No hard math here, but a lot of fun.

Freeman, Mae B. and Ira, *Fun with Figures,* Random House, Inc., New York, 1946.

Following the "do-it-yourself" idea, the authors show in this book how to have fun with geometric figures. Not much math, but worth the price for fun.

Fujii, *An Introduction to the Elements of Mathematics,* John Wiley & Sons, Inc., New York, 1961.

Presents the fundamental ideas underlying mathematics for the reader with the nontechnical background. It also discusses history of mathematics and gives an elementary treatment of analysis and problem solving.

Gaines, *Cryptanalysis—Codes and Ciphers,* Dover Publications, New York, 1956.

This introductory intermediate-level text is the best book in print on cryptograms and their solutions.

Gamow, *One, Two, Three . . . Infinity,* Mentor Books, New York, 1947.

Problems of mathematics, physics, and astronomy clarified for the layman. This paperback should be available at your corner store. Also available in hardback edition from The Viking Press, Inc., New York.

Gamow and Stern, *Puzzle-Math,* The Viking Press, Inc., New York, 1958.

Interesting brain-twisters and puzzles based on everyday situations that can be untangled by mathematical thinking.

Gardner, *Mathematical Puzzles of Sam Loyd,* vols. I and II, Dover Publications, New York, 1960.

A delightful collection of puzzles by one of the greatest puzzle makers of all time. A classic.

Gardner, *Mathematics, Magic and Mystery*, Dover Publications, New York, 1957.

Another interesting, inexpensive paperback by the mathematical editor of *Scientific American*. Well worth the price.

Gardner, *Scientific American Book of Mathematical Puzzles & Diversions*, Book 1, 1959; Book 2, 1961.

Two of the finest collections of mathematical recreations available. Highly recommended.

Guilbaud, G. T., *What Is Cybernetics?*, Criterion Books, New York, 1959.

A clear exposition of the new field—updated, inclusive, and intended for informed laymen (economists, psychologists, and sociologists as well as all scientists) who find their work area influenced by cybernetics.

Heafford, *The Math Entertainer*, Emerson Books, Inc., New York, 1959.

Mathematical teasers designed to puzzle the reader. Answers and complete explanations are given for all problems.

Hilbert and Cohn-Vosser, *Geometry and the Imagination*, Chelsea Publishing Company, New York.

One of the world's outstanding mathematicians shows how to gain insight and intuitive understanding and how to use this understanding to obtain mathematical results.

Huff, *How to Take a Chance*, W. W. Norton & Company, Inc., New York, 1959.

Entertaining but soundly exact discussions of various aspects of chance, probability, and error, especially as applied to everyday life.

Huff, *How to Lie With Statistics*, W. W. Norton & Company, Inc., New York, 1955.

Humorous, but penetrating and authoritative explanation of the basic conceptions and misconceptions of statistics. Vivid illustrations in cartoon style fully capture and even extend the content.

Hunter, *Figurets: More Fun with Figures*, Oxford University Press, New York, 1958.

More fascinating mathematical puzzles, these problems for the adult layman are cast in the form of entertaining anecdotes.

Johnson, *Paper Folding for the Mathematics Class*, NCTM, Washington, D.C., 1957.

Illustrated directions for folding the basic constructions, geometric concepts, circle relationships, products and factors, polygons, knots, polyhedrons, symmetry, conic sections, recreations.

Johnson and Glenn, *Exploring Mathematics on Your Own*, Webster Publishing Company, St. Louis, 1960, 1961.

"Adventures in Graphing," "Computing Devices," "Fun with Mathematics," "Invitation to Mathematics," "Number Patterns," "Pythagorean Theorem," "Sets—Sentences and Operations," "Short-Cuts in Computing," "Topology," "Understanding Numeration Systems," "World of Measurement," "World of Statistics."

Jones, *Elementary Concepts of Mathematics*, The Macmillan Company, New York, 1947.

Designed for the non-math student, this text clarifies such concepts of everyday importance as compound interest, averages, probability, games of chance, graphs, etc.

Kasner and Newman, *Mathematics and the Imagination*, Simon and Schuster, Inc., New York, 1940.

An outstanding book which can be read and enjoyed by all. Not merely a "puzzle book"; this volume contains some excellent mathematical ideas.

Kemeny, Snell, and Thompson, *Introduction to Finite Mathematics*, Prentice-Hall, Inc., Englewood Cliffs, N.J., 1957.

One of the most modern of the new college freshman texts. An excellent book for good mathematics students. Well written and sound.

Khinchin, A. Y., *Three Pearls of Number Theory*, Graylock Press, 1952.

Brief history and complete, elementary proofs of three famous theorems of additive number theory. Delightful and profitable reading for all who love number theory—from amateur to expert.

Kline, *Mathematics in Western Culture*, Oxford University Press, New York, 1953.

This book gives a remarkably fine account of the influence mathematics has exerted on the development of philosophy, the physical sciences, religion, and the arts in Western life.

Kojima, *The Japanese Abacus*, Charles E. Tuttle Co., Inc., New York, 1961.

Describes the Japanese abacus and its use together with appropriate problems. This instrument is standard in the Orient and is amazing for the speed and accuracy it affords in competition.

Kraitchik, *Mathematical Recreations*, Dover Publications, New York, 1953.

One of the most thorough compilations of recreational mathematical problems. Highly recommended.

Kramer, *The Mainstream of Mathematics*, Oxford University Press, New York, 1951.

A historical treatment of mathematical thought from primitive number to relativity.

Levinson, *The Science of Chance: From Probability to Statistics*, Holt, Rinehart, and Winston, New York, 1950.

Compact, highly readable survey of chance and statistics, covering many forms of speculation and risk in business, as well as the odds in games of chance.

Lieber, Lillian R. and Hugh G., *The Education of T. C. Mits*, W. W. Norton & Company, Inc., New York, 1944.

A delightful, easy to read book that contains some interesting philosophy as well as mathematics.

Lieber, Lillian R., with drawings by Hugh G. Lieber, *Non-Euclidean Geometry*, Galois Institute, Long Island, New York, 1940.

Various geometries needed for different surfaces, treated postulationally, making Euclidean and non-Euclidean geometries easier and interesting. Very readable.

Logsdon, *A Mathematician Explains*, University of Chicago Press, Chicago, 1961.

This book tells informally what a student or interested layman should know about mathematics.

Maxwell, *Fallacies in Mathematics*, Cambridge University Press, New York, 1959.

Some mathematical fallacies traced in depth, often unexpectedly, to the source of error, with a will to please as well as to instruct.

McCracken, *Digital Computer Programming*, John Wiley & Sons, Inc., New York, 1957.

An excellent book on advanced computer programming using a mythical machine TYDAC.

Menger, *You Will Like Geometry*, Museum of Science and Industry, Chicago, 1961.

An especially interesting booklet which discusses unusual curves and amusements of geometry. Believe it or not, you *will* like geometry if you catch the infectious spirit of the guide book which was published as an adjunct to the Illinois Institute of Technology's permanent geometry exhibit at the Museum of Science and Industry.

Meserve, *Fundamental Concepts of Geometry*, Addison-Wesley Publishing Company, Reading, Mass., 1955.

A book giving the main concepts behind much of classical geometry.

Meyer, *Fun With Mathematics*, Premier Books, Fawcett, New York.

A collection of puzzles, problems, number facts, and curiosities.

Mosteller, Rourke, Thomas, *Probability: A First Course*, Addison-Wesley Publishing Company, Reading, Mass., 1961.

An introduction to probability and some statistics for a one-semester course; requires only two years of high school algebra.

Mott-Smith, *Mathematical Puzzles*, Dover Publications, New York, 1954.

Another collection of 188 interesting mathematical puzzles in an inexpensive edition.

Newman, *World of Mathematics*, Simon and Schuster, Inc., New York.

A well-known four-volume work.

Niven and Zuckerman, *An Introduction to the Theory of Numbers*, John Wiley & Sons, Inc., New York, 1960.

A substantially complete introduction to the theory of numbers, using an analytical (not historical) approach. The basic concepts are covered in the first part of the book, followed by more specialized material in the final three chapters.

Niven, *Numbers: Rational and Irrational*, Random House, Inc., New York, 1961.

This book deals with the number system, one of the basic

structures in mathematics. It is concerned especially with ways of classifying numbers into various categories.

Ore, *Number Theory and Its History*, McGraw-Hill Book Company, Inc., New York, 1948.
One of the most readable books on elementary number theory, with many interesting historical references.

Peck, Lyman C., *Secret Codes, Remainder Arithmetic, and Matrices*, NCTM, Washington, D.C., 1961.
A sprightly enrichment pamphlet using fun with secret codes to introduce ideas from modern mathematics. Many problems, with answers.

Polya, *Mathematics and Plausible Reasoning*, Princeton University Press, Princeton, N.J., 1954.
This is a guide to the practical art of plausible reasoning; the first volume deals with induction and analogy in mathematics, the second with patterns of plausible inference. The most suitable books your panel could find on these important topics.

Polya, *Mathematical Discovery*, John Wiley & Sons, Inc., New York, 1962.
Presents ways and means which lead to the discovery of the solution of problems. It is a continuation of Polya's two earlier works, *How to Solve It* and *Mathematics and Plausible Reasoning*.

Rademacher and Toeplitz, *The Enjoyment of Mathematics*, Princeton University Press, Princeton, N.J., 1957.
Probably the most outstanding "popular" mathematics book. Each chapter starts with simple observations easily within the grasp of all, and smoothly catapults the reader into the heart of a genuine mathematical research-type problem. Highly recommended for good students.

Ringenberg, *A Portrait of 2*, NCTM, Washington, D.C., 1956.
Written to enlarge the reader's concept of number, this pamphlet discusses the number 2 as an integer, a rational number, real number, and a complex number. Lays an excellent foundation for modern mathematical ideas.

Salkind, *The Contest Problem Book*, Random House, Inc., New York, 1961.
Composed of problems from the annual high school contests of the Mathematical Association of America, and cosponsored

by the Society of Actuaries. The interest on the part of these organizations in developing and administering the high school contests is based on the firm belief that one way of learning mathematics is through selective problem solving.

Sawyer, *Math Patterns in Science*, American Education Publishers, New York, 1960.
Shows how math and science are interrelated by employing a series of simple illustrated exercises that tie algebra patterns into everyday general science applications.

Sawyer, *A Concrete Approach to Abstract Algebra*, W. H. Freeman and Company, Golden Gate, San Francisco, 1961.
Gives excellent introductory material for preparation in modern abstract algebra.

Sawyer, *Mathematician's Delight*, Penguin Books, Inc., Baltimore, Md., 1943.
A well-written popular volume, dispelling the fear that surrounds mathematics.

Sawyer, *Prelude to Mathematics*, Penguin Books, Inc., Baltimore, Md., 1955.
A delightful account of some stimulating and surprising branches of mathematics. Highly recommended.

Schaaf, *Basic Concepts of Elementary Mathematics*, John Wiley & Sons, Inc., New York, 1960.
Discusses the nature of number and enumeration, the logical structure of arithmetic.

Steinhaus, *Mathematics Snapshots*, Oxford University Press, New York, new edition, 1960.
Pictures help visualize mathematics; the simple text and clear illustrations make this a book to be enjoyed by anyone with a knowledge of algebra. Many interesting suggestions for models and projects.

Turnbull, H. W., *The Great Mathematicians*, New York University Press, New York, 1961.
A readable and fascinating biographical history of mathematics from the early Egyptians to the great men of the twentieth century. The ideas and lives of the men who have dedicated themselves to the first and most exacting of man's skills.

van der Waerden, *Science Awakening: Egyptian, Babylonian and Greek Mathematics*, Oxford University Press, New York, 1961.

A history of mathematics for the general reader from the Egyptians as the "inventors" of geometry through the decline and final decay of Greek mathematics.

Whitehead, *An Introduction to Mathematics*, Oxford University Press, New York, 1958.

A highly recommended book by a well-known English mathematician.

Wilder, *Foundations of Math*, John Wiley & Sons, Inc., New York, 1952.

A fine book which stresses set theory and logic. Not always easy to read, but well worth the effort.

Williams, *The Compleat Strategyst*, McGraw-Hill Book Company, Inc., New York, 1954.

An excellent book on the theory of game strategy. Although written in a light vein, it provides a sound introduction to this fast growing field of modern mathematics.

Wylie, *101 Puzzles in Thought and Logic*, Dover Publications, New York, 1957.

Brand new problems you need no special knowledge to solve. Introduction with simplified explanation of general scientific method and puzzle solving. A fine book to stimulate logical thinking.

C. R. C. *Standard Mathematical Tables*, Chemical Rubber Co. Publications.

Long established as a time-saving reference aid. It contains a wealth of comprehensive, up-to-date mathematical data for use in algebra, geometry, calculus, trigonometry, statistics, differential equations, finance and investment, statistical analysis, and in virtually any problematic area.

24th Yearbook of National Council of Teachers of Mathematics, *The Growth of Mathematical Ideas Grades K-12*, NCTM, Washington, D.C., 1959.

An excellent book not only for teachers but also for students interested in modern concepts.

Answers and Hints

Chapter 1
Problem Set 1-6
1. $f(2) = 0$, $f(-7) = 45$, $f(1+x) = x^2 + 2x - 3$, $f(2 + \Delta x) = \Delta x^2 + 4\Delta x$.

3. $\dfrac{\pi d^2}{4}$.

7. $\tfrac{71}{9}$; 0.

9. $A = \dfrac{15 + n}{12}$, where n equals the number of months.

11. $\dfrac{-21(2t - \Delta t)}{t^2(t + \Delta t)^2}$.

13. $\dfrac{\Delta t + 2t}{3}$.

17. Yes. The domain would consist of all dates from the date of founding to the present.

Problem Set 1-10
1. 7, −5.   3. 0, 8.   5. $\tfrac{3}{7}$.   7. 0.   9. −10.
13. Approximately 119 turns.

Problem Set 1-11
3. $x^2 + \dfrac{B}{A}x + \dfrac{C}{A} = 0$, yes, $A\left(x^2 + \dfrac{B}{A}x + \dfrac{C}{A}\right) = 0$.   5. $\tfrac{2}{7}$, −8.
9. $\tfrac{1}{3}(2z - 1 \pm \sqrt{7z^2 - 13z + 4})$.   11. $z = 5$; $z = 2$.
12. $\tfrac{1}{6}(5i - 2 \pm \sqrt{4i - 57})$.   13. $b = \tfrac{1}{2}(-5 \pm \sqrt{65})$.

303

15. $-2, 3$.   **17.** $\frac{1050}{137} \approx 7.7$.   **19.** 9 in. by 12 in.

20. (a) $\dfrac{5x+2}{(2x+\frac{5}{4})^2 - (\frac{11}{4})^2}$, (b) $\dfrac{3}{\sqrt{\left(\frac{3}{\sqrt{2}}\right)^2 - \left(\sqrt{2}\left[x - \frac{3}{2}\right]\right)^2}}$,

(c) $\dfrac{3x+1}{(x+1)^2 + (\sqrt{2})^2}$.

21. (a) $\dfrac{20x+12}{(2x+1)^2 + (\sqrt{3})^2}$, (b) $\dfrac{13t}{(3t-2)^2 + (4)^2}$, (c) $\dfrac{-15}{(3x-1)^2 - (1)^2}$,

(d) $\dfrac{196a}{(2ax+b)^2 + (\sqrt{4ac-b^2})^2}$ if $b^2 \leq 4ac$, $\dfrac{196a}{(2ax+b)^2 - (\sqrt{b^2-4ac})^2}$

if $b^2 \geq 4ac$.

25. $|k| < 12$.   **27.** $\pm \dfrac{\sqrt{6}}{2}, \pm 2i$.   **28.** $-2, 4, -3, 5$.   **29.** $-1.3, 1.9$.

31. $\frac{1}{4}(-3 \pm j\sqrt{1591})$.   **33.** $4, t = -4 + \sqrt{16+2s}$.   **35.** $4\frac{2}{9}$ hr.

Problem Set 1-12

1. 7.   **2.** $-1, \frac{1}{3}$.   **3.** 10.   **4.** No solution.   **5.** $\frac{1}{2}(13 \pm \sqrt{161})$.

6. 3.   **7.** 0.   **8.** $-\frac{1}{2}$.   **9.** ± 6.   **10.** $\dfrac{41 \pm \sqrt{141}}{70} \approx \begin{cases} .76 \\ .42 \end{cases}$.

11. No solution.   **13.** $(x=2, y=5), (x=\frac{14}{5}, y=\frac{23}{5})$.

15. $-\sqrt{x-3} = x-5$ has solution $x=4$. $-\sqrt{3-2x}+5=0$ has solution $x=-11$. $-\sqrt{w+4}+w-2=0$ has solution $w=5$.
$-4\sqrt{2x+3}+5=0$ has solution $x = -\frac{23}{32}$.
$-\sqrt{3x^2+7}+2=0$ has solutions $x = \pm i$.

16. No solution.   **17.** $x = \dfrac{3 - 2\sqrt{2}}{50} = \dfrac{(\sqrt{2}-1)^2}{50}$.   **19.** -1.

21. $x = 4$.   **23.** No solution.   **25.** 9.

Problem Set 1-13

1. $2x\,\Delta x - 7\Delta x + \Delta x^2$.   **3.** $-\dfrac{2x + \Delta x}{x^2(x+\Delta x)^2}$.

5. $\dfrac{6\Delta x + 15x^2\,\Delta x + 15x\,\Delta x^2}{[2 - 5(x+\Delta x)^2][2 - 5x^2]}$.   **7.** $-\dfrac{2\Delta y}{y(y+\Delta y)}$; $-\dfrac{2\Delta t}{2t(2t+\Delta t)}$.

9. $\dfrac{1}{\sqrt{t+\Delta t - 3} + \sqrt{t-3}}$.   **11.** $\dfrac{6}{[3(x+\Delta x)+1][3x+1]}$.

Problem Set 1-14. Self-Test

(a), (b) [number line diagrams]

(c) All points on the axis except the point 2.

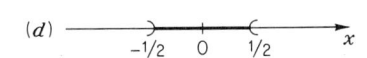

(d) [number line diagram]

5. $-\dfrac{7\Delta t}{2t(t + \Delta t)}.$

7. 0.

Chapter 2
Problem Set 2-1

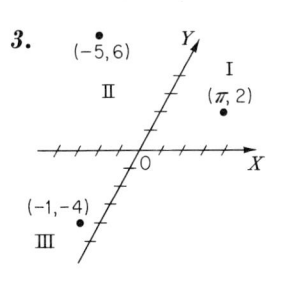

1. Points: $(-3, 2)$ in II, $(3,1)$ in I, $(2,-5)$ in IV.

3. Points: $(-5, 6)$ in II, $(\pi, 2)$ in I, $(-1, -4)$ in III.

4. On the x axis.

5. On a line, $x = -2$, parallel to the y axis and 2 units to the left of the y axis.

6. On a line, $y = 5$, parallel to the x axis and 5 units above the x axis.

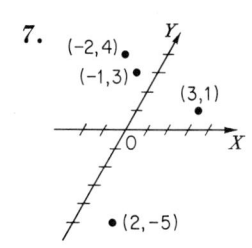

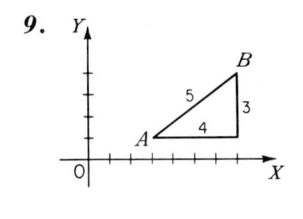

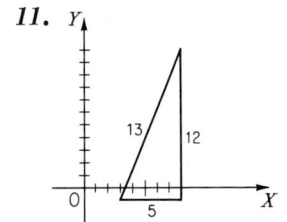

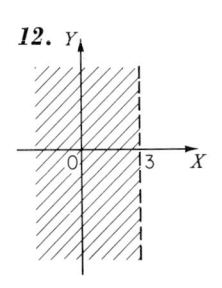

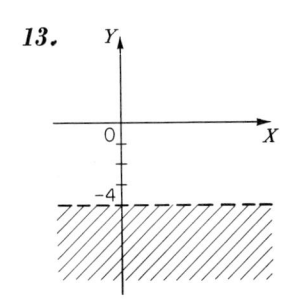

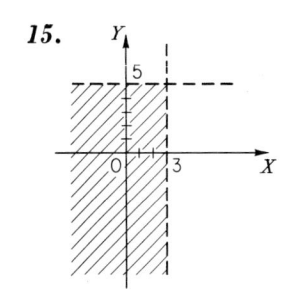

7. Points $(-2, 4)$, $(-1, 3)$, $(3, 1)$, $(2, -5)$ collinear.

9. Right triangle with legs 4, 3 and hypotenuse 5.

11. Right triangle with legs 5, 12 and hypotenuse 13.

12. Region $x \geq 3$.

13. Region $y \leq -4$.

15. Region $x \geq 3$ and $y \leq 5$.

Answers and Hints

17.

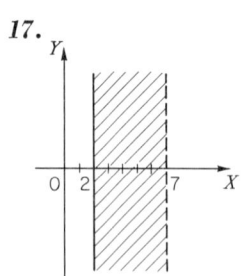

19.

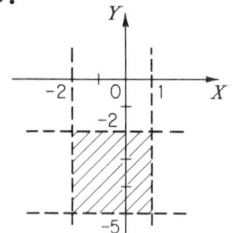

21.

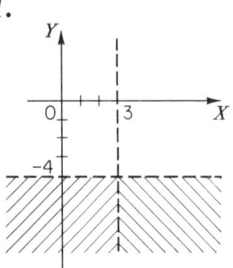

23.

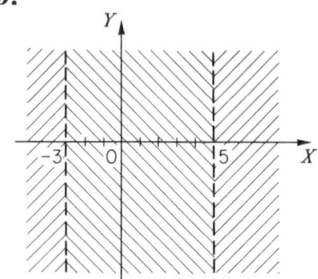

25.

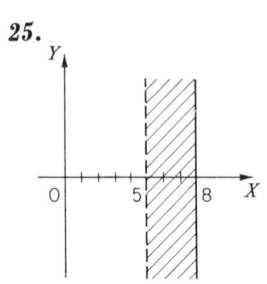

27.

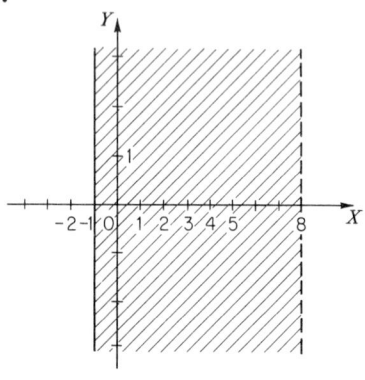

29.

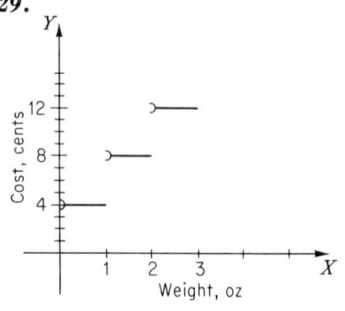

31.

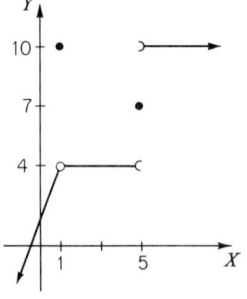

Problem Set 2-2
2. $\sqrt{157} \approx 12.5$.   **3.** $\sqrt{37} \approx 6.1$.   **4.** $2\sqrt{26}$.   **5.** $3\sqrt{2}$.
6. $\sqrt{(h-3)^2 + (k-1)^2}$.   **7.** $\sqrt{16 + (r-s)^2}$.   **9.** $\sqrt{61}$.
11. $3\sqrt{10}$.   **13.** $\sqrt{17}, \sqrt{29}, \sqrt{2}$.   **15.** $3\sqrt{2}, \sqrt{37}, \sqrt{85}$.

21. The point (h,k) will lie on a circle having center $(3,1)$ and radius 2.
25. The point (r,s) will lie on a circle having center $(-1,-4)$ and radius 3.
27.   **29.** 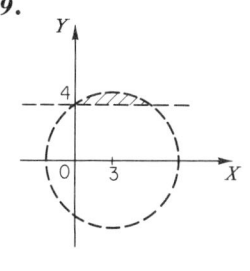   **31.** $(x-3)^2 + (y+2)^2 < 25$.
33. $d = x + y$.

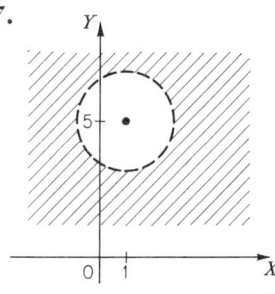

35. $p = \frac{1}{15}\sqrt{(x_2 - x_1)^2 + 225(y_2 - y_1)^2}$.

Problem Set 2-3
1.   **2.** $x^2 + y^2 - 8x - 6y = 0$.   **3.**

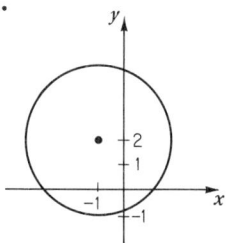

$(x+1)^2 + (y-2)^2 = 9$
or
$x^2 + y^2 + 2x - 4y - 4 = 0$.

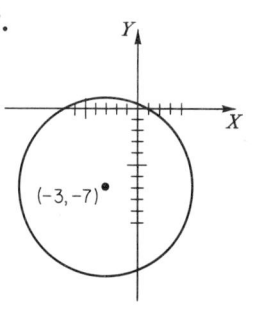

$x^2 + y^2 + 6x + 14y - 6 = 0$.

5. $x^2 + y^2 - 2hx - 2ky + h^2 + k^2 - 4 = 0$.   **7.**

6. $(x - h)^2 + (y - k)^2 = r^2$ or
$x^2 + y^2 - 2hx - 2ky + (h^2 + k^2 - r^2) = 0$.

$x + 4y - 16 = 0$.
8. $x - y + 3 = 0$.

Answers and Hints

9.

10. $2x_1 x + 2y_1 y = x_1^2 + y_1^2.$

11. $2x - 8y + 37 = 0.$

13. $x = 5.$

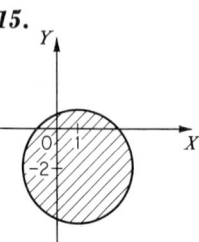

15.

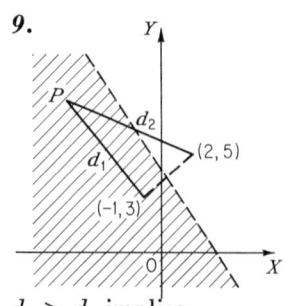

$d_2 > d_1$ implies
$-6x - 4y + 19 > 0$ or
$6x + 4y - 19 < 0.$
All points P with $d_2 > d_1$
lie below the line
$6x + 4y - 19 = 0.$

17. $y = 2.$ **19.** $y = x.$ **21.** $y = 0.$

23. $y^2 = (\sqrt{(x-1) + (y-3)^2})^2$ or $y = \dfrac{x^2}{6} - \dfrac{x}{3} + \dfrac{5}{3}.$ **25.** $(\tfrac{4}{5}, \tfrac{19}{5}).$

27. Line parallel to the y axis and 7 units to the right of the y axis.

29. Line parallel to the y axis and 4 units to the left.

31. Line parallel to the y axis and $\dfrac{\pi}{3}$ units to the right.

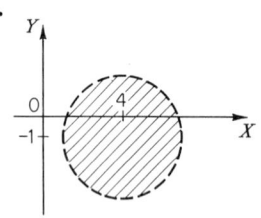

32.

All points inside the circle $(x-4)^2 + (y+1)^2 = 9$ having center $(4,-1)$ and radius 3.

33. No real locus. Sometimes called an imaginary circle.

35. Circle with center $(0,0)$ and radius 5.

37.

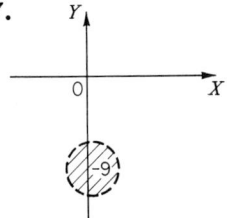

All points inside the circle
$$x^2 + (y+9)^2 = 4$$
having center $(0, -9)$ and radius 2.

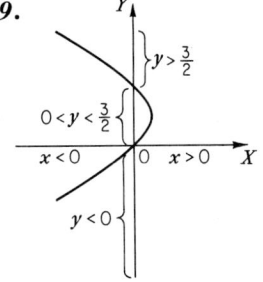

39.

Write $x = -2y(y - \frac{3}{2})$. Then $x > 0$, if $-2y(y - \frac{3}{2}) > 0$ or $y(y - \frac{3}{2}) < 0$ implies $0 < y < \frac{3}{2}$. Also $x = 0$ when $y = 0$ and $y = \frac{3}{2}$. Finally $x < 0$ when $y(y - \frac{3}{2}) > 0$, i.e., when $y < 0$ and $y > \frac{3}{2}$.

Problem Set 2-4

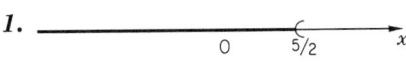

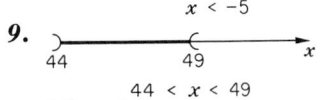

1. $x < 5/2$

3. $x < 1$

5. $x < -5$

7. $-5 < x < 9$

9. $44 < x < 49$

11. $x < -4, x > 10$

13. All real x values.

14. No x value satisfies.

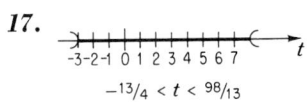

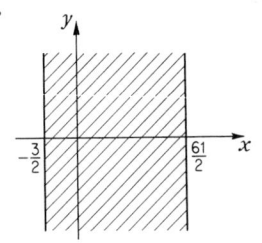

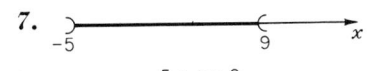

15. $x > 73/14$

17. $-13/4 < t < 98/13$

19. $t > -1/2$

21.

The shaded region shown including the boundaries.

23. $|x| > 5$.

25. All real y; all real x.

27. The given conditions imply $xy \geq 0$. If x and y are of opposite algebraic signs, the inequality will not be satisfied.

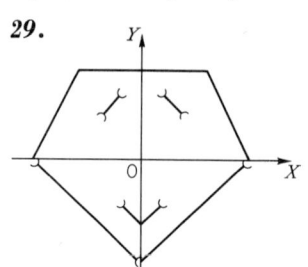

29.

31. (a) $x < 0$, $x > 6$.
(b) $0 \leq x < 1$, $3 < x < 4$, $4 < x < 5$, $5 < x \leq 6$.
(c) $1 < x < 2$, $2 < x \leq 3$, $x = 4$.
(d) $x = 2$.
(e) $x = 1$, $x = 5$.

Problem Set 2-5

1. $m = \dfrac{5 - (-3)}{1 - 2} = -8$. **3.** No slope. **4.** $m = 0$. **5.** $m = 4$.

7. Illustrations: slope $GB = \dfrac{10}{6} = \dfrac{5}{3}$. Slope $DA = \dfrac{8}{-6 - 6 - 8} = -\dfrac{2}{5}$.

9. Intercepts: $(-\tfrac{2}{7}, 0)$, $(0, \tfrac{2}{3})$. $m = \dfrac{\tfrac{2}{3} - 0}{0 - (-\tfrac{2}{7})} = \dfrac{7}{3}$.

11.

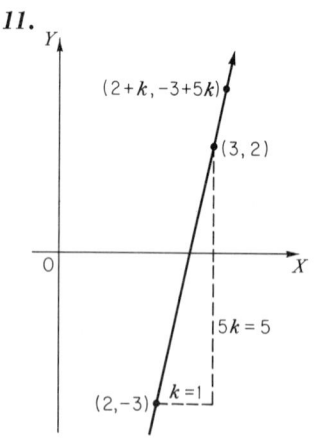

$m = \dfrac{5k}{k} = 5$, $k \neq 0$. Note that any point on the line is given by

$$(x = 2 + k,\ y = -3 + 5k).$$

If the *parameter* k is eliminated, the equation of the line is found:

$$y = -3 + 5(x - 2) \quad \text{or} \quad 5x - y - 13 = 0.$$

Answers and Hints   311

13.

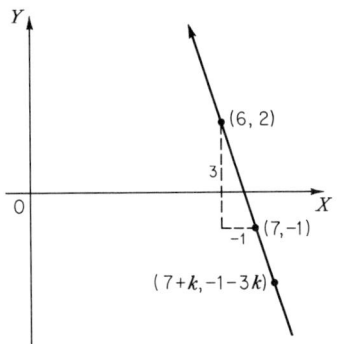

$$m = \frac{-3k}{k} = -3, k \neq 0.$$
$(x = 7 + k, y = -1 - 3k).$
$y = -1 - 3(x - 7)$ or
$3x + y - 20 = 0.$

15.

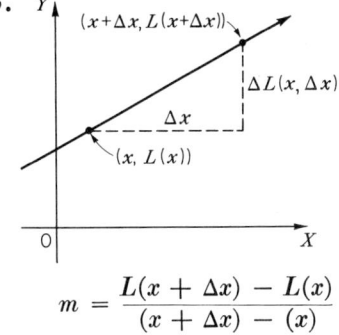

$$m = \frac{L(x + \Delta x) - L(x)}{(x + \Delta x) - (x)}$$
$$= \frac{\Delta L(x, \Delta x)}{\Delta x}.$$

16. Illustration no. 8: $y = -\frac{3}{2}x + \frac{5}{2}$. (a) $y + \Delta y = -\frac{3}{2}(x + \Delta x) + \frac{5}{2}$.
(b) $\Delta y = -\frac{3}{2}\Delta x.$ (c) $m = \frac{\Delta y}{\Delta x} = -\frac{3}{2}.$

17. $m = 0.$ **18.** No defined slope.

21. See figure for Prob. 15. Since $\Delta L(x, \Delta x)$ and Δx have the same sign, their ratio $m = \dfrac{\Delta L(x, \Delta x)}{\Delta x}$ is positive.

22. Here $\Delta L(x, \Delta x)$ and Δx are opposite in sign and their ratio is negative.

23.

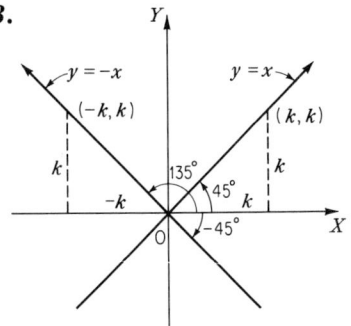

25. See (2) of no. 23.
26. $m = -1.$

(1) $m = \dfrac{k}{k} = 1,$

(2) $m = -\dfrac{k}{k} = -1, k \neq 0.$

27.

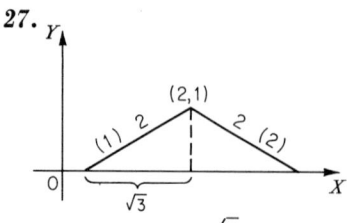

(1) $m = \dfrac{1}{\sqrt{3}} = \dfrac{\sqrt{3}}{3}$.

(2) $m = -\dfrac{\sqrt{3}}{3}$.

Problem Set 2-6

1. $y - 6 = 2(x - [-1])$ or $2x - y + 8 = 0$. 3. $x + y - 1 = 0$.
5. $y - [-2] = -(\frac{1}{3})(x - 4)$ or $x + 3y + 2 = 0$. 7. $m = 0$, $y + 3 = 0$.
9. $m = \frac{7}{2}$, $7x - 2y + 14 = 0$. 10. $m = -\dfrac{b}{a}$.
11. $y = -3x + 5$, $m = -3$, $b = 5$. 13. $y = -\frac{1}{4}x + \frac{1}{3}$, $m = -\frac{1}{4}$, $b = \frac{1}{3}$.

15. $y = 3x + 5$, $m = 3$, $b = 5$.
16. $m = -4$. Parallel lines have the same slope.

17. $m = -\frac{1}{2}$, $y - 1 = -\frac{1}{2}(x - 2)$. 19. $m = -\frac{4}{3}$, $y = -(\frac{4}{3})(x + 1)$.
21. $m = 0$, $y + 7 = 0$. 23. $x + 3 = 0$.
25. $m = -1$, $y + 6 = -(x - 4)$. 27. $y = mx$. 29. $y = 0$.
31. $m = \dfrac{24}{11}$, $y = \dfrac{(24x - 45.8)}{11}$. 33. $m = -\dfrac{79}{91}$, $y = \dfrac{(-79x + 1320.3)}{91}$.

35. Use Prob. 10. $a = \dfrac{3}{2}$, $b = 7$, $\dfrac{2x}{3} + \dfrac{y}{7} = 1$.

37. $(x = -1, y = 0)$, $(x = 0, y = -\frac{3}{7})$. 39. $y = -8$. 41. $x = 5$.

43. Intercepts: $A(-3,0)$, $B(0,8)$. Mid-point of AB is $M\left(\dfrac{-3+0}{2}, \dfrac{0+8}{2}\right)$.
Use Prob. 7, Set 3-10. Slope of line OM: $m = -\frac{8}{3}$. Apply Prob. 27 and obtain desired line $y = -\dfrac{8x}{3}$.

44. Use $m = -2$.
45. $m = -\dfrac{2}{5}$, $y = -\dfrac{2x}{5} + b$, $-3 = -\dfrac{2(4)}{5} + b$, $b = -\dfrac{7}{5}$. $y = \dfrac{(-2x - 7)}{5}$.

47. The solution ($x = 1$, $y = 3$) of the first and second equations must satisfy the third equation. $c = -9$.

49.

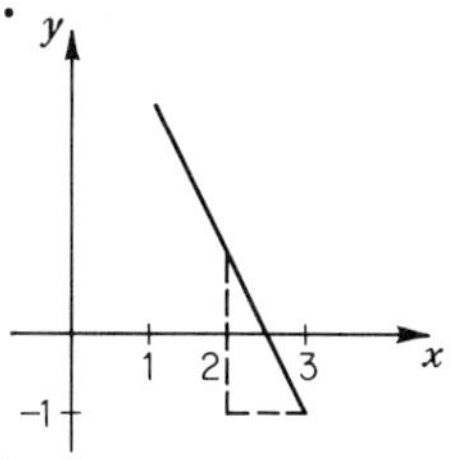

51. (a) $x = 2$.
(b) 4 units.
(c) 10.
(d) 20 square units.

53.

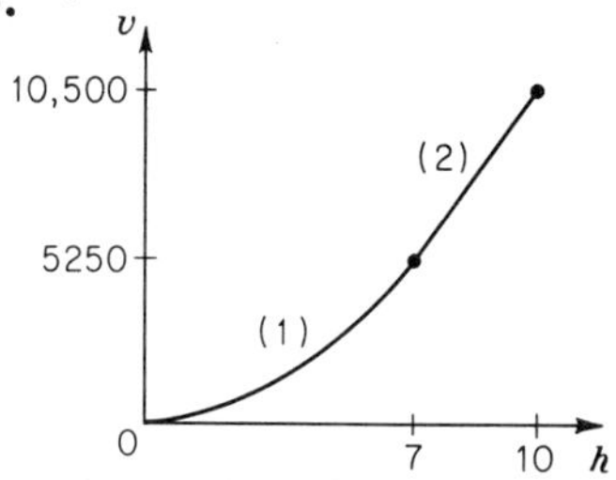

(1) $V = \frac{1}{2}(h)\left(\frac{75h}{7}\right)(20) = \frac{50h^2}{7}$, $0 \leq h \leq 7$;

(2) $V = \frac{1}{2}(7)(75)(20) + (h-7)(75)(20)$
$= 1{,}500h - 5{,}250$, $7 \leq h \leq 10$.

Problem Set 2-7. Self-Test

1. $y - 6x + 17 = 0$.

3. $2 < x < 8$

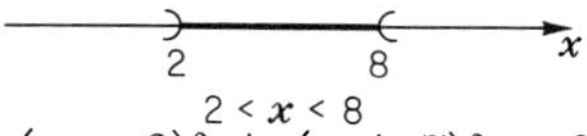

5. $(x-2)^2 + (y+7)^2 = 9$.

7.

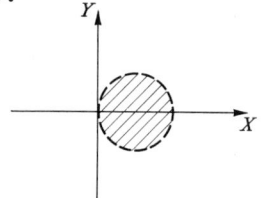

Chapter 3
Problem Set 3-1

3. $m = 2$.

4. Right triangle OPA is similar to right triangle OMP. $\frac{|MP|}{|AM|} = \frac{|OM|}{|MP|}$, $|MA| = \frac{16}{3}$, $|OA| = \frac{25}{3}$. The slope of the tangent line AP is $m = -\frac{3}{4}$. Note that this is the negative reciprocal of the slope of the radius OP.

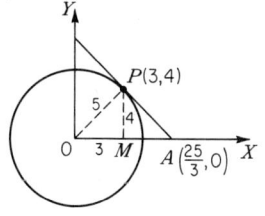

314 *Answers and Hints*

5. See Prob. 4.
9. $y = 2|x|$ or $y = 2x$, $x \geq 0$; $y = -2x$, $x \leq 0$. $m = 2$ is the slope at $(3,6)$. $m = -2$ is the slope at $(-2,4)$. At $(0,0)$ no definite slope is determined.
15. $3x + 4y - 25 = 0$. The line and the circle meet in two coincident points.

Problem Set 3-3

1. 45. **2.** 27.

3. (1) $f(x)$ is continuous at $x = 5$. Why? (2) $f(x)$ is not continuous at $x = 3$. Why?
5. Neither are continuous at $x = 4$. Why? Both are continuous at $x = 6$. Why?
7. Choose $\delta \leq .5$ and $|x - 3| < \delta \leq .5$. Then $|4x - 12| < 2$.
9. Yes. No. **11.** Yes. **13.** 4.
15. $\lim\limits_{\Delta x \to 0} \dfrac{-3}{1 + \Delta x} = -3$. **17.** (a) Yes. (b) Yes. (c) Yes.
19. $y = \dfrac{4}{x - 4}$.

21. HINT: $|f_1(x) - L| < \epsilon$ for all x satisfying $0 < |x - b| < 8$. But, for all such x, $f_1(x) = f_2(x)$. Therefore?

Problem Set 3-4

1.

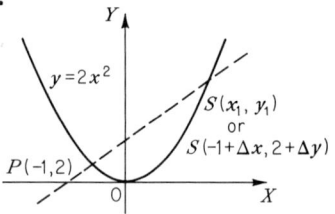

Slope $PS = \dfrac{y_1 - 2}{x_1 + 1} = \dfrac{2x_1^2 - 2}{x_1 + 1}$
$= 2(x_1 - 1)$, $x_1 \neq -1$.
Slope of tangent line $= \lim\limits_{x_1 \to -1} 2(x_1 - 1)$
$= -4$.

3. $2 + \Delta y = 2(-1 + \Delta x)^2$, $\Delta y = -4\Delta x + 2\Delta x^2$.
Slope $PS = \dfrac{\Delta y}{\Delta x} = \dfrac{-4\Delta x + 2\Delta x^2}{\Delta x} = -4 + 2\Delta x$, $\Delta x \neq 0$.
Slope of tangent line at P is $\lim\limits_{\Delta x \to 0} \dfrac{\Delta y}{\Delta x} = \lim\limits_{\Delta x \to 0}(-4 + 2\Delta x) = -4$.
5. $P(2,3)$, $S(2 + \Delta x, 3 + \Delta y)$. Slope $PS = \dfrac{\Delta y}{\Delta x} = \dfrac{-3\Delta x}{\Delta x (2 + \Delta x)} = \dfrac{-3}{2 + \Delta x}$,
$\Delta x \neq 0$. Slope of tangent line at $P = \lim\limits_{\Delta x \to 0} \dfrac{\Delta y}{\Delta x} = -\dfrac{3}{2}$.

Answers and Hints

7. $P(1,5)$, $S(1 + \Delta x, 5 + \Delta y)$.

Slope $PS = \dfrac{\Delta y}{\Delta x} = \dfrac{10\Delta x + 5\Delta x^2}{\Delta x} = 10 + 5\Delta x$, $\Delta x \neq 0$.

Slope of tangent line at $P = 10$. Equation of tangent line: $10x - y - 5 = 0$.

9. $P(2,1)$, $S(2 + \Delta x, 1 + \Delta y)$. $\dfrac{\Delta y}{\Delta x} = -\dfrac{1}{2 + \Delta x}$, $\Delta x \neq 0$.

Slope of tangent line $= -\frac{1}{2}$. Equation of tangent line: $x + 2y - 4 = 0$.

11. Slope of tangent line $= -10$. Equation of tangent line: $10x + y + 9 = 0$.

13. $m = -3$; tangent line $3x + y - 2 = 0$.

15. $m = 8$, $8x - y - 3 = 0$. **16.** $m = 10$; tangent line $10x - y - 2 = 0$.

17. $m = 5$, $5x - y - 7 = 0$.

19. $P(2,14)$, $S(2 + \Delta x, 14 + \Delta y)$.

$\dfrac{\Delta y}{\Delta x} = \dfrac{36\Delta x + 18\Delta x^2 + 3\Delta x^3}{\Delta x} = 36 + 18\Delta x + 3\Delta x^2$, $\Delta x \neq 0$.

$m = 36$, $36x - y - 58 = 0$.

20. Use $\lim\limits_{\Delta x \to 0} \dfrac{\Delta y\, (x_1, \Delta x)}{\Delta x}$.

Problem Set 3-5

1. $m = \lim\limits_{\Delta x \to 0} \dfrac{\Delta y}{\Delta x} = \lim\limits_{\Delta x \to 0} (10x_1 + 5\Delta x) = 10x_1$.

3. $\dfrac{\Delta y}{\Delta x} = \dfrac{-12}{x(x + \Delta x)}$, $\Delta x \neq 0$; $m = \lim\limits_{\Delta x \to 0} \dfrac{\Delta y}{\Delta x} = -\dfrac{12}{x^2}$. $-12, -3, -\frac{4}{3}, -\frac{1}{3}$.

5. $-\dfrac{12}{x^2} = -3$. $(-2, -6)$, $(2, 6)$.

6. $\Delta y = f(x + \Delta x) - f(x) = \Delta f(x, \Delta x)$, $y = f(x)$.

7. Slope at $(0,0) = 0$. Tangent line is $y = 0$. Slope at $(2,8) = 12$. Tangent line is $12x - y - 16 = 0$.

8. If the curve is rising as x increases, Δx and Δy have the same sign and $\dfrac{\Delta y}{\Delta x}$ is positive. Therefore $\lim\limits_{\Delta x \to 0} \dfrac{\Delta y}{\Delta x} > 0$.

9. If the curve is falling as x increases, Δx and Δy have opposite signs and $\dfrac{\Delta y}{\Delta x}$ is negative. Therefore $\lim\limits_{\Delta x \to 0} \dfrac{\Delta y}{\Delta x} < 0$.

11. $3x_1^2 = 12$. The tangent line at $(-2, -8)$ is $12x - y + 16 = 0$. The tangent line at $(2, 8)$ is $12x - y - 16 = 0$.

13. $14x - 6 = 2$. The tangent line at $(\frac{4}{7}, -\frac{8}{7})$ is $14x - 7y - 16 = 0$.

15. Slope at $(a,b) = 200 - 120a + 12a^2$.

316   *Answers and Hints*

16. $12x^2 - 120x + 200 = 12\left(x - 5 + \frac{5\sqrt{3}}{3}\right)\left(x - 5 - \frac{5\sqrt{3}}{3}\right)$. Slope is zero when $\left(x = 5 - \frac{5\sqrt{3}}{3} \cong 2.1, \ y \cong 192.4\right)$, $\left(x = 5 + \frac{5\sqrt{3}}{3} \cong 7.9, \ y \cong -1614.4\right)$. y is increasing when $12x^2 - 120x + 200 > 0$. That is, when $x < 5 - \frac{5\sqrt{3}}{3}$ or $x > 5 + \frac{5\sqrt{3}}{3}$.

17. $\lim\limits_{\Delta x \to 0} \frac{\Delta y}{\Delta x} = -4x$. The slope of the tangent line at $(2, -4)$ is -8.

18. $3x^2 - 4x - 7 = 3(x + 1)(x - \frac{7}{3})$. Slope is zero at $(-1, 9)$ and $(\frac{7}{3}, -\frac{257}{27})$. y is increasing when $3x^2 - 4x - 7 > 0$. $x < -1, \ x > \frac{7}{3}$.

19. $\lim\limits_{\Delta x \to 0} \frac{\Delta y}{\Delta x} = 2x - 10$. At minimum point $(5, 2)$ the slope is zero.

Problem Set 3-6. Self-Test

1. (a) $2x + 3y + 5 = 0$.
    (b) $2x + 3y - 5 = 0$.
    (c) $2x - 3y = 0$.

2. (a) $y = -\frac{3}{4}x + \frac{5}{4}$.
    (b) $\frac{x}{3} + \frac{y}{-4} = 1$.

3. (a) $m = -7$ at $(0,0)$.
    (b) $y + 7x = 0$.

4. $m = -\frac{1}{2}$ at $(2,1)$.

5. $(2,8); (2,8); (-4,-64)$.

6. 12.

7. $\lim\limits_{x \to 1} \frac{x^2 + 2x - 3}{x - 1} = 4$.

8. Not defined at $x = 1$.

Chapter 4
Problem Set 4-2

3. $-2x$.

5. $\frac{\Delta f(x, \Delta x)}{\Delta x} = \frac{-2x \Delta x - \Delta x^2}{(x + \Delta x)^2 x^2 \Delta x}$, $f'(x) = -\frac{2}{x^3}$.

7. $\frac{\Delta f(x, \Delta x)}{\Delta x} = \frac{21x^2 \Delta x + 21x \Delta x^2 + 7\Delta x^3}{\Delta x}$, $f'(x) = 21x^2$.

9. 0.

11. $\frac{\Delta y (x, \Delta x)}{\Delta x} = \frac{0}{\Delta x}$, $y' = 0$.

13. $\frac{\Delta y (x, \Delta x)}{\Delta x} = \frac{\Delta x}{(\sqrt{x + \Delta x - 5} + \sqrt{x - 5}) \Delta x}$, $y' = \frac{1}{2\sqrt{x - 5}}$.

14. See 15.

15. $\frac{\Delta f (t, \Delta t)}{\Delta t} = \frac{-2\Delta t}{(\sqrt{7 - 2t - 2\Delta t} + \sqrt{7 - 2t}) \Delta t}$, $f'(t) = -\frac{1}{\sqrt{7 - 2t}}$.

17. $6x - 6$.

21. $(y + \Delta y)^2 = 2(x + \Delta x) + 3$, $(2y + \Delta y) \Delta y = 2\Delta x$, $\dfrac{\Delta y}{\Delta x} = \dfrac{2\Delta x}{(2y + \Delta y) \Delta x}$,

$\lim\limits_{\Delta x \to 0} \dfrac{\Delta y}{\Delta x} = \dfrac{1}{y}$.

23. Let $A = \sqrt[3]{4x^2 - 1}$, $B = \sqrt[3]{4(x + \Delta x)^2 - 1}$. Then

$$\dfrac{\Delta g(x, \Delta x)}{\Delta x} = \dfrac{A - B}{AB \, \Delta x} = \dfrac{(A - B)(A^2 + AB + B^2)}{AB(A^2 + AB + B^2) \, \Delta x} = \dfrac{A^3 - B^3}{AB(A^2 + AB + B^2) \, \Delta x}$$

$$= \dfrac{-8x \, \Delta x - 4\Delta x^2}{AB(A^2 + AB + B^2) \, \Delta x},$$

$$g'(x) = \dfrac{-8x}{3(\sqrt[3]{4x^2 - 1})^4}.$$

25. Verify that (1), (2), (3) of definition in 5-5 hold.

Problem Set 4-4

1. $36x^8$.   **3.** $6x$.   **5.** $7x^6$.   **7.** 0.   **9.** $\dfrac{5x}{\sqrt{5x^2}}$.

11. $y'(x) = 3x^2$, $m = y'(2) = 12$.   **13.** $y'(x) = 15x^4$, $m = y'(3) = 1{,}215$.

14. $y'(x) = 15x^2$, $x^3 = 8$, $m = 60$.

15 $y'(x) = 21x^2 = \frac{21}{25}$. $(-\frac{1}{5}, -\frac{7}{125})$, $(\frac{1}{5}, \frac{7}{125})$.

19. $y = 3x$, $y = 3x + 4$.

21. $\left(\dfrac{15 + 5\sqrt{3}}{3}, \dfrac{-9000 + 8000\sqrt{3}}{9}\right)$, $\left(\dfrac{15 - 5\sqrt{3}}{3}, \dfrac{9000 - 8000\sqrt{3}}{9}\right)$.

$x < \dfrac{15 - 5\sqrt{3}}{3}$ and $x > \dfrac{15 + 5\sqrt{3}}{3}$.

23. $(-1, 9)$, $(\frac{7}{3}, -\frac{257}{27})$.   **25.** $(2, 16)$, $(-4, -128)$.

Problem Set 4-5

1. $20x^3 - 6x - 2$.

3. $y' = 3x^2 - 4x + 3 = 2$. Tangent line at $(\frac{1}{3}, -\frac{113}{27})$ is $54x - 27y - 131 = 0$. Tangent line at $(1, -3)$ is $2x - y - 5 = 0$.

5. $y'(x) = 35x^6 + 3$, $y - 10{,}944 = 25{,}518(x - 3)$.

7. $y'(x) = 2x - 3$, $y'(1) = -1$, $y'(2) = 1$.

9. $x^2 - 5x + 6 = 0$. $y'(2) = -1$, $y'(3) = 1$.

11. $y' = 3x^2 \geq 0$ for all x. y never decreases as x increases.

13. $(\frac{2}{3}, -\frac{4}{3})$.   **15.** $p'(0) = -12$, $p'(\pm 2\sqrt{3}) = 24$.

17. Points of intersection: $(1, 12)$, $(-\frac{7}{3}, 12)$. $y'(1) = 10$, $y'(-\frac{7}{3}) = -10$. Tangent lines: $y - 12 = 10(x - 1)$, $y - 12 = -10(x + \frac{7}{3})$.

19. $y - 77 = 30(x - 4)$. **21.** $y = 9$. **23.** $y - 216 = 3(\sqrt[3]{230})^2(x - \sqrt[3]{230})$.

25. Points of intersection: $(-\frac{3}{5}, -\frac{36}{5})$, $(2, 11)$. $y'(-\frac{3}{5}) = -6$, $y'(2) = 20$.
Tangent lines: $y + \frac{36}{5} = -6(x + \frac{3}{5})$, $y - 11 = 20(x - 2)$.

27. Slope of secant is -5. Point: $(-\frac{3}{2}, \frac{9}{4})$.

29.

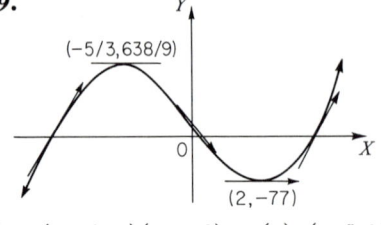

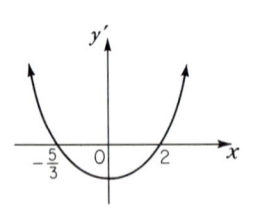

(−5/3, 638/9)

(2, −77)

$y' = (3x + 5)(x - 2)$. (a) $(-\frac{5}{3}, \frac{638}{9})$, $(2, -77)$. (b) y increases as x increases when $y' > 0$. $x < -\frac{5}{3}$, $x > 2$. (c) y decreases as x increases when $y' < 0$. $-\frac{5}{3} < x < 2$.

30. See Prob. 31.

31. $\dfrac{\Delta y\ (x, \Delta x)}{\Delta x} = \dfrac{3\Delta x}{(x + \Delta x)x\ \Delta x}$. $y' = \dfrac{3}{x^2} = 12$. Points: $(-\frac{1}{2}, 6)$, $(\frac{1}{2}, -6)$.

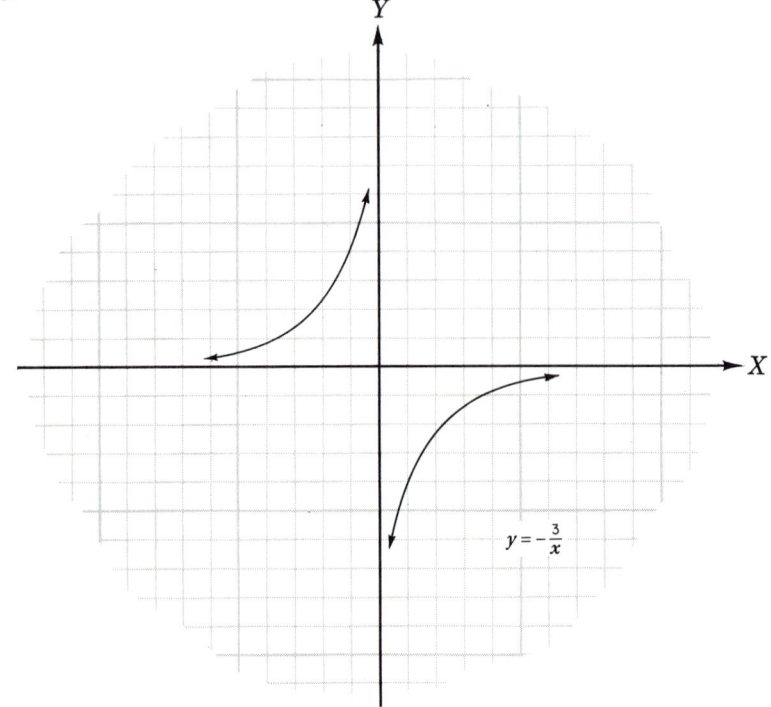

33.

$y = -\dfrac{3}{x}$

35. For a function which is continuous and differentiable it is geometrically apparent that for any two numbers x_1 and x_2 there is at least one number $x_1 < x < x_2$ such that the slope of the tangent line to the curve at $(x, h(x))$ is equal to the slope of the secant through $(x_1, h(x_1))$ and $(x_2, h(x_2))$. That is, $h'(x) = \dfrac{h(x_2) - h(x_1)}{x_2 - x_1}$. Since $h'(x) = 0$ for all x by hypothesis, $h(x_2) = h(x_1)$ for all x_1 and x_2. Hence, $h(x)$ is constant.

37. If a polynomial is neither a constant function nor a linear function, then it has at least one term of the form cx^n with $n > 1$. Therefore, the derivative of the polynomial has a term ncx^{n-1} with $n - 1 > 0$. Hence, the value of the slope changes as x changes and is not constant.

39. $24x - y - 57 = 0$. **40.** $4x + y - 7 = 0$.

41. (1) For each value of k the equation $7x + 11y - 12 + k(2x - 9y + 13) = 0$ represents a line through the point of intersection of the two given lines. Since the desired line passes through $(5, 7)$,

$$7(5) + 11(7) - 12 + k[2(5) - 9(7) + 13] = 0 \quad \text{and} \quad k = \tfrac{5}{2}.$$

The desired line is

$$7x + 11y - 12 + \tfrac{5}{2}(2x - 9y + 13) = 0 \quad \text{or} \quad 24x - 23y + 41 = 0.$$

(2) **Second Solution.** The point of intersection of the given lines is $(-\tfrac{7}{17}, \tfrac{23}{17})$. The slope of the desired line is $\tfrac{24}{23}$. Its equation is $24x - 23y + 41 = 0$.

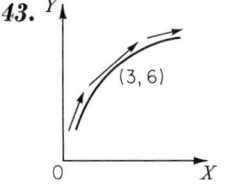

43.

$p'(x)$ is positive and decreasing as x increases through $x = 3$ since $p''(3)$ is negative. The curve is below the tangent line at $(3, 6)$.

45. $3x^2 + 2x - 21 = 0$. Points: $(-3, -1)$, $(\tfrac{7}{3}, -\tfrac{1787}{27})$. Tangent lines: $2x - y + 5 = 0$, $y + \tfrac{1787}{27} = 2(x - \tfrac{7}{3})$.

Problem Set 4-6

1. $y'(x) = 6(x - 1)$.

x	y	y'	Conclusions
$x < 1$		$-$	
1	2	0	Minimum
$x > 1$		$+$	$y = 2$ at $x = 1$.

3.

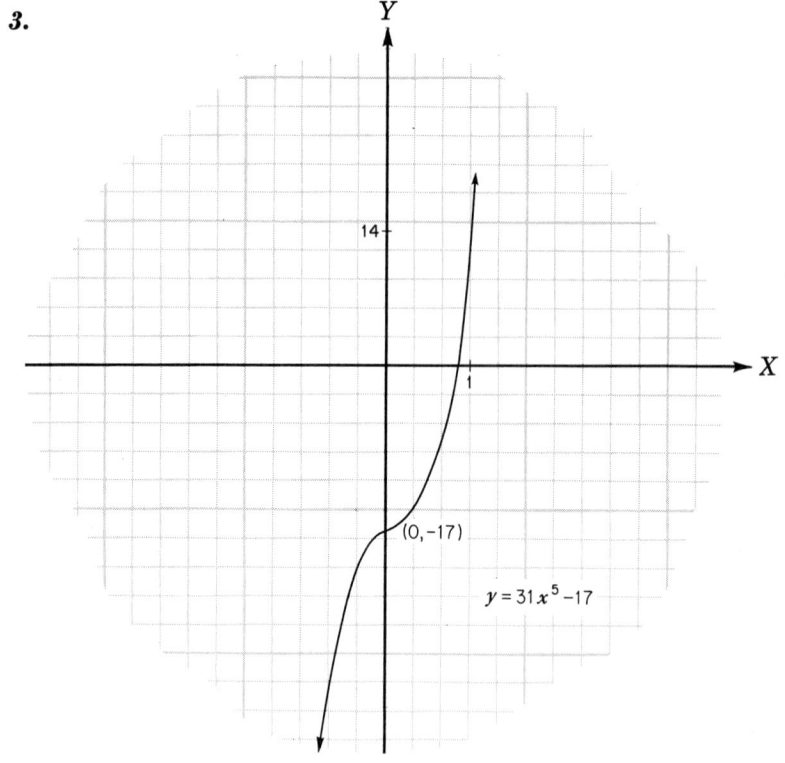

When $x < 0$, $y' > 0$. At $(0, -17)$, $y' = 0$. When $x > 0$, $y' > 0$. Hence, $(0, -17)$ is neither a max nor a min.

5.

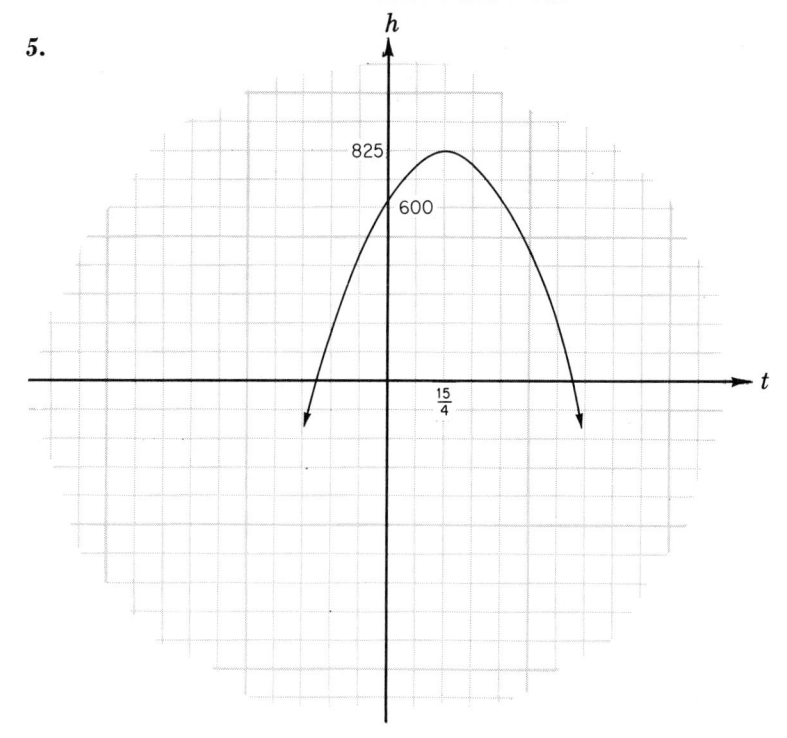

$$h = 600 + 120t - 16t^2.$$

$y' = 120 - 32t$. When $t < \frac{15}{4}$, $y' > 0$. At $(\frac{15}{4}, 825)$, $y' = 0$. When $t > \frac{15}{4}$, $y' < 0$. The point is a max.

7. $y' = 4(x^3 - 8)$. Min at $(2, -42)$.
9. $y' = 8(x - \frac{7}{8})$. Min $y = -\frac{17}{16}$ at $x = \frac{7}{8}$.
11. $h' = -2(t - 60)$. Max $h = 4{,}000$ at $t = 60$.
13. $y' = 4(x + \frac{5}{12})$. Min $y = -\frac{241}{72}$ at $x = -\frac{5}{12}$.
15. $y' = \frac{3}{2}(x - 2)^2$. Neither max nor min. Inflection point: $(2,4)$.
17. $y' = 4x(x - \sqrt{6})(x + \sqrt{6})$. Min at $(-\sqrt{6}, -24)$, $(\sqrt{6}, -24)$. Max at $(0, 12)$. Locus symmetric to y axis.

Problem Set 4-7

1. $y' = 3x^2 - 16x + 30 = 25$. Points: $(\frac{1}{3}, \frac{58}{27})$, $(5, 68)$.

3. Let the dimensions of the box be x ft $\times$ $2x$ ft $\times \dfrac{36}{x^2}$ ft. The surface area

$$A = 4x^2 + \frac{216}{x} \text{ sq ft} \qquad A' = \frac{8(x^3 - 27)}{x^2}.$$

If $0 < x < 3$, then $A' < 0$. If $x = 3$, then $A' = 0$. If $x > 3$, then $A' > 0$.
Min $A = 108$ sq ft when $x = 3$ ft. Dimensions: 3 ft × 6 ft × 4 ft.

5. $A = 8r^2 + \dfrac{250}{r}$. $A' = \dfrac{(16r^3 - 250)}{r^2}$. Min A when $r = \dfrac{5}{2}$ in. and $h = \dfrac{20}{\pi}$ in.

7. $I = (500 - x)(300 + x)$. Max I when $x = 100$. Monthly rate $4.00.

9. $50.

11. $A = x(200 - 4x)$. Max $A = 2{,}500$ sq ft when $x = 25$.

13. Max area $2{,}250$ sq ft.

Problem Set 4-8

1. $y' = (x^3 - 1)(8x - 2) + (4x^2 - 2x + 1)(3x^2) = 20x^4 - 8x^3 + 3x^2 - 8x + 2$.
3. $y'(x) = (3x^2 - 2x + 5)(8x - 3) + (4x^2 - 3x)(6x - 2) = 48x^3 - 51x^2 + 52x - 15$, $y'(2) = 269$.
5. $y'(x) = 4x^3 + 6x^2 + 14x + 10$, $y'(-2) = -26$.
7. $y'(x) = 10x^9 + 7x^6 + 35x^4 + 12x^3 - 15x^2 + 14x + 3$, $y'(1) = 66$.
9. $y'(x) = 60x^4 - 32x^3 - 30x + 10$, $y'(5) = 33{,}360$.
11. $y' = \dfrac{2x}{x-2} + (x^2 + 5)\dfrac{d\left(\dfrac{1}{x-2}\right)}{dx} = \dfrac{2x^2 - 4x - 1}{(x-2)^2}$.
13. $y' = (2x - 5)(2x^2 - 10x + 13)$; $(\tfrac{5}{2}, -\tfrac{99}{16})$.
15. $y' = 2a^2 x$. Point: $(0, -b^2)$.
17. $(15x^2 + 21x + 8)(3x + 2)(x - 1) = y'$. Points: $(-\tfrac{2}{3}, \tfrac{11}{27})$, $(1, -32)$.
19. $y'(0) = 11$.

Problem Set 4-9

1. $5(x^3 - 5x)^4 (3x^2 - 5)$. 3. $4(x^2 - 5x + 2)^3 (2x - 5)$.
5. $10(7x^2 - 4x + 3)^9 (14x - 4) = 20(7x^2 - 4x + 3)^9 (7x - 2)$. 7. $8(x - 3)^7$.
8. $6(t^2 - 2t + 3)^2 (t - 1)$.

9.

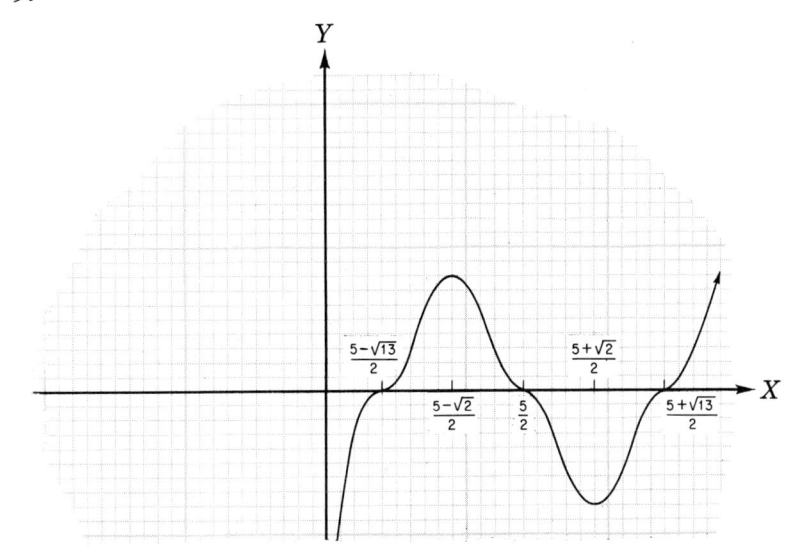

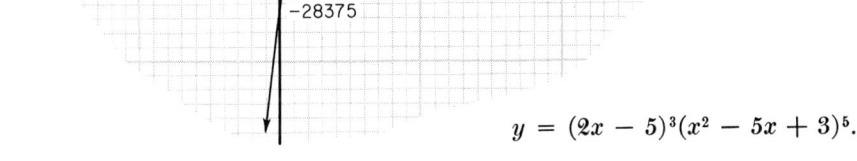

[Graph showing $y = (2x - 5)^3(x^2 - 5x + 3)^5$ with x-intercepts labeled $\frac{5-\sqrt{13}}{2}$, $\frac{5-\sqrt{2}}{2}$, $\frac{5}{2}$, $\frac{5+\sqrt{2}}{2}$, $\frac{5+\sqrt{13}}{2}$, and minimum value -28375.]

$y = (2x - 5)^3(x^2 - 5x + 3)^5.$

$y' = 13(2x - 5)^2(x^2 - 5x + 3)^4(2x^2 - 10x + 11).$

Max: $\left(\dfrac{5 - \sqrt{2}}{2}, \dfrac{161051}{512}\sqrt{2}\right).$

Min: $\left(\dfrac{5 + \sqrt{2}}{2}, -\dfrac{161051}{512}\sqrt{2}\right).$

Points of inflection: $\left(\dfrac{5 \pm \sqrt{13}}{2}, 0\right)$, $(\tfrac{5}{2}, 0).$

Problem Set 4-10

1. $8t(24t^2 - 19).$

3. $-\dfrac{3(2 + xy^5)}{5x^2y^4}.$

5. $\dfrac{dy}{dx} = \dfrac{8xy - 2xy^5 - 3}{5x^2y^4 - 4x^2}$
$= -\tfrac{15}{4}.$

7. $405x^8 - 837(x^9 - 3x^3 + 5x - 18)^{26}(9x^8 - 9x^2 + 5) - 3,075(x^{19} - 3\sqrt{5}\,x^{23} + 5)^{74}(19x^{18} - 69\sqrt{5}\,x^{22})$.

9. $\dfrac{37 - 2xy^8}{135y^2 - 8x^2y^7}$.

Problem Set 4-11

1. 6, 6. **3.** See Prob. 3, Set 4-7.

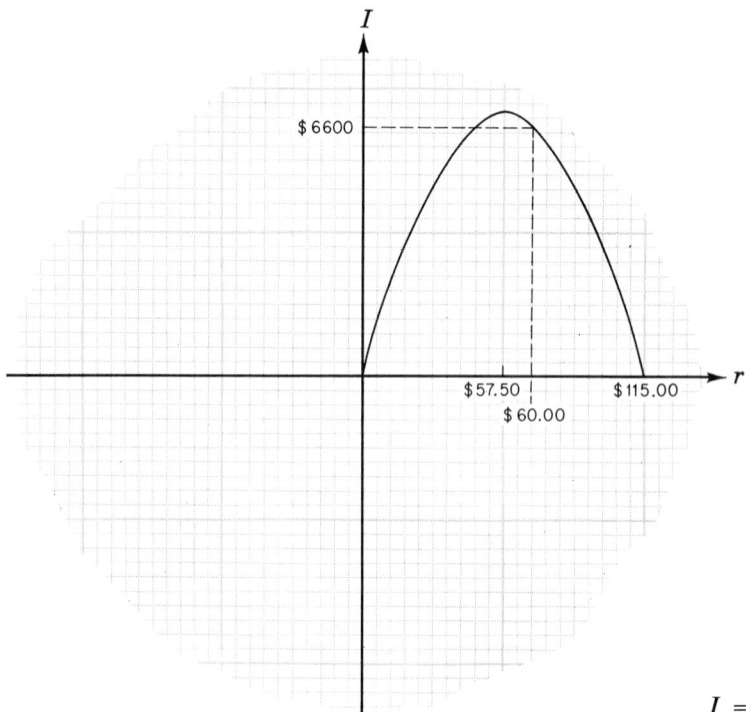

5.

$I = r(230 - 2r)$.

(a) Let I = total income, r = rent per apartment. $I = r(245 - 2r)$. For max I, $r = \$61$.

(b) From the graph, it is seen that I has a max value at $r = \$57.50$ and then decreases continuously as r increases. Since r cannot be less than $\$60$, max I is given by $r = \$60$.

(c) Similar to (b). $\$60$.

7. Consider the number $-\tfrac{1}{2}$.

9. Let the base be $2b$. Then $A = b\sqrt{10{,}000 - b^2}$. Max $A = 500$ sq cm when $b = 50\sqrt{2}$ cm.

11. (a) $V = \dfrac{x^2 + 25}{x}$, $V =$ value of fraction. There is no greatest or least value.

$$= (x^2 + 25) \cdot \dfrac{1}{x}.$$

(b) There is no greatest value. Min $V = 10$ when $x = 5$.

13. (a) $S = 0$ when $t = \tfrac{5}{2}$ sec.

15. $y - 4x + 4 = 0$, $y + 4x + 4 = 0$.

(b) $\dfrac{\sqrt{33} - 5}{4}$ sec.

17. (a) Let $C =$ cost of running the steamer for 27 miles at v miles per hour.

$$C = \dfrac{27}{v}\left(\tfrac{1}{50}v^3 + 135\right)$$

Min C is given by $v = 15$ mph.

(b) $v = 15$ mph.

19. $I = (500 - x^2)(40 + 5x)$.

(a) $90.

(b) 80 per cent.

(c) The integers from 0 to 22 inclusive.

21. $S = \pi r^2 + \dfrac{(6 - \pi r)^2}{4}$.

(a) $\dfrac{6}{4 + \pi}$ ft from one end.

(b) No cut should be made. The entire wire should be bent into a circle.

23. $S = kd^3\sqrt{1 - d^2}$; k is a constant, d is the depth. S is max when $b = \tfrac{1}{2}$ ft, $d = \dfrac{\sqrt{3}}{2}$ ft.

25. $\dfrac{dy}{dx} > 0$ when $x < \tfrac{17}{10}$ or $x > 2$.

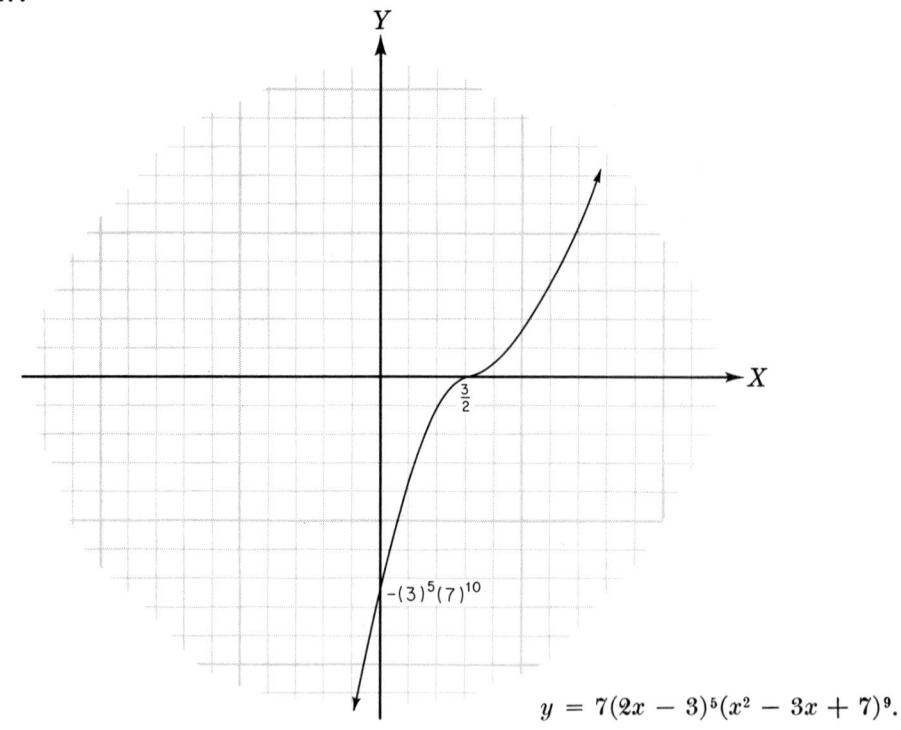

27.

$-(3)^5(7)^{10}$

$\frac{3}{2}$

$y = 7(2x - 3)^5(x^2 - 3x + 7)^9.$

29.

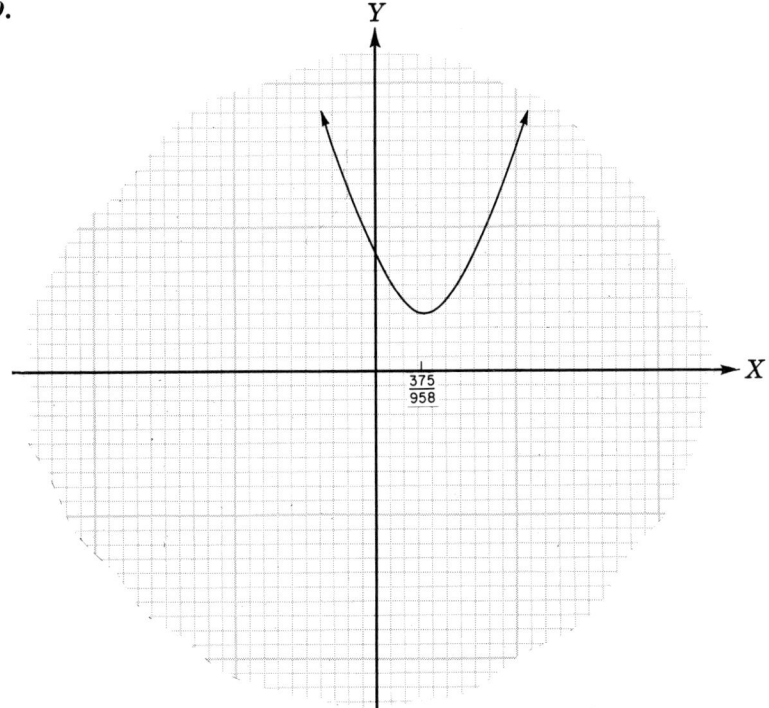

$\frac{375}{958}$

$$y = (479x^2 - 375x + 2{,}193)^{71}.$$

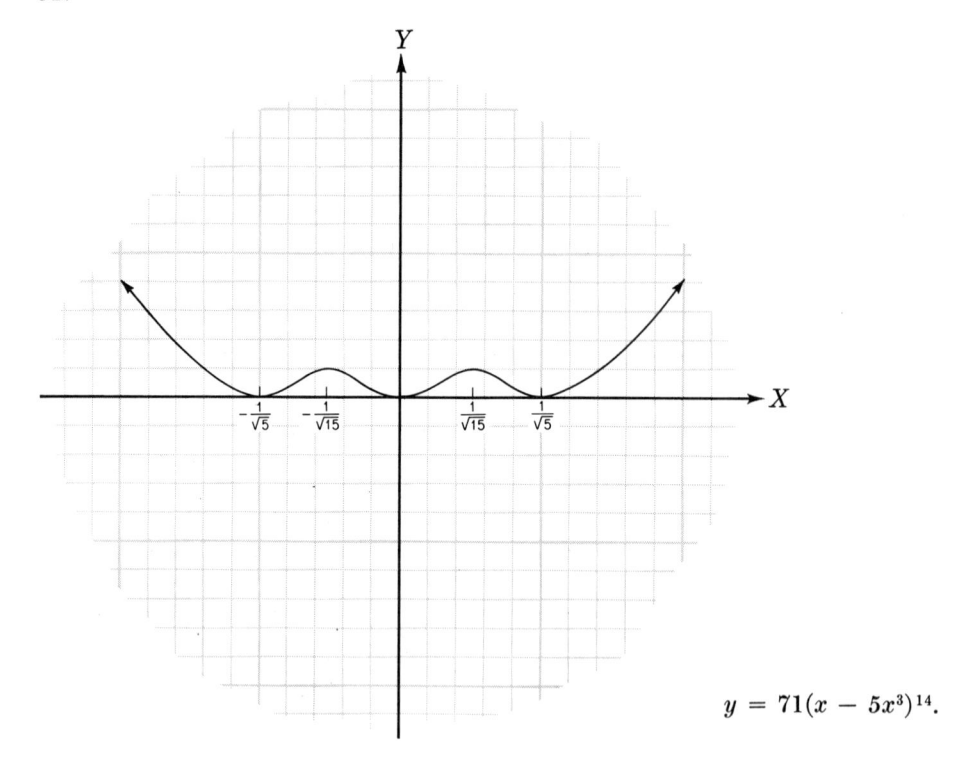

31.

$y = 71(x - 5x^3)^{14}$.

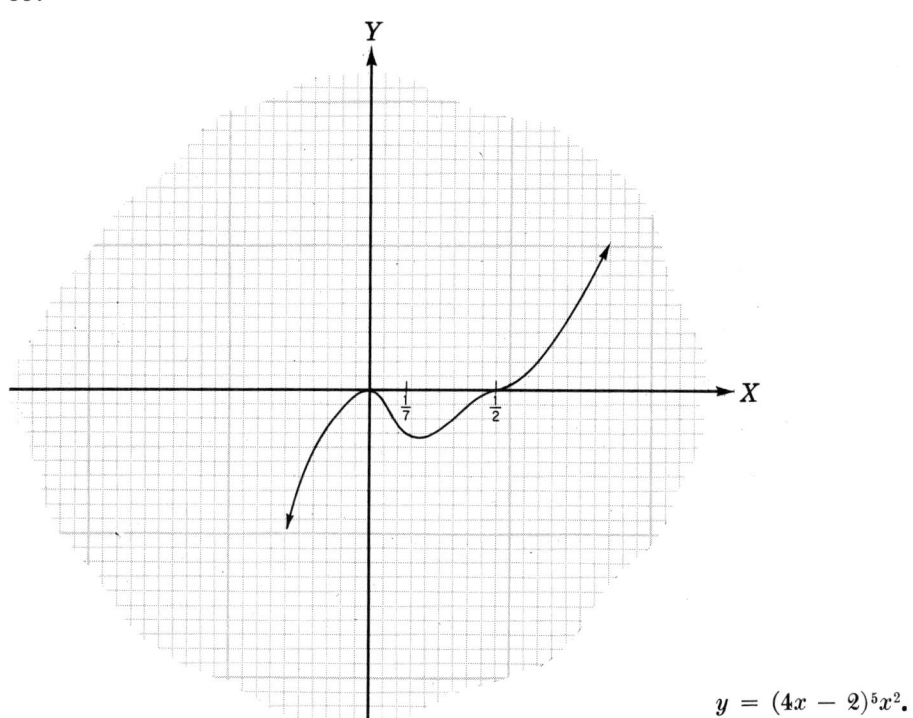

33.

$y = (4x - 2)^5 x^2.$

35.

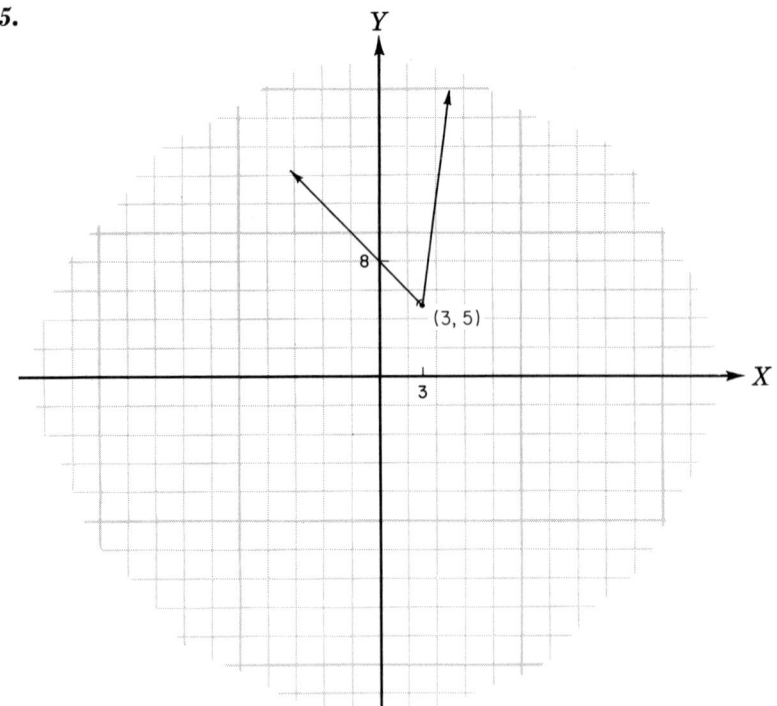

(3, 5)

$$y = \begin{cases} 4x - 7 \text{ if } x \geq 3 \\ 8 - x \text{ if } x < 3. \end{cases}$$

Problem Set 4-12. Self-Test

1. $\Delta y = \frac{3}{2}\Delta x$.
$\lim\limits_{\Delta x \to 0} \frac{\Delta y}{\Delta x} = \frac{3}{2}$.

3.

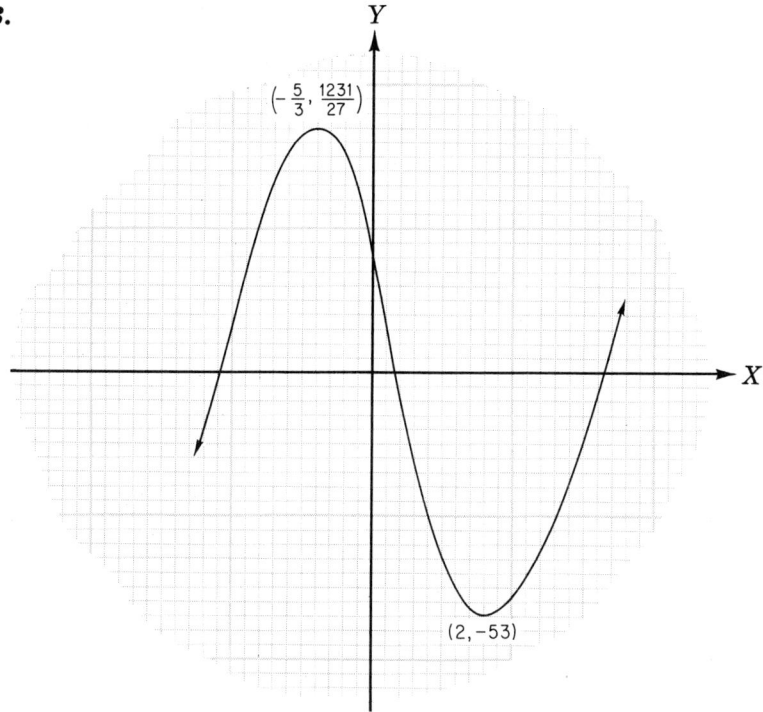

Graph showing $y = 4x^3 - 2x^2 - 40x + 3$ with labeled points $\left(-\frac{5}{3}, \frac{1231}{27}\right)$ and $(2, -53)$.

5. $S = 260$ ft. $v = 32\sqrt{10}$ ft.

7. Distance from (x,y) to $(3,-1)$ is $D = \sqrt{(x-3)^2 + (x^2+1)^2}$. (x,y) is approximately $(.7, .49)$ for min D.

Chapter 5

Problem Set 5-1

1. $h(1) = 11$, $h(4) = 51.75$, $h(0)$ is meaningless (Why?), $h(3) = \dfrac{274 + 3\sqrt{3}}{9}$.

3. $\dfrac{y + x^2 y^3}{xy + x^2}$ **11.** ± 2.   **13.** 4.   **15.** $-\frac{1}{2}$.

17. No solution.   **19.** $\dfrac{(3 - 2\sqrt{2})}{50}$.   **21.** 1, 6.   **23.** 7.   **25.** $-3, -1$.

Problem Set 5-2
1. 3.5×10^4.   3. 5.8×10^{-3}.   5. $y = 3.6 \times 10^{-2}$.   11. $x = 4.4 \times 10^{-3}$.
13. 2.7×10^{-5}.   21. 11: 4.5×10^{-3}
    13: 2.7×10^{-5}.

Problem Set 5-3
1. $-3(4x^3 - 5x + 6)^{-4}(12x^2 - 5)$.
3. $14(x^2 - 4x + 3)^{14}(2x^3 - 5x + 1)^{-9}(30 + 13x - 55x^2 + 36x^3 - 6x^4)$.
5. $8x + 15x^{-4} + 2 + 55x^{-6}$.   7. $-4(x^2 - 5x + 3)^{-5}(2x - 5)$.
9. $-\dfrac{4}{x^5}$.   11. (a) No.
    (b) Yes, when $x < 0$.
13. $(2x - 5)(2x + 3)^{-1} - 2(2x + 3)^{-2}(x^2 - 5x + 1)$.
15. $(0,6)$, $(4,0)$.

Problem Set 5-4
1. $\dfrac{3}{2\sqrt{3x - 7}}$.   3. $\tfrac{8}{5}(3x^2 - 7x + 5)^{-\frac{3}{5}}(6x - 7)$.
5. $-12t^2$.   7. $\dfrac{(4t - 3)(120t^2 - 76t + 51)}{3(5t^2 - 3t + 2)^{\frac{3}{3}}}$.
9. $572(4x - 7)^{12}$.
11. $A = 8r^2 + \dfrac{250}{r}$. $A' = \dfrac{(16r^3 - 250)}{r^2}$.   Min A when $r = \dfrac{5}{2}$ in. and
$h = \dfrac{20}{\pi}$ in.
13. $N = \dfrac{k}{s^3}$. $T = N(s - 3) = k(s^{-2} - 3s^{-3})$. $T' = -\dfrac{k(2s - 9)}{s^4}$.   Max T
when $s = \$4.50$.
15. $\dfrac{25T_0}{12}$.   17. Point: $(1,1)$. Min distance: $\sqrt{5}$.

19. $48.
21. $V = 3\pi r^2(8 - r)$.   Max V given by $r = \tfrac{16}{3}$ in., $h = 8$ in.
23. 5 in. by 10 in.
25. $C = 40x + 50\sqrt{16 + (5 - x)^2}$. Verify that the graph of this function has a min at $(-\tfrac{1}{3}, 320)$ and increases continuously on either side of that point. Hence, the most economical route will be straight from the house to the point.

27. (a) $I = k\left[\dfrac{1}{x^2} + \dfrac{1}{6(6-x)^2}\right]$. I is the intensity x ft from point A. Min I is given by $x = \dfrac{6\sqrt[3]{6}}{\sqrt[3]{6}+1}$.

(b) At a point 6 in. from A.

29. $\dfrac{1-2x}{x^2(x-1)^2}$. **31.** $\dfrac{24-4x}{(2x+7)^2(4x-5)^{1/2}}$. **33.** $\dfrac{10x+107}{6(2x+9)^{4/3}(5x+7)^{1/2}}$.

Problem Set 5-5. Self-Test

2. (a) $\dfrac{2}{\sqrt{4x-3}}$.

(b) $\frac{1}{2}(2x-3)^{1/4}(x^2-5x+1)^{1/2}(17x^2 - 73x + 50)$.

(c) $-\dfrac{5}{2x^2}$.

4.

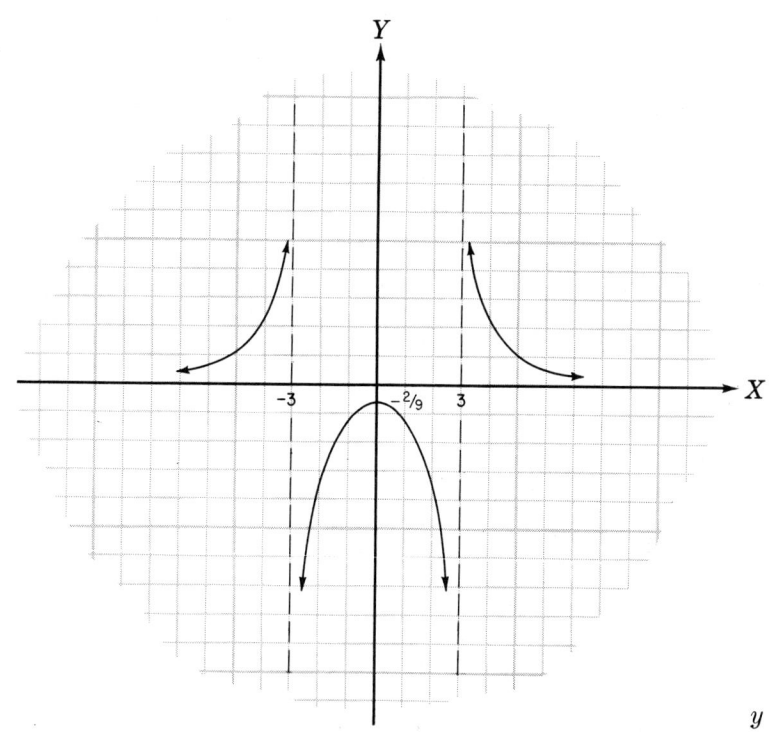

$$y = \dfrac{\sqrt{4+x^2}}{x^2-9}.$$

5. $\dfrac{199}{8}$.

Chapter 6
Problem Set 6-2
1. 32 ft/sec.
3. 80 ft/sec for third second.
5. $\Delta s\ (2,5) = 470$, $\dfrac{\Delta s\ (2,5)}{5} = 94$.
7. If D is the distance one way, then the time required for the round trip is $\dfrac{D}{30} + \dfrac{D}{60} = \dfrac{D}{20}$. Now $R\left(\dfrac{D}{20}\right) = 2D$ and $R = 40$ mph, the average rate.

Problem Set 6-3
1. $v(t) = 2t - 3$, $a(t) = 2$. $v(7) = 11$, $a(7) = 2$. $v(11) = 19$, $a(11) = 2$. If $-100 \le t < \frac{3}{2}$, then $v(t) < 0$. If $t = \frac{3}{2}$, then $v(\frac{3}{2}) = 0$. If $\frac{3}{2} < t \le 100$, then $v(t) > 0$. When $t = -100$, $s = 10{,}305$, $v = -203$. As t increases from $t = -100$ to $t = \frac{3}{2}$ the particle moves from $s = 10{,}305$ to $s = \frac{11}{4}$. As t increases from $t = \frac{3}{2}$ to $t = 100$ the particle moves from $s = \frac{11}{4}$ to $s = 9705$.
3. $v(t) = 64 - 32t$, $a(t) = -32$. $v(0) = 64$, $v(5) = -96$, $v(10) = -256$. As t increases from $t = -100$ to $t = 2$ the particle moves from $s = -166{,}400$ to $s = 64$. As t increases from $t = 2$ to $t = 100$ the particle moves from $s = 64$ to $s = -153{,}600$.
5. $v = 0$, $a = 0$. The particle is at rest at $s = 11$.
7. $36t^2 - 3 = 6$. $s(-\frac{1}{2}) = 5$, $s(\frac{1}{2}) = 5$.
9. $18t = 36$. $s(2) = 12$.
11. At rest when $72 - 6t = 0$. $s(12) = 432$. Interval: $0 \le t \le 12$.
13. $v(t) = -32t + 160$, $v(15) = -320$ ft/sec.
15. $v(t) = -16t - 20$. $-8t^2 - 20t + 100 = 96$, $t = \dfrac{-5 + \sqrt{33}}{4}$.
$v\left(\dfrac{-5 + \sqrt{33}}{4}\right) = -4\sqrt{33} \cong -23.0$ ft/sec.
17. $v(3) = -6$ ft/sec.
19. $v(t) = 10t + 7$. $5t^2 + 7t = 160$.
$v\left(\dfrac{-7 + \sqrt{3249}}{10}\right) = \sqrt{3249} = 57$ ft/sec $\cong 39$ mph.

Problem Set 6-4
1. $V = \dfrac{4\pi R^3}{3}$, $\dfrac{dR}{dt} = \dfrac{\frac{dV}{dt}}{4\pi R^2}$, $\dfrac{dR}{dt}\bigg]_{R=5} = \dfrac{400}{4\pi(5)^2} = \dfrac{4}{\pi}$ in./min.

Answers and Hints 335

3. $\dfrac{4\pi R^3}{3} = 288\pi$, $R = 6$ in., $\dfrac{dR}{dt}\bigg]_{R=6} = \dfrac{600}{4\pi(6)^2} = \dfrac{25}{6\pi}$ in./min.

5. $S = 4\pi R^2$, $\dfrac{dS}{dt} = 8\pi R \dfrac{dR}{dt}$, $\dfrac{dS}{dt}\bigg]_{R=6} = 8\pi(6)\left(\dfrac{25}{6\pi}\right) = 200$ sq in./min.

7. $\dfrac{dh}{dt} = \dfrac{\dfrac{dV}{dt}}{96h}$, $\dfrac{dh}{dt}\bigg]_{h=10} = \dfrac{500}{96(10)} = \dfrac{25}{48}$ in./min.

9.

$y = 3x^2 - 4x + 2x^{-1}$, $y' = \dfrac{2(3x^2 + x + 1)(x - 1)}{x^2}$.

x	y	y'	Conclusions
$x < 0$		−	
$0 < x < 1$		−	
1	1	0	Min $y = 1$ at $x = 1$
$x > 1$		+	

If $x \searrow L^-$, the $y \nearrow L^+$. If $x \nearrow 0$, then $y \searrow L^-$. Curve crosses x axis between $x = -1$ and $x = 0$. If $x \searrow 0$, then $y \nearrow L^+$. If $x \nearrow L^+$, then $y \nearrow L^+$.

11.

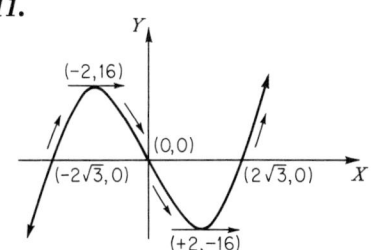

$y = x(x^2 - 12)$, $y' = 3(x^2 - 4)$.
Intercepts: $(0,0)$, $(\pm 2\sqrt{3}, 0)$.
Max $y = 16$ at $x = -2$.
Min $y = -16$ at $x = 2$.

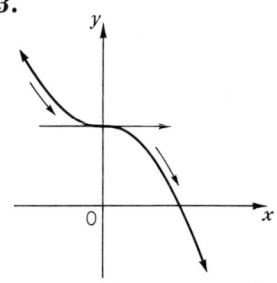

13.

$y = 12 - x^3$, $y' = -3x^2$.

x	y	y'	Conclusions
$x < 0$		−	
$x = 0$	12	0	neither
$x > 0$		−	

15. $y' = 2(x - 1)$. Min $y = -8$ at $x = 1$. Intercepts: $(1 \pm 2\sqrt{2}, 0)$.
17. $P = (100 + 20w)(500 - 25w)$. $w = 7\tfrac{1}{2}$ weeks.

336 Answers and Hints

19. Let P = profit, s = selling price per article, N = number of articles sold. $P = N(s - 1)$ with $N = \dfrac{k}{s^{\frac{3}{2}}}$.

Max, P when $s = \$\frac{5}{3} \approx \$1.67$.

21. $\dfrac{40\pi}{3}$ min, $\dfrac{1}{\pi}$ in./min, $\dfrac{1}{4\pi}$ in./min, $\dfrac{25}{\pi(12.3)^2}$ in./min.

23. $y' = 3(x + 1)(x - 5)$. Consider x increasing. If $x < -1$, then $y' > 0$ and y is increasing. If $x = -1$, $y = 16$, then $y' = 0$. If $-1 < x < 5$, then $y' < 0$ and y is decreasing. If $x = 5$, $y = -92$, then $y' = 0$. If $x > 5$, then $y' > 0$ and y is increasing.

25. See Prob. 5, Set 4-7.

27. $V = \frac{3}{4}\pi h^3$.

$12 = \dfrac{9\pi}{4} h^2 \dfrac{dh}{dt}$

$\dfrac{dh}{dt} = \dfrac{4}{15\pi}\left(\dfrac{9\pi^2}{20}\right)^{\frac{1}{3}}$ when $V = 100$ cu ft.

29. $\dfrac{dV}{dt} = 4\pi r^2 \dfrac{dr}{dt}$, $V = \dfrac{4}{3}\pi r^3$.

31. $\dfrac{dW}{dS} = 4H^3 + 12H^2 S \dfrac{dH}{dS} = 4H^3 + 72H^2 S^2 = 4(3S^2 - 5)^2(21S^2 - 5)$,

$\dfrac{dH}{dS} = 6S$.

33. $\dfrac{dH}{dS} = 18$, $\dfrac{dW}{dS} = 356{,}224$.

Problem Set 6-5. Self-Test

1. (a) 2 sec later 40 ft from its initial position.
(b) $0 \leq t \leq 2$.

2. (a) 95 ft.
(b) 159 ft.
(c) 96 ft/sec.
(d) $\dfrac{8 + \sqrt{159}}{4}$ sec later.

3. $V = \dfrac{48}{\sqrt{3}} h^2$

$\dfrac{dh}{dt} = \dfrac{25\sqrt{3}}{24}$ in./min.

Chapter 7
Problem Set 7-1

1. 140. **3.** 31. **5.** 55. **7.** 11.25.
9. 106. **11.** 26. **13.** $n = 8$. **15.** 15.

Answers and Hints

17. HINT: $\sum_{x=1}^{n} [x^3 - (x-1)^3] = n^3 = \sum_{x=1}^{n} [3x^2 - 3x + 1]$

$$= 3 \sum_{x=1}^{n} x^2 - 3 \sum_{x=1}^{n} x + n.$$

19. Disprove by showing the two expressions to be unequal when $n = 2$.

Problem Set 7-3

1. $\dfrac{5}{n} \left[\dfrac{1}{n} + \dfrac{2}{n} + \cdots + \dfrac{n-1}{n} \right] < A < \dfrac{5}{n} \left[\dfrac{1}{n} + \dfrac{2}{n} + \cdots + n \right].$

$\dfrac{5}{2} \cdot \dfrac{n-1}{n} < A < \dfrac{5}{2} \cdot \dfrac{n+1}{n}.$

$A = \dfrac{5}{2}.$

3. $\dfrac{6}{n} \sum_{K=0}^{n-1} \left[5\left(\dfrac{6k}{n}\right) + 4 \right] < A < \dfrac{6}{n} \sum_{K=1}^{n} \left[5\left(\dfrac{6k}{n}\right) + 4 \right].$

$A = 114.$

5. $\dfrac{3}{n^3} \sum_{K=1}^{n-1} k^2 < A < \dfrac{3}{n^3} \sum_{K=1}^{n} k^2.$

$A = 1.$

7. Notice that the total area, A, is twice the area between 0 and $\tfrac{1}{2}$.

$\dfrac{1}{2n^2} \sum_{K=1}^{n-1} k - \dfrac{1}{4n^3} \sum_{K=1}^{n-1} k^2 < A < \dfrac{1}{2n^2} \sum_{K=1}^{n} k - \dfrac{1}{4n^3} \sum_{K=1}^{n} k^2.$

$A = \tfrac{1}{6}.$

Problem Set 7-4

1. $0 \leq W_1 \leq 1$, $1 \leq W_2 \leq 2$, $2 \leq W_3 \leq 3$, $3 \leq W_4 \leq 4$, $4 \leq W_5 \leq 5$.

$10 \leq \sum_{K=1}^{5} W_k \leq 15.$

Problem Set 7-6

1. $\lim\limits_{x \to \infty} \sum_{i=1}^{n} \left[3\left(\dfrac{i}{n}\right) + 5 \right] \dfrac{1}{n} = \dfrac{13}{2}.$

3. $\lim\limits_{n \to \infty} \sum_{i=1}^{n} 5\left(\dfrac{4i}{n}\right)\left(\dfrac{4}{n}\right) = 40.$

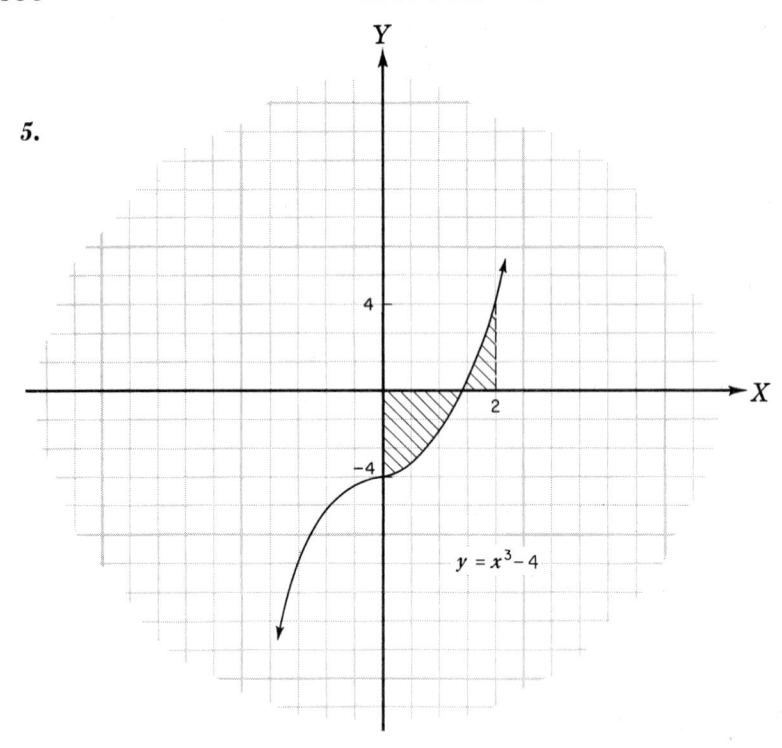

5.

$$\lim_{n\to\infty}\sum_{i=1}^{n}\left[\left(\frac{2i}{n}\right)^3-4\right]\frac{2}{n}=\lim_{n\to\infty}\left[\frac{16}{n^4}\cdot\frac{n^4+2n^3+n^2}{4}-8\right]=-4.$$ Notice that for $0\leq x\leq \sqrt[3]{4}$, $f(x)\leq 0$ so that summing over that interval in the direction of increasing x leads to a negative value whereas summing over the interval $\sqrt[3]{4}\leq x\leq 2$ leads to a positive value. Hence $\int_0^2(x^3-4)\,dx$ gives

the algebraic sum of the two areas, one of which is negative, the other positive.

7. $\lim\limits_{n\to\infty} \sum\limits_{i=1}^{n} \left(4 + \dfrac{i}{n}\right)\dfrac{1}{n} = \dfrac{9}{2}.$

9. $\lim\limits_{n\to\infty} \sum\limits_{i=1}^{n} \left[2 + \dfrac{3i}{n}\right]^3 \dfrac{3}{n} = \lim\limits_{n\to\infty} \left\{\dfrac{3}{n}\sum\limits_{i=1}^{n} 8 + \dfrac{108}{n^2}\sum\limits_{i=1}^{n} i + \dfrac{162}{n^3}\sum\limits_{i=1}^{n} i^2 + \dfrac{81}{n^4}\sum\limits_{i=1}^{n} i^3\right\}$
$= \dfrac{609}{4}.$ See Fig. 7-8.

11. $\lim\limits_{n\to\infty} \sum\limits_{i=1}^{n} \dfrac{k}{n} = k.$

13. $\lim\limits_{n\to\infty} \sum\limits_{i=1}^{n} \left(\dfrac{ki}{n}\right)^3 \dfrac{k}{n} = \dfrac{k^3}{3}.$

15. $\lim\limits_{n\to\infty} \sum\limits_{i=1}^{n} \left(\dfrac{ki}{n}\right)^4 \dfrac{k}{n} = \dfrac{k^5}{5}.$

17. Use the definition of definite integral and Theorem 3, Sec. 7-1.
19. Same as Prob. 17 with Theorem 2, Sec. 7-1.

Problem Set 7-7

3. $9\displaystyle\int_0^1 x^2\,dx + 9\int_1^2 x^2\,dx - 4\int_0^1 x\,dx - 4\int_1^2 x\,dx = 16.$

5. 728.

7. HINT: $\sum\limits_a^c f(\xi)\,\Delta x = \sum\limits_a^b f(\xi)\,\Delta x + \sum\limits_b^c f(\xi)\,\Delta x,$ and $\lim(H + B) = \lim H + \lim B.$

9. See Prob. 19, Set 7-6.

Problem Set 7-8

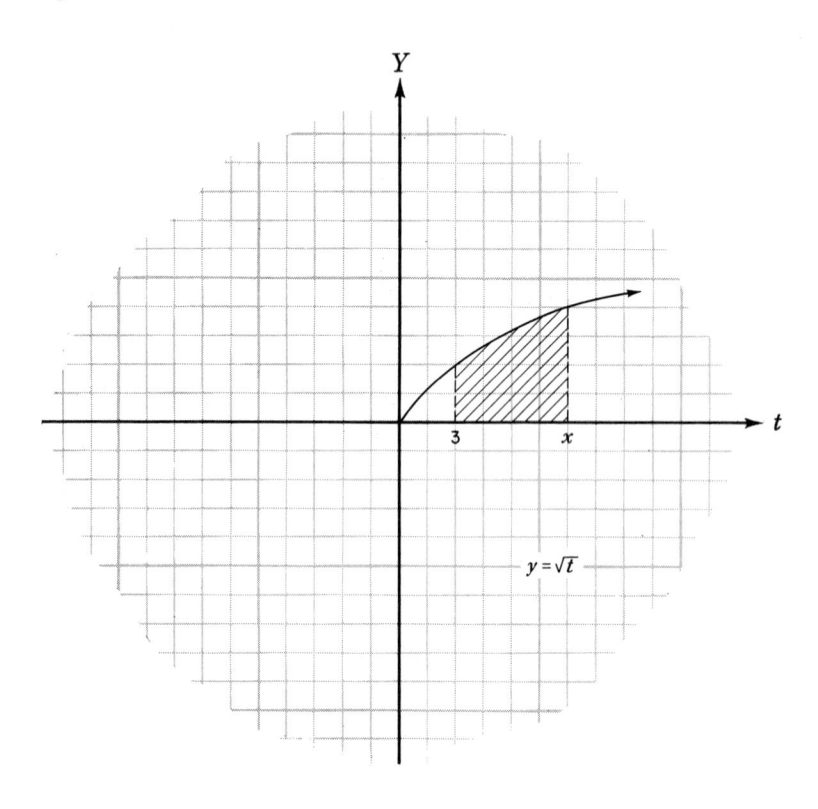

1 a

1b

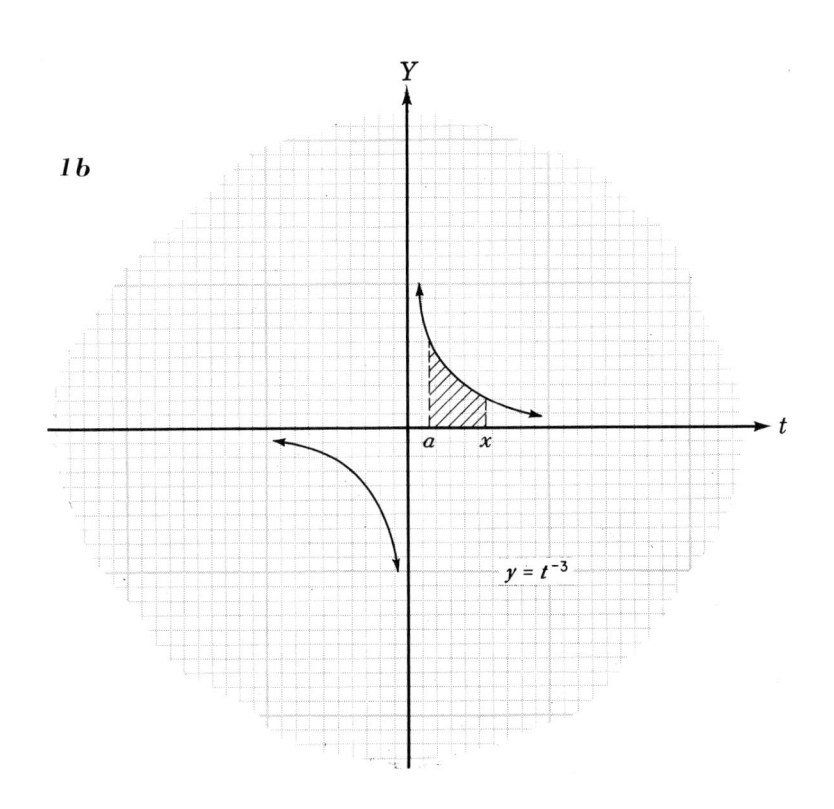

$y = t^{-3}$

3. $\dfrac{x^2}{2} - \dfrac{2}{3}x^{3/2} + C.$

5. $-\dfrac{1}{2}x^{-2} + \dfrac{2}{3}x^{3/2} - 5x + C.$

7. $\dfrac{x^3}{3} + \dfrac{x^2}{2} + C.$

9. $\dfrac{x^3}{3} - \dfrac{5}{2}x^2 + C.$

342 Answers and Hints

11.

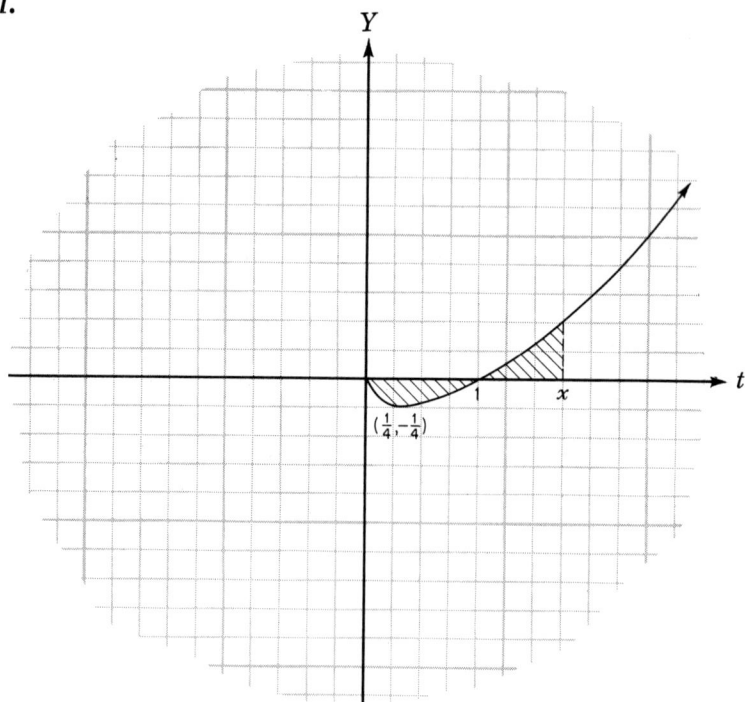

$y = (t - \sqrt{t})$.

Answers and Hints 343

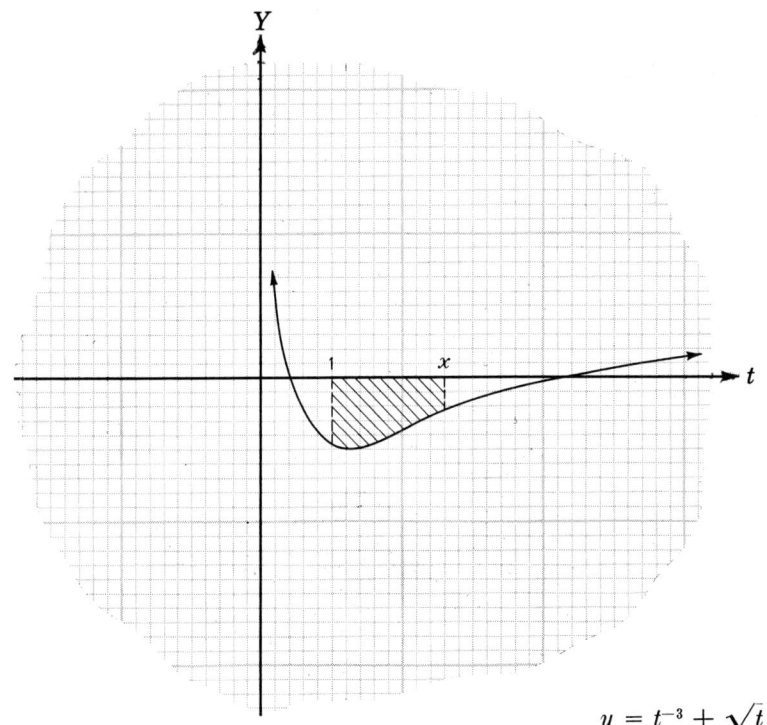

$y = t^{-3} + \sqrt{t} - 5$
$y = $ min when $t = \sqrt[7]{36}$.
point of inflection
at $t = 2\sqrt[7]{18}$.

Problem Set 7-9

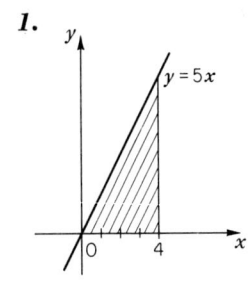

1.

$y = 5x$.
$\int_0^4 5x\ dx = \dfrac{5x^2}{2} + c \Big|_0^4 = 40.$
$A = \tfrac{1}{2} \cdot 4 \cdot 20 = 40.$

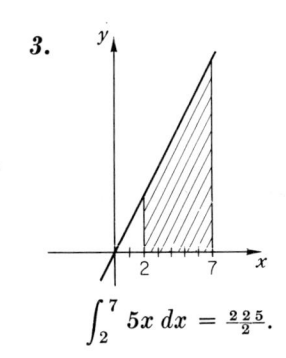

3.

$\int_2^7 5x\ dx = \tfrac{225}{2}.$

344   *Answers and Hints*

5.

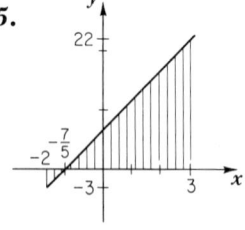

$$\left|\int_{-2}^{-\frac{7}{5}} (5x + 7)\, dx\right| + \left|\int_{-\frac{7}{5}}^{3} (5x + 7)\right| dx$$
$$= \left|-\tfrac{9}{10}\right| + \left|\tfrac{484}{10}\right| = 49.3.$$

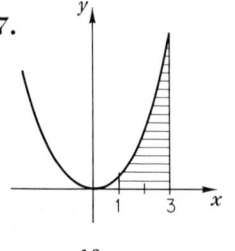

7.

$$\int_{1}^{3} 3x^2\, dx = 26.$$

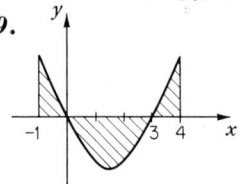

9.

$$\int_{-1}^{0} (x^2 - 3x)\, dx + \left|\int_{0}^{3} (x^2 - 3x)\, dx\right|$$
$$+ \int_{3}^{4} (x^2 - 3x)\, dx = \tfrac{49}{6}.$$

11.

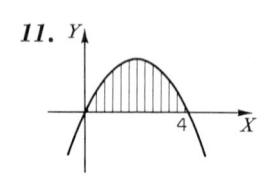

$$\int_{0}^{4} (-x^2 + 4x)\, dx = \tfrac{32}{3}.$$

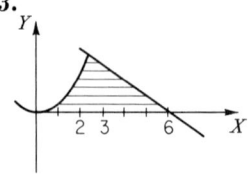

13.

$$\int_{0}^{2} x^2\, dx + \int_{2}^{3} (6 - x)\, dx = \tfrac{37}{6}.$$

15.

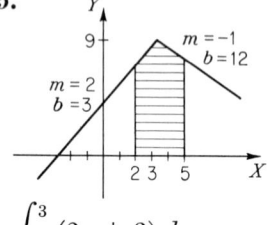

$$\int_{2}^{3} (2x + 3)\, dx$$
$$+ \int_{3}^{5} (-x + 12)\, dx = 24.$$

17. $I_0 = \int_{0}^{a} 2\pi r^3\, dr = \dfrac{\pi a^4}{2}.$

19. $\tfrac{1}{2}x^6 - \tfrac{2}{3}x^{\frac{3}{2}} + C.$

21. The integrand is undefined at $t = 0$.

25. $\int_{1}^{3} \dfrac{dx}{x} > \int_{11}^{13} \dfrac{dx}{x}.$

29. $g(x) = \dfrac{1}{4 - x}.$

Problem Set 7-11

1. 20.   **3.** 6,435.   **5.** $\tfrac{16}{7}$.   **7.** 148.

9. $\tfrac{1879}{160}$.   **13.** $\tfrac{670}{3}$.   **15.** $\tfrac{1}{11}[8'' - 3'']$.

Problem Set 7-12. Self-Test
1. 372.
2. For each $\epsilon > 0$, there exists an N such that when $x > N$, $|f(x) - 2 > \pi|$ $< \epsilon$.
3. $39 < A < 102$. 4. Take $\xi = 1 + \dfrac{3i}{n}$. Ans.: $\frac{267}{4}$.
5. (a) 10. 6. (b) and (c) are meaningless.
(b) $\frac{73}{6}$.
(c) 240.
7. $\frac{98}{3}$. 8. 0.

Chapter 8
Problem Set 8-1
1. $h(t) = -16t^2 + 32t + 560$ if $0 \leq t \leq 7$. $v(7) = -192$ ft/sec.
3. $h(t) = -16t^2 + 96t$. Reaches highest point when $v(t) = 0$, i.e., at $t = 3$ sec. $h(3) = 144$ ft. At ground level $h(t) = 0$, $t = 0$, $t = 6$. $v(6) = -96$ ft/sec.
5. Ball 1. $v_1(t) = -32t$ Ball 2. $v_2(t) = -32t + 30$
$h_1(t) = -16t^2 + 45$. $h_2(t) = -16t^2 + 30t$
If the balls meet, $h_1(t) = h_2(t)$, i.e., $t = \frac{3}{2}$.
$v_1(\frac{3}{2}) = -48$ ft/sec. $v_2(\frac{3}{2}) = -18$ ft/sec.
$h_1(\frac{3}{2}) = 9$ ft above ground. $h_2(\frac{3}{2}) = 9$ ft above ground.
The student should note that *both* balls are going down when they meet.
7. $h(50) = 40,000$ ft, max ht. Remains in air 100 sec. $v(100) = -1,600$ ft/sec, which is dangerous.
8. $h = 16(\frac{9}{2})^2 = 324$ ft.   9. $v = 32(\frac{9}{2}) = 144$ ft/sec $\cong 98$ mph.
11. (a) $D(t) = \frac{5}{2}t^2 + v_0 t$. $D(4) = 40 + 4v_0 = 60$, $v_0 = 5$ ft/sec.
(b) $D(t) = \frac{5}{2}t^2 + 5t = \frac{60}{3}$, $t = 2$, $v(2) = 15$ ft/sec. Note that although $\frac{1}{2}$ the time has elapsed, only $\frac{1}{3}$ the distance is covered.
(c) $D(t) = \frac{5}{2}t^2 + 5t = \frac{60}{2}$, $t = -1 + \sqrt{13} \cong 2.6$ sec.
(d) $D(2) = 20$ or $\frac{1}{3}$ the length of the chute in $\frac{1}{2}$ the time.
12. $a = \dfrac{v^2}{2s}$. $a = 21,600$ miles/hour/hour.
13. $v = -kt + v_0$. $D = -kt^2/2 + v_0 t$.
$D(.01) = -k(.01)^2/2 + v_0(.01) = .75$ ft $\Big\}$ $k = 15,000$. $v_0 = 150$ ft/sec at
$v(.01) = -k(.01) + v_0 = 0$ moment of impact.

15. $v = -12t + v_0$. $D = -6t^2 + v_0 t$. When $v = 0$, $D = 3750$.
$$\left.\begin{array}{r}0 = -12t + v_0 \\ 3750 = -6t^2 + v_0 t\end{array}\right\} v_0 = 300 \text{ ft/sec}.$$

17. (a) $\int_0^3 4x\,dx = 2x^2 \big|_0^3 = 18$ in.-lb. (b) 72 in.-lb. (c) 288 in.-lb. (d) $W = \int_7^{13} 4x\,dx = 240$ in.-lb.

19. $v = -32t + v_0$. $h = -16t^2 + v_0 t$. $h(1) = -16 + v_0 = 80$ ft, $v_0 = 96$ ft/sec. Hence $v(t) = -32t + 96 = 0$ at $t = 3$ and $h(3) = 144$ ft.

23. Assuming $F = ma$ is valid on the planet, $a = -50$ ft/sec². The braking power needed is approximately 0.8 of the power available.

25. $16\sqrt{129}$ ft/sec. **27.** $F(x) = 2x^2 - 24$.

31.

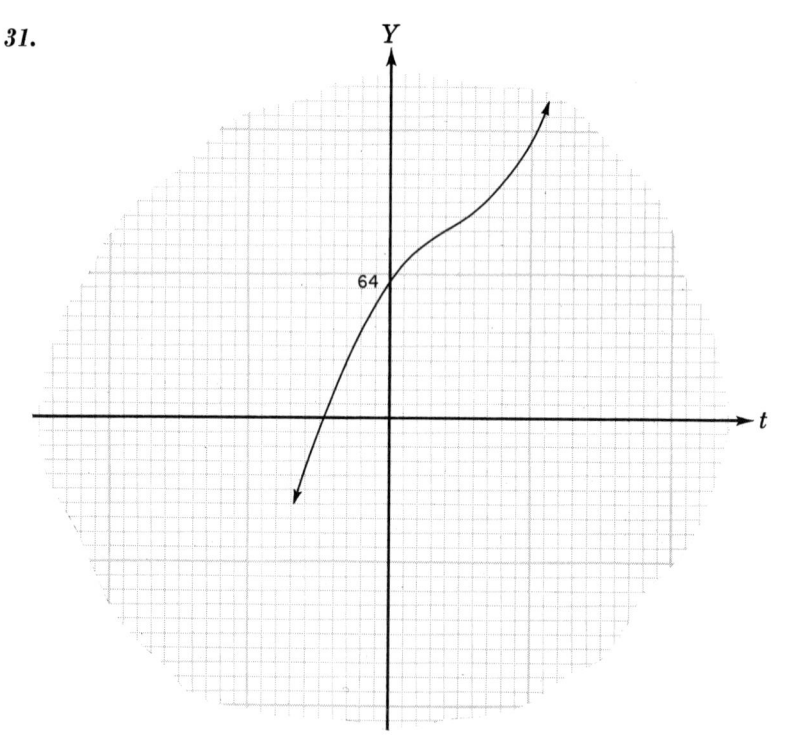

$y = t^3 - \tfrac{7}{2}t^2 + 6t + 64$ point of inflection at $(\tfrac{7}{6}, \tfrac{7325}{108})$.
$y = t^3 - \tfrac{7}{2}t^2 + 6t + 64$.

33. $\dfrac{dy}{dx} = c$, then $y = cx + k$.

35.

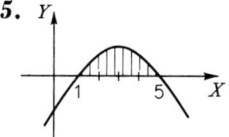

$$A = \int_1^5 (-x^2 + 6x - 5)\, dx = \tfrac{32}{3}.$$

37.

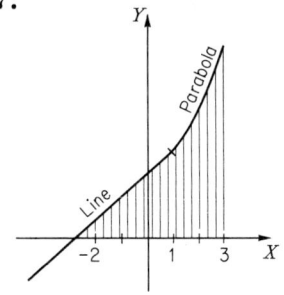

$$A = \int_{-2}^{1} (2x + 5)\, dx + \int_{1}^{3} (6x^2 + 1)\, dx = 66.$$

Problem Set 8-2

1. The area is evaluated using vertical strips.

3. $\int_0^2 (2x^2 - x^3)\, dx.$ **5.** $\int_2^4 \left(\dfrac{y}{2}\right)^{\frac{1}{3}} dy.$ **7.** 2,272 ft-lb.

9. Yes.   **11.** $\dfrac{4b}{a} \int_0^a \sqrt{a^2 - x^2}\, dx.$ **13.** $\lim\limits_{a \to \frac{\pi}{2}} \int_0^a \tan x\, dx.$

15. $4 \int_0^a (a^{\frac{2}{3}} - x^{\frac{2}{3}})^{\frac{3}{2}}\, dx.$ **17.** $2 \int_0^{\frac{-1+\sqrt{17}}{2}} [\sqrt{4 - y^2} - y^2]\, dy.$

19. $100 \int_0^5 h \sqrt{25 - h^2}\, dh.$

Problem Set 8-3

1. $\dfrac{113\pi}{12}$ in.-lb.

3. (a) $500\pi.$
(b) $165\pi.$

5. $\dfrac{2\pi}{3} \int_0^{3\sqrt{3}} (108 - 4y^2)\, dy = 144\pi \sqrt{3}.$

7. $\pi \int_6^7 (49 - y^2)\, dy = \dfrac{20\pi}{3}.$

9. (a) 144π cu in.
(b) $\dfrac{236\pi}{3}$ in.-lb.

11. $249.6\pi \int_0^4 y\, dy = 1996.8\pi$ ft-lb.
13. (a) 36 ft-lb.
(b) 60 ft-lb.
(c) 193 ft-lb.

Problem Set 8-4. Self-Test
1. $s = -16t^2 + 128t$
(a) 220 ft.   (b) Upwards.   (c) 256 ft.   (d) 128 ft/sec.

2. See Prob. 12, Set 8-1.   **3.** 40 ft-lb; 80 ft-lb.   **4.** 3,300 ft-lb.

5. $\int_{-1}^{2} (2 + x - x^2)\, dx = 4\frac{1}{2}$ square units.   **6.** 108π cubic units.

7. $249.6\pi \int_0^3 (8 + y)\, dy = 7113.6\pi$ ft-lb.   **8.** $\int_{30}^{31} 10^x\, dx$.

Chapter 9
Problem Set 9-1

1. If $u < 0$, then $-u > 0$ and $\dfrac{d\ln(-u)}{dx} = \dfrac{1}{-u} \cdot \dfrac{d(-u)}{dx} = \dfrac{1}{u}\dfrac{du}{dx}$. Also

$\int \dfrac{du}{u} = \int \dfrac{d(-u)}{-u} = \ln(-u) + C$.

3. $15x^{-1}$.   **5.** $\dfrac{-3x^2}{1-x^3}$.

7. $\dfrac{6x^2 - 34x + 41}{(x^2 - 5x + 3)(2x - 7)}$.   **11.** $\dfrac{23}{(3x-4)(2x+5)}$.

13. (a) $\dfrac{3(\ln x)^2}{x}$.   (b) $\dfrac{3}{x}$.   **15.** $\dfrac{7\ln x - 7}{(\ln x)^2}$.

17. $\ln(x^2 + 5) + C$.   **19.** $\frac{1}{2}\ln|x^2 - 5| + C$.
21. $\frac{3}{2}\ln|y^3 - 3| + C$.   **23.** $\frac{7}{2}z^2 - 5\ln|z| + C$.
25. $\frac{1}{2}t^2 - t + \frac{1}{4}\ln|t| + C$.

Problem Set 9-2
3. 0.34657.   **5.** $\frac{1}{2}\ln 3 = 0.54930$.   **7.** 0.81093.

9. $2.77260k$.   **11.** $10\sqrt{21} + 8\ln\left(\dfrac{5 - \sqrt{21}}{2}\right)$.

Problem Set 9-3

1. $21e^{7x}$.
3. $(4x^3 - 3)e^{x^3-5} + 3x^2(x^4 - 3x)e^{2x^3-10}$.
5. 148.41.
7. $-e^{-t} + c$.
9. $x = ce^{kt}$. With the given boundary conditions, $x = 2^{-\frac{t}{1500}}$.
11. Write the equation of the tangent line at the point (a, e^{-a}) and show that its intercepts are $(1 + a, 0)$ and $[0, e^{-a}(1 + a)]$. Hence, $A = \frac{1}{2}e^{-a}(1 + a)^2$ and max A occurs with vertices $(2, 0)$ and $(0, 2e^{-1})$.
15. Use the result from Prob. 14 that $\cosh^2 x - \sinh^2 x = 1$.
17. $(x^2 - 5)^4(4x + 9)^{12}(12x + 7)^{-\frac{2}{3}}(1120x^3 + 1760x^2 - 2570x - 2{,}000)$.

Problem Set 9-4. Self-Test

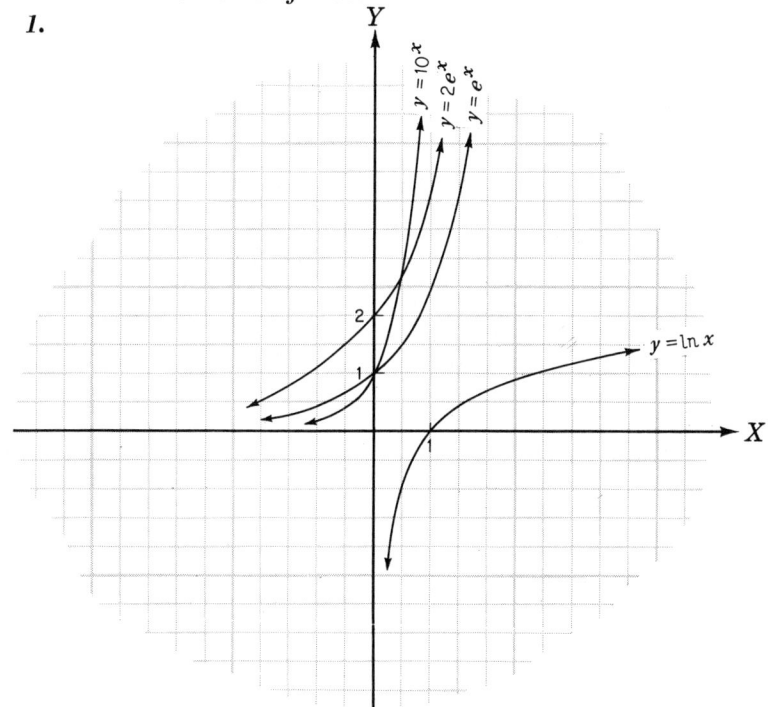

1.

Coordinates of intersections with $x = 2$ are $(2, 0.69315)$, $(2, 7.3891)$, $(2, 100)$, $(2, 14.7782)$.

2. $y = e^{-1}x$; $m = 0.4$; y-intercept $= 0$.

3. (a) $\frac{5}{3}e^{3x} + c$. (b) $\ln(x^2 + 1) + c$. (c) $4x + \ln x + \frac{3}{x} + c$.

4. (a) $y\left[\dfrac{12}{4x+3} + \dfrac{4}{5(2x+7)}\right]$.

(b) $\dfrac{4}{4x-3} - \dfrac{2x+1}{x^2+x+1}$.

(c) 0.

5. (a) 6.07764. (b) -5.90346. (c) 69.55983.

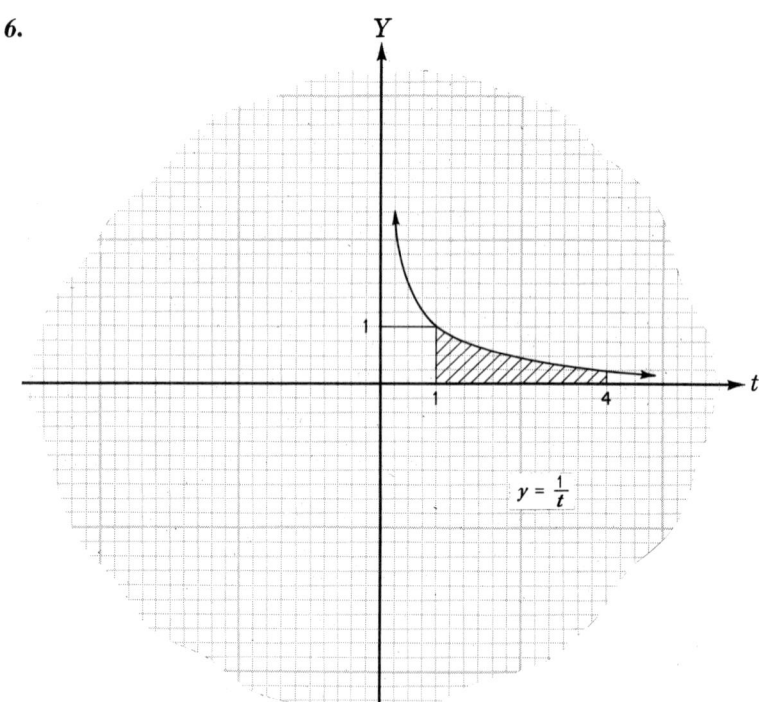

6.

[Graph of $y = \frac{1}{t}$ with shaded region between $t=1$ and $t=4$]

7. See Probs. 12 and 14, Set 9-3.   8. Show that $\pi > \ln \pi^e$ so $e^\pi > \pi^e$.

Chapter 10
Problem Set 10-2

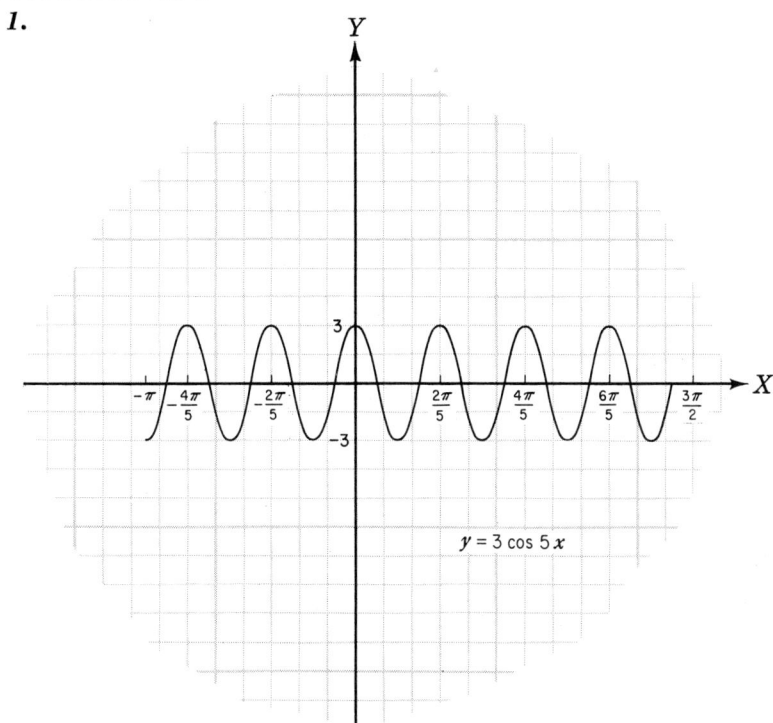

1.

$y = 3 \cos 5x$

3. 0.   **5.** −1.   **7.** $\dfrac{\sqrt{3}}{2}$.

9.

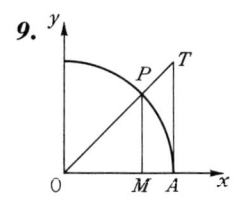

$2\triangle OMP < 2 \text{ sector } OAP < 2\triangle OAT.$

10.

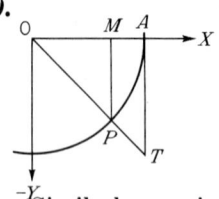

Similarly: $-\sin t \cos t < -t < -\tan t$

$\therefore \sin t \cos t > t > \tan t$.

11. HINT: The order of an inequality is reversed when dividing by a negative number.

13.

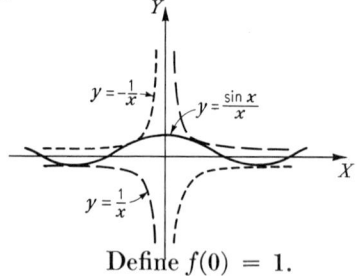

Define $f(0) = 1$.

Problem Set 10-3

1. $-6 \sin (2x - 5)$. **3.** $2xe^{x^2} - 5 \cos x$. **5.** $-\cos y \sin (\sin y)$.

7. $\sin x + x \cos x$. **9.** $\dfrac{\cos t}{2\sqrt{1 + \sin t}}$.

11. $27x^2 \cos (x^3) \cos (x^2) - 18x \sin (x^3) \sin (x^2)$.

13. $e^{-x}[\cos x - \sin x]$. **15.** $y = x$.

17. $\dfrac{d(\cot x)}{dx} = \dfrac{d\left(\dfrac{\cos x}{\sin x}\right)}{dx} = -\csc^2 x$. **19.** $-\csc x \cot x$.

25. $\dfrac{dy}{dx} = [x]e^x$, $x \neq$ an integer. $y(3) = 3e^3$. $y'(3)$ does not exist.

27. (a) $\displaystyle\int \cot x \, dx = \int \dfrac{\cos x}{\sin x} \, dx = \ln |\sin x| + c$.

(b) Both a and b must be within the same open interval, $n\pi < a, b < (n + 1)\pi$, for any integer n.

29. Both, when x is such that the logarithms are defined, since $\sec x = \dfrac{1}{\cos x}$.

Problem Set 10-5

1. Multiply the integrand by $\dfrac{\csc x - \cot x}{\csc x - \cot x}$ and let $v = \csc x - \cot x$.

3. Use the identity $\frac{1}{2}(1 + \cos 2x) = \cos^2 x$ and Theorem 2, Sec. 7-7.

5. HINT: $\dfrac{1}{\cos^2 t} = \sec^2 t$.　　7. HINT: $\cot^2 \theta = \csc^2 \theta - 1$.　　11. 0.3.

13. Both since $\tan\left(\dfrac{\pi}{4} + \dfrac{x}{2}\right) = \sec x + \tan x$ is an identity.

15. Yes since $\csc u - \cot u = \tan \dfrac{u}{2}$ is an identity.

17. $\int \sec^6 t \, dt = \int \sec^2 t \, dt + 2\int \sec^2 t \tan^2 t \, dt + \int \sec^2 t \tan^4 t \, dt$
$= \tan t + \tfrac{2}{3} \tan^3 t + \tfrac{1}{5} \tan^5 t + C.$

25. Meaningless. See Sec. 7-10.

29. $4 \ln \left| \dfrac{4 - \sqrt{16 - x^2}}{x} \right| + \sqrt{16 - x^2} + c.$

Problem Set 10-6. Self-Test

1. Read Sec. 10-1.
2. Use an identity for $\cos(A + B)$ and argue that $\cos x < \cos(x + 6)$.
3. $a = 0, b = \dfrac{\pi}{2}$; $a = \dfrac{\pi}{4}, b = \dfrac{3\pi}{4}$; $a = -\pi, b = 0$. See Sec. 7-10.
4. $f(0) = 3$. HINT: $\lim\limits_{\alpha \to 0} \dfrac{\sin \alpha}{\alpha} = 1$.
5. (a) $y - .5420x + .1595 = 0$.　(b) $-.551$.
6. $\ln |\sec x| + C$.
7. (a) $\dfrac{x}{2} + \tfrac{1}{4} \sin 2x + C$.

(b) $-\tfrac{5}{3} \cos 3x + C$.

(c) $\ln |\csc x - \cot x| + C$.

8. Consider the area bounded by the x axis and $y = x \sin x$, $2 \leq x \leq 5$. Approximate this area by circumscribed rectangles using abscissas $2, \dfrac{5\pi}{6}, \pi, \dfrac{7\pi}{6}, \dfrac{5\pi}{4}, \dfrac{4\pi}{3}, \dfrac{3\pi}{2}, 5$. $\int_2^5 x \sin x = -4.1$ to 2 significant digits.

Chapter 11
Problem Set 11-3

1. $x \sin x + \cos x + C$.

3. Use technique of example 3. $\dfrac{e^x}{2}[\sin x - \cos x] + C$.

5. See Example 2.   7. $\ln|x + \sqrt{1 + x^2}| + C$.

9. Let $x = 3\sin\theta$. $\tfrac{1}{3}\ln\left|\dfrac{3 - \sqrt{9 - x^2}}{x}\right| + C$.

11. π.

15. Let $v = 16 + y^2$ and show that the integral becomes $\tfrac{1}{2}\int(v^{\frac{1}{2}} - 16v^{-\frac{1}{2}})\,dv$.
Ans.: $\tfrac{1}{5}(16 + y^2)^{\frac{5}{2}} - \tfrac{16}{3}(16 + y^2)^{\frac{3}{2}} + C$.

Problem Set 11-4. Self-Test

1. $x \sin x + \cos x + C$.   2. $\ln\left|\dfrac{x + \sqrt{16 + x^2}}{4}\right| + C$.

3. $\dfrac{e^x}{2}[\sin x - \cos x] + C$.

4. $8\left(\text{the principal angle whose sine is }\dfrac{x}{4}\right) + \tfrac{1}{2}x\sqrt{16 - x^2} + C$.

5. Use integration by parts with $u = \tan^{n-1} x$, $dv = \tan x \sec x\, dx$ to show that $\int \tan^n x \sec x\, dx = \dfrac{1}{n}\left[\sec x \tan^{n-1} x - (n-1)\int \tan^{n-2} x \sec x\, dx\right]$.
Let $x = 5\sec\theta$ in the given integral and use this formula to derive the answer.

$\dfrac{1}{2500}x(x^2 - 25)^{\frac{3}{2}} - \dfrac{3}{200}x(x^2 - 25)^{\frac{1}{2}} + \tfrac{3}{8}\ln\left|\dfrac{x + \sqrt{x^2 - 25}}{5}\right| + C$.

6. 28.5 sq units.   7. $x \log x - x + C$.

8. Approximate by a rectangle of base 1.5, altitude $-\dfrac{2}{3\pi}$ to obtain $-.32$.

List of Symbols

ϵ, 2
$x \epsilon X$, 2
$f(x)$, 3, 6
$\{x\}$, 4
Δ, 5
$\neq$, 7, 8
$<, >, \leq, \geq$, 8, 178
$|b|$, absolute value, 8
$\frac{5}{0}$, 13
$\frac{0}{0}$, 13
$\frac{0}{5}$, 13
Σ, 21, 167, 182
Δf, 27
$P(x,y)$, 34
$*$, 42n.
$\lim_{x \to b} f(x)$, 70
ϵ, 70
δ, 70
$\frac{\Delta y}{\Delta x}$, 82, 86, 92, 94
$D_x F$, 91
$f'(x)$, 91
y', 91, 92
$\frac{df(x)}{dx}$, 91, 173
$(x + \Delta x)^n$, 99
$y = P(x)$, 110
$\frac{dy}{dx}$, 125

$\frac{dy}{dt}$, 129
$\lim_{\Delta t \to 0} \frac{\Delta y}{\Delta t}$, 130
$a^{m+n}, \frac{a^m}{a^n}, (a^n)^k, (a \cdot b)^n$, 138
$\sqrt{x}$, 139
$0^0, 0^{-|k|}$, 139n.
$\frac{\Delta s}{\Delta t}$, 156, 214
$\frac{dv(t)}{dt}$, 159, 214
$\lim_{n \to \infty} \frac{k}{n}$, 173
x_i, 183
$\lim_{n \to \infty} \sum_{i=1}^{n} f(\zeta_i) \Delta x_i$, 184
$\int$, 185
$\int_a^b |f(t)| \, dt$, 191
$I(b)$, 199
$\int \frac{dx}{x}$, 239, 243
$\frac{dL(x)}{dx}$, 240
$L(ax)$, 240
e, 242
e^x, 250
$P(C_t, S_t)$, 256
$\lim_{t \to 0} \frac{\sin t}{t}$, 264
π, 271

355

Index

Abscissa, 33
Absolute maximum, 105
Absolute minimum, 106
Absolute value, 8
Acceleration, 159
Algebra, basic properties of, 10
Analytic geometry, 31
Applications, of integral, 214
   of minima and maxima, 118
Area, 173, 175
Auxiliary equation, 23
Average rate of change, 156
Average velocity, 156
Axes, 31

Basic properties of algebra, 10
Begle, *Introductory Calculus*, 109
Bell, E. T., *Men of Mathematics*, 33
Bisector, perpendicular, 44
Bounds, lower and upper, 182

Calculus, integral, 195
Cartesian coordinate system, 31
Characteristic of $\log_{10}$ and $\log_e$, 248
Circle, unit, 47
Circuit, electric, 21
Complex numbers, 11, 18
Composite function, derivative of, 129
Conditional equation, 15
Conditional inequality, 49

Constant, derivative of, 97
Continuity, 72
Continuous function, 72
Coordinate system, cartesian, 31, 33
Correspondence, 1
Cos u, 264
Cosh x, 252
Cot t, 256
Csc t, 265
Curve, slope of, at a point, 100

Definite integral, 184, 185
   Riemann, 206
Definition, domain of, 2, 3
   trigonometric, 255
Delta, 5, 71
   f, 27
Delta process, 92
Denominator, zero, 257
Derivative, 91, 156
   $a \cdot e^x$, 251
   of a composite function, 129
   of a constant, 97
   of cos u, 262, 265
   of cot u, 268
   of csc u, 268
   of kx^n, n, complex, 150
      n, integral, 98, 126, 148
      n, rational, 147, 149, 150
      n, real, 150
   of a polynomial, 100, 101

357

Index

Derivative, rate of change, 156
   of sec u, 268
   of sin u, 262, 264, 265
   of sum, 97
   of tan u, 268
Derivatives, list of, 268
Descartes, René, 31
Differentiation, of a constant, 97
   fundamental notion of, 151
   of kx^n, 98
   of a power of function, 126
   of a product, 123
   of a sum, 97
   theorems on, 97
   of u^n, 126
   of $u \cdot v$, 123
Digits, significant, 142
Distance between two points, 39
Division, meaning of, 12, 13
   by zero, 13
   zero result, 13
Domain of definition, 2, 3

Epsilon, 70
Equation, auxiliary, 23
   conditional, 15
   equivalent, 15
   of a line, 57
   locus of, 43
   point-slope, 58
   slope y-intercept, 59
   solution set of, 14
   two-point, 58
Equivalent equation, 15
Exponents, 138
   fractional, 139
   negative, 139
   zero, 139
Extraneous value, 24

Family of lines, 61
Formula, for function, 4, 16
   general, 190
   quadratic, 16, 17
Formulating problems, 219
Fractional exponents, 139

Function, 1, 2, 3
   composite, 129
   continuous, 72
   formula for, 4, 16
   hyperbolic cosine, 252
   hyperbolic sine, 252
   inverse, of ln x, 250
   limit of, 69
   logarithmic and exponential, 238
   maximum value of, 105
   minimum value of, 106
   multiple-valued, 4
   nonnegative, 199
   notation, 5
   polynomial, 100
Fundamental law of logarithms, 240
Fundamental notion of differentiation, 151
Fundamental theorem of integral calculus, 195

General formula, 190
General rate of change, 162
Geometry, analytic (coordinate), 31
Graphs, of cos x, 257–260
   of sec x, 257–260
   sin x, 260
   of tan x, 257–260

Identity, 15
   trigonometric, 271
Increments, 86
Indefinite integral, 200
Inequalities, 8
   of first degree, solutions of, 49
Inequality, conditional and unconditional, 49
Integrals, 184
   applications of, 214
   $\dfrac{dt}{t}$, 243
   $e^u \dfrac{du}{dx}\, dx$, 251
   $\dfrac{1}{u} \dfrac{du}{dx}\, dx$, 245

Index

Integrals, applications of, $t^k dt$, 197
  of csc u, 270
  definite, 184, 185
  fundamental theorem of, 195
  indefinite, 200
  Riemann definite, 206
  of sec u, 270
  of tan u, 270
  theorems on, 189
  trigonometric, 270
Integration, 198
  by parts, 123, 277, 280
  techniques of, 211
  of trigonometric functions, 268
Introductory Calculus, Begle, 109
Inverse function of ln x, 250

Limit, 71, 172
  of a function, 69
  notation, 173
  remarks on, 184
  of trigonometric functions, 260
Line, equation of, 57
  point-slope equation of, 58
  slope of, 53
  tangent to curve at a point, 66
Lines, family of, 61
  parallel, 60
Ln x, 241, 247
  inverse function of, 250
Locus of an equation, 43
Logarithm, common, 241
  ln x, 241
  naperian, 247
  natural, 243
  of zero, 245
Logarithmic and exponential functions, 238
Logarithms, fundamental law of, 240
  rules of, 242
Lower bounds, 182

Matrices, 12
Maxima, applications of, 118

Maximum, absolute and relative, 105
Maximum points, tests for, 110
Men of Mathematics, E. T. Bell, 33
Minima, applications of, 118
Minimum, 105
  absolute and relative, 106
Minimum points, tests for, 110
Motion, 214
Multiple-valued function, 4

Negative exponents, 139
Notation, function, 5
  limit, 173
  scientific, 142
  sigma, 167
Number as area, 268
Number system, structure of, 10
Numbers, complex, 16, 17, 18
  quaternions, 11
  rational, 11
  real, 11

One, 12
Ordered pairs, 3
Ordinate, 33
Origin, 31

Pairs, ordered, 3
Parabola, 46
Parallel lines, 60
Parts, integration by, 123, 277, 280
Perpendicular bisector, 44
Point-slope equation of a line, 58
Points, tests, for maximum, 110
  for minimum, 110
Polygon, 174
Polynomial, derivative of, 100, 101
Power of a function, differentiation of, 126
Problems, formulating, 219
  (*See also* problems at ends of sections)
Process, delta, 92
Product, differentiation of, 123

Quadratic formula, 16, 17
Quaternions, 11

Rate of change, average, 156
  general, 162
  of slope of $y = f(x)$, 267
  of velocity, 159
Rational numbers, 11
Real numbers, 11
Regular polygon, 174
Relation, 3, 4
Relative maximum, 105
  conditions for, 108
Relative minimum, 106
  conditions for, 108
Riemann definite integral, 206
Rules of logarithms, 242

Scientific notation, 142
Sec u, 256
  derivative of, 268
  slope of, 82
Sech x, 253
Self-tests, explanation of, 30
Sigma notation, 167
Significant digits, 142
Sin u, 264
Sinh x, 252
Slope, 53
  of line, 53
  no, 53
  rate of change of $y = f(x)$, 267
  of secant, 82
  of tangent line, 82
Slope y-intercept equation, 59
Slope zero, 53, 54
Solution set of an equation, 14
Speed, 159
Spring problem, 179
Structure of the number system, 10
Subset, 3
Substitution, trigonometric, 280
Sum, 167
Symbols, list of, 355
System, cartesian coordinate, 31
  number, 10

Tables of ln x, use of, 247
Tan u, 256
  derivative of, 268
Tangent, 66
  slope of, 82
Tanh x, 253
Techniques of integration, 211
Tests for maximum and minimum points, 110
Theorems, on differentiation, 97
  on integrals, 189
Trigonometric definitions, 255
Trigonometric functions, 255
  integration of, 268
  limits of, 260
Trigonometric identities, 271
Trigonometric integrals, 270
Trigonometric substitution, 280
Two-point equation of a line, 58

Unconditional inequality, 49
Unit circle, 47
Upper bounds, 182
Use of tables of ln x, 247
Useful trigonometric identities, 271
Useful trigonometric integrals, 270

Value, absolute, 8
  extraneous, 24
Velocity, 156, 158
  average, 156
  rate of change of, 159
  at time t_1, 158

Work, 180
  definition of, 183

Zero, 12, 245
  division by, 13
  logarithm of, 245
  slope, 53, 54
Zero exponents, 139

Logarithms of Numbers from 1.00 to 9.99

N	0	1	2	3	4	5	6	7	8	9
1.0	0.0000	0.004321	0.008600	0.01284	0.01703	0.02119	0.02531	0.02938	0.03342	0.03743
1.1	0.04139	0.04532	0.04922	0.05308	0.05690	0.06070	0.06446	0.06819	0.07188	0.07555
1.2	0.07918	0.08279	0.08636	0.08991	0.09342	0.09691	0.1004	0.1038	0.1072	0.1106
1.3	0.1139	0.1173	0.1206	0.1239	0.1271	0.1303	0.1335	0.1367	0.1399	0.1430
1.4	0.1461	0.1492	0.1523	0.1553	0.1584	0.1614	0.1644	0.1673	0.1703	0.1732
1.5	0.1761	0.1790	0.1818	0.1847	0.1875	0.1903	0.1931	0.1959	0.1987	0.2014
1.6	0.2041	0.2068	0.2095	0.2122	0.2148	0.2175	0.2201	0.2227	0.2253	0.2279
1.7	0.2304	0.2330	0.2355	0.2380	0.2405	0.2430	0.2455	0.2480	0.2504	0.2529
1.8	0.2553	0.2577	0.2601	0.2625	0.2648	0.2672	0.2695	0.2718	0.2742	0.2765
1.9	0.2788	0.2810	0.2833	0.2856	0.2878	0.2900	0.2923	0.2945	0.2967	0.2989
2.0	0.3010	0.3032	0.3054	0.3075	0.3096	0.3118	0.3139	0.3160	0.3181	0.3201
2.1	0.3222	0.3243	0.3263	0.3284	0.3304	0.3324	0.3345	0.3365	0.3385	0.3404
2.2	0.3424	0.3444	0.3464	0.3483	0.3502	0.3522	0.3541	0.3560	0.3579	0.3598
2.3	0.3617	0.3636	0.3655	0.3674	0.3692	0.3711	0.3729	0.3747	0.3766	0.3784
2.4	0.3802	0.3820	0.3838	0.3856	0.3874	0.3892	0.3909	0.3927	0.3945	0.3962
2.5	0.3979	0.3997	0.4014	0.4031	0.4048	0.4065	0.4082	0.4099	0.4116	0.4133
2.6	0.4150	0.4166	0.4183	0.4200	0.4216	0.4232	0.4249	0.4265	0.4281	0.4298
2.7	0.4314	0.4330	0.4346	0.4362	0.4378	0.4393	0.4409	0.4425	0.4440	0.4456
2.8	0.4472	0.4487	0.4502	0.4518	0.4533	0.4548	0.4564	0.4579	0.4594	0.4609
2.9	0.4624	0.4639	0.4654	0.4669	0.4683	0.4698	0.4713	0.4728	0.4742	0.4757
3.0	0.4771	0.4786	0.4800	0.4814	0.4829	0.4843	0.4857	0.4871	0.4886	0.4900
3.1	0.4914	0.4928	0.4942	0.4955	0.4969	0.4983	0.4997	0.5011	0.5024	0.5038
3.2	0.5051	0.5065	0.5079	0.5092	0.5105	0.5119	0.5132	0.5145	0.5159	0.5172
3.3	0.5185	0.5198	0.5211	0.5224	0.5237	0.5250	0.5263	0.5276	0.5289	0.5302
3.4	0.5315	0.5328	0.5340	0.5353	0.5366	0.5378	0.5391	0.5403	0.5416	0.5428
3.5	0.5441	0.5453	0.5465	0.5478	0.5490	0.5502	0.5514	0.5527	0.5539	0.5551
3.6	0.5563	0.5575	0.5587	0.5599	0.5611	0.5623	0.5635	0.5647	0.5658	0.5670
3.7	0.5682	0.5694	0.5705	0.5717	0.5729	0.5740	0.5752	0.5763	0.5775	0.5786
3.8	0.5798	0.5809	0.5821	0.5832	0.5843	0.5855	0.5866	0.5877	0.5888	0.5899
3.9	0.5911	0.5922	0.5933	0.5944	0.5955	0.5966	0.5977	0.5988	0.5999	0.6010
4.0	0.6021	0.6031	0.6042	0.6053	0.6064	0.6075	0.6085	0.6096	0.6107	0.6117
4.1	0.6128	0.6138	0.6149	0.6160	0.6170	0.6180	0.6191	0.6201	0.6212	0.6222
4.2	0.6232	0.6243	0.6253	0.6263	0.6274	0.6284	0.6294	0.6304	0.6314	0.6325
4.3	0.6335	0.6345	0.6355	0.6365	0.6375	0.6385	0.6395	0.6405	0.6415	0.6425
4.4	0.6435	0.6444	0.6454	0.6464	0.6474	0.6484	0.6493	0.6503	0.6513	0.6522
4.5	0.6532	0.6542	0.6551	0.6561	0.6571	0.6580	0.6590	0.6599	0.6609	0.6618
4.6	0.6628	0.6637	0.6646	0.6656	0.6665	0.6675	0.6684	0.6693	0.6702	0.6712
4.7	0.6721	0.6730	0.6739	0.6749	0.6758	0.6767	0.6776	0.6785	0.6794	0.6803
4.8	0.6812	0.6821	0.6830	0.6839	0.6848	0.6857	0.6866	0.6875	0.6884	0.6893
4.9	0.6902	0.6911	0.6920	0.6928	0.6937	0.6946	0.6955	0.6964	0.6972	0.6981
5.0	0.6990	0.6998	0.7007	0.7016	0.7024	0.7033	0.7042	0.7050	0.7059	0.7067
5.1	0.7076	0.7084	0.7093	0.7101	0.7110	0.7118	0.7126	0.7135	0.7143	0.7152
5.2	0.7160	0.7168	0.7177	0.7185	0.7193	0.7202	0.7210	0.7218	0.7226	0.7235
5.3	0.7243	0.7251	0.7259	0.7267	0.7275	0.7284	0.7292	0.7300	0.7308	0.7316
5.4	0.7324	0.7332	0.7340	0.7348	0.7356	0.7364	0.7372	0.7380	0.7388	0.7396